GUARDIANS

GUARDIANS

Strategic Reconnaissance Satellites

Curtis Peebles

LONDON

IAN ALLAN LTD

First published 1987

ISBN 0 7110 17654

Published by Ian Allan Ltd, Shepperton, Surrey; and printed in the United
States of America.

Contents

———————————

Introduction

Every moment of every day, year in and year out, a watch is being kept. The subject of this attention may be a missile test site in northwestern Russia, an air base in California, a naval exercise off the coast of Virginia, a new radar site in Siberia, or a terrorist camp in a Middle Eastern country. These and tens of thousands of other places and things are being watched. It is the latest chapter in a story older than recorded history. Ever since humans first formed themselves into organized groups, they have wondered what the other group over the hill is planning and have taken action to find out.

Today's spies are different from their shadowy predecessors. They are mechanical—coasting through the cold silence of space. They may skim just above the atmosphere or go about their work partway to the moon. Their attention is turned toward the earth and the many activities taking place on its surface. The missions these satellites undertake are varied. Some take pictures, others listen, still others follow the movements of ships. Early-warning and nuclear-detection satellites keep watch for missile launches and nuclear tests.

Like spies of old, military satellites live in the shadows. In most cases, something as basic as what they look like is secret. Yet if the satellites remain hidden, their impact is not. Because of the satellites, the world is a safer place. Through their constant watch, both sides know the number, location, and status of the other's weapons. And both sides know both sides know. New threats can be identified and countered. A nation can act from knowledge rather than from fear and ignorance. Surprise and bluff are no longer useful tactics. In this way, military satellites represent a stabilizing influence—acting as guardians of whatever peace exists in the world.

This book seeks to be a guide to the world's military space systems—their development, history, technology, and place in the larger world. The book is organized by mission and subdivided by generations. The would-be analyst faces many problems; the most obvious one is the layers of secrecy. Such secrecy is meant to prevent an enemy from learning any useful information about intelligence–gathering sources. The secrecy is understandable and necessary but frustrating for the writer. One is forced to sort through the available bits and pieces; such things as launch vehicle, orbital characteristics and inclination, behavior, and even the radio frequency of the satellite's telemetry are all valuable clues.

No claim is made that this is the definitive history. There are too many gaps in the available information. No doubt there are large parts of the story still secret; some parts may remain so forever. Still, one can see at least the broad outline. All sources are open and reliable. Often, however, they are obscure and difficult for the general public to obtain. Where available, official documents have been used.

I would like to thank the people who helped: Phil Henderson; Sue Henderson; Peter Campbell; Larry R. Strawderman; Gay Haran; Harry Waldron; Brig. Gen. Donald W. Goodman, USAF; John H. Wright; W. M. McDonald; Lyna Gamma; Brig. Gen. Russell Berg, USAF (Ret.); and David C. Humphrey.

A special thanks to James E. Oberg for the information on Plesetsk, "Ferret-D," and the military Salyut. I also wish to recognize the help of Harry N. Miller and Mil-Air Books as well as the staffs of the Lyndon B. Johnson Presidential Library, the U.S. Air Force Historical Research Center at Maxwell Air Force Base, the Edwards Air Force Base History Office, the Central Intelligence Agency, the Defense Intelligence Agency, the U.S. Air Force, Lockheed Aircraft, San Diego Aerospace Museum, and the San Diego State University and California State University Long Beach libraries.

CHAPTER 1

The Fearful Darkness

At the end of World War II, the United States and the Soviet Union stood astride the globe eyeing each other warily. They, of all the combatants, had emerged stronger than when they entered the war. They alone had the power to shape the postwar world. This inevitably brought the two superpowers increasingly into conflict. Their vastly different natures, philosophies, interests, goals, and views of the world offered no grounds for agreement; they indeed were mutually exclusive. Soviet behavior in Poland, the communist coup in Czechoslovakia, and, ultimately, the Berlin blockade hardened the U.S. attitude. A bare 3 years after U.S. and Soviet soldiers had joyously shaken hands on the Elbe River, an iron curtain stretched across central Europe and World War III seemed only an incident away. Looking across the cold war landscape, Western intelligence saw only a dark void. Virtually nothing was known about Russia. The Air Force assembled target lists from material in the Library of Congress, but even things as trivial as rationing, economic conditions, and public morale were beyond reach. Russia was a closed nation. The Central Intelligence Agency (CIA) called it "denied territory." Diplomats were isolated in Moscow. Their travel was restricted to the city itself. Offices and apartments were bugged and the diplomats were subjected to blatant surveillance and provocation. They were reduced to watching military parades and flyovers or looking at photos of party officials to see who had been purged.[1]

What little information the West had on Russia came from refugees, Soviet soldiers who had defected or been captured during the war and not returned, and ex–German POWs who were released in the years following the end of the war. Most of their information was trivial; the rest was spotty, confused, or out-of-date. Further confusing the picture were "paper mills" which sold second-hand or fraudulent intelligence, trying both to fill the information vacuum and their operators' pockets.

By 1948, even these limited sources were drying up. At the same time, the need for information on Russia was becoming greater with the blockading of Berlin. A Soviet attack seemed imminent—indeed only a matter of time. Today it is difficult to ascertain the desperation of those years: the U.S. was deeply frustrated and worried about its inability to learn what the Russians were planning. It was a near-wartime atmosphere of "do something, do anything." What the U.S. decided to do was send in agents.

Agents

The agents would perform various missions. The most important was to watch airfields and rail lines for any step-up in activities that would signal preparations for the attack many expected. Another was to provide information on specific targets—the location of military units and aircraft; whether a particular airfield was equipped to handle nuclear weapons, security, and living conditions; and, most importantly, future Soviet plans.[2] Finally, the agents were to provide links with the anti-Soviet resistance movements in the Ukraine and Baltic. Some areas, such as the Carpathian Mountains, were actually controlled by these nationalists.[3] The value of such resistance movements for sabotage and disruption, in the event of war, was obvious.

The agents themselves were displaced persons or defectors. Some were ethnic Russians, but most were from the various nationalities of the Soviet empire—Balts, Ukrainians, Armenians, or Georgians. Their motives varied—some wanted to fight communism, others sought independence for their homeland. What they had in common was toughness, competency, self-assurance, and courage. They were willing to undertake a long-term, extremely dangerous mission. If caught, they could expect only death.

Once a potential candidate was spotted, he was quietly asked if he would be willing to go back to Russia on a mission. If he agreed, he was taken to a safe house for a series of medical and psychological tests to determine his stability, coordination, and learning ability. The candidate would also go over his life history—every place he had been, everything he had ever done, and everybody he knew or who could recognize him. The individual's reaction to stress was also tested by subjecting him to trick questions, interrogations, and lie detector tests. The latter served to weed out any bogus emigrés sent by the Russians. The final selection was by

the case officers who would train the agents.

The training lasted nearly a year. It covered such areas as physical conditioning; aircraft recognition; life in the Soviet Union (prices, new regulations, etc.); and practicing coding, decoding, and transmitting messages. Specific exercises involved watching an airfield and estimating the level of activity, runway length, and building layouts. Another exercise was photography and a third was parachute jumps. The most important, as the agent's survival would depend on it, was going over the fake life history of the agent and the documents that would support it. This was a mixture of both the agent's actual life and invented details and experiences to cover the time spent in the West. The agent had to know it as if it truly was his life story.

The agents operated either singly or in two-man teams. As the targets they were to watch were deep inside Russia, they were sent in by aircraft. Unmarked C-47s were used, flown by Czech and Hungarian crewmen. The drops were at night during the dark of the moon. Although all the flights were tracked by the Russians, and some were fired upon, no aircraft was lost in the four years of operations. The first airdrop was made on 5 September 1949; it was a two-man team into the Carpathian Mountains. Other missions soon followed. The agent was dressed in Soviet and East German clothing and carried, for trade or bribery, several watches and $5 bills (the latter were available to Soviet troops in East Germany). The agent carried an RS-1 radio concealed in his suitcase. Once he landed, the agent would bury his parachute, then make his way to a nearby railroad station and catch a train to the target area. Once there, he would send a brief radio message to indicate his safe arrival. The agent would settle down, get a job, and try to blend into the background. Reports were sent by either radio or letters with invisible ink messages.

Once inside Russia, the agent was utterly alone; there was nobody to help him, no place to hide, and no escape from his mission. He faced the KGB [*] and the most extensive police state the world had ever seen. The KGB had the power of life and death over every Soviet citizen from a Politburo member to the lowliest peasant. Any suspicious individual who attracted their attention would be investigated. Just as dangerous to the agent were ordinary Soviet citizens. The recurring spy scare campaigns convinced them Western agents were rampant. This created a population of citizens who would readily spy on their neighbors, report strangers, and betray fellow workers. If persuasion did not work, fear would. Any stranger asking for help could only be a KGB provocation. Even a friend or relative would turn in an agent to protect themselves from a police trap. Such a climate of paranoia made even a casual conversation on a train dangerous. Any slip or even irrational suspicion was enough to destroy the agent.

Another danger to agents was document checks by the police. Having looked at thousands of individual documents, the police would readily spot a forged or altered one. A Soviet citizen had to carry a host of papers. Most important was the internal passport, which was the primary identity document. It was valid for five years and was different for each Soviet Republic. Another was the labor book, which showed every place an individual worked. Each entry had to be signed by the factory's personnel manager. Other documents that might be needed included a military status book, party card, officer's identity book, or work and travel permits.

The task of forging the necessary documents was a monumental one. The documents had to be filled out correctly and the signatures had to be by the proper issuing officials. Specific documents varied from one part of a Soviet Republic to another. Even the ink had to be correct. The CIA had great difficulty producing the necessary documents. For the early years of the airdrop program, the CIA had to use captured German forgeries originally printed during World War II. It was not until 1951 that the CIA finally learned the techniques used to print Soviet documents.

Once the forms were in hand, it was necessary to fill them out. This required vast amounts of precise information—birth dates, names of employees and landlords, addresses of factories and homes, personnel and histories of military units, and much more. Ironically, the widespread destruction of birth, housing, employment, and military records during World War II was an advantage. It was possible to create a life story that could not be checked, simply because the people, places, and things listed no longer existed. To counter this, the Soviets, between 1949 and 1953, undertook a reregistration campaign. Each person had to be identified by a living witness before his passport would be re-registered. The difficulties one faced from this were obvious.[5]

Given all this, it is not surprising that virtually all agents were soon captured. Each agent, just before the drop, was given a danger

---

[*] From 1946 until 1954, the Soviet political police was called the MGB. On 13 March 1954, it was renamed the KGB. For simplicity, the more familiar name is used.[4]

sign that was to be placed in a message if he was under KGB control. In case after case, the sign would appear in their messages. The KGB tried to deceive the West by "playing back" the agents, that is, forcing him to send false information. The CIA went along to keep the agent alive. The Russians were often sloppy. In one case, handwriting analyses indicated an agent's messages were written by seven different people. Despite the best efforts of the CIA and the courage of the agents themselves, the West soon had virtually no sources remaining inside Russia.[6]

During this same time, the need for information was growing. In September 1949, as the first agents were being sent into Russia, a U.S. aircraft picked up the fallout from the Soviet Union's first atomic bomb explosion. The Soviets also had a copy of the U.S.

B-29 (TU-4) to carry the bomb; several B-29s had been forced to land in Russia after being damaged on bomb runs over Japan. In August 1953, a few months after Stalin's death, the Soviets detonated their first hydrogen bomb. Clearly, it would be only a matter of time before they had a stockpile. It would not be the agents who would provide information on the buildup, however. Because the tremendous effort and lives expended for so little information could not be justified, the airdrop agent program was terminated in 1954.

Tinker, Tailor, Tourist—Spy

Even as the airdrop operations were ending, a new possibility presented itself. After

A 1958 "tourist" photo of the Soviet M-50 Bounder, a supersonic long-range bomber, at an airfield near Moscow. Because a telephoto lens was used, the fuselage appears short. In reality, the M-50 was 185 feet (56.2 meters) long with an 83 feet (25.3 meters) wingspan. The aircraft was a failure and never went beyond prototype stage. *Eisenhower Library Photo.*

"Tourist" photo of a Soviet P-4 class PT boat. Fifty-five feet long, it is armed with machine guns, torpedoes, and depth charges. The North Vietnamese Navy used similar torpedo boats in their attack on the USS *Maddox* on August 4, 1964, which became known as the Tonkin Gulf incident. *Lyndon B. Johnson Library Photo.*

Stalin's death, many of the controls on foreign visitors were eased. Visitors were no longer limited to a few closely controlled, utterly loyal fellow travelers. Now even ordinary tourists could enter the Soviet Union. Such visitors could be used to provide reports and photos. These tourist-spies would not try to steal secret documents or recruit Soviet citizens, they would just keep their eyes open and take a picture or two. They were a varied lot coming from America, Western Europe, and the Third World. Their backgrounds were equally diverse—government officials, businessmen, diplomats, chess players, churchmen, athletes, and students.[7] The CIA would approach them and ask if they were willing to volunteer. Many were happy to do so. The CIA would then put together a secret itinerary of targets to observe or photograph. The tourists were briefed on their assignments. The planning had to allow for the fact that these amateur agents usually had little technical background. To compensate, they were told what to look for—the color of smoke coming out of a factory or something equally simple.

They were also warned about what actions to avoid and the counterintelligence activities they would face. So prepared, these unlikely agents went forth to match wits with the KGB.

Despite their amateur status, they brought back thousands of photos—subjects not associated with tourists' snapshots. These included photos of a missile assembly plant and storage area; the first shots of several new types of Soviet submarines, including the first nuclear-powered one; a missile-carrying destroyer; more than a dozen new surface-to-air missile sites spotted from airliners; information on long-range bombers, manned spaceflight and missile-support facilities, bacteriological warfare, early work on antiballistic missiles, factories, rail yards, port installations, and similar military information. Tourists' reports on aircraft serial numbers caused a doubling of production estimates for one particular type. The amateur spies also bought Soviet electronic equipment and merchandise. From this and other sources, production figures, numbers of machine tools, alloys, and other facets of Soviet industrial capacity could be found.

Some of the agents were very resourceful—a female East German, studying in Moscow, was able to collect plant, water, and soil samples from the restricted zone around the Mozhaysk atomic facility. She also sent a Christmas package of "hot" pinecones to her contact.

Tourists were not the only information sources the U.S. had. A spin-off was "mounted operations." Unlike the tourists' missions, these were specifically planned and organized by the CIA. The agents, predominantly American, were recruited to go into Russia and report in depth on a specific installation or area of expertise. These agents were much more qualified than the tourist-spies; they were knowledgeable about Russia and usually able to speak the language. Not surprisingly, many were graduate students or professors who had specialized in studies of the Soviet Union. They could undertake much more complex missions than could be given to an ordinary tourist. The CIA provided them with an all-expense-paid visit, along with a week or more of training.[8,9] Their professional expertise gave them access to Soviet laboratories and personnel. They could examine Soviet technology firsthand. Most importantly, they could ask questions—what is this, what does it do, how does it work? The agents' expertise with Western equipment allowed an assessment of Soviet technology to be made.

At the other end of the spectrum was a "mole." He was Lt. Col. Pyotr Popov of Soviet military intelligence—the GRU. He was the most productive source for the CIA in the 1950s. Popov offered to help the CIA in November 1952. Within a few months, he fulfilled one of the Pentagon's highest priorities by turning over a copy of the 1951 edition of the Soviet army field regulations—the Soviets' basic manual of tactics—a document for which the U.S. had been willing to pay $500,000. In the six years Popov worked for the CIA, he provided information on GRU operations and agents in both Europe and the U.S., new Soviet tanks, amphibious vehicles and armored personnel carriers, details on tactical missile systems and nuclear submarines, as well as information on Soviet tactics for the use of nuclear weapons. In all, his information saved the U.S. a minimum of half a billion dollars in research and development. Inevitably, Popov came under suspicion. He was recalled to Moscow in November 1958 and arrested. The Soviets tried to use him as a double agent, but Popov was able to warn his case officer in Moscow. After a meeting in October 1959, the CIA officer, who had diplomatic immunity, was expelled and Popov was shot.[10,11]

The climate of fear that prevailed during the agent drops had eased by this time. Still the Soviet Union remained a "denied area." At the same time the tourists' operations were underway and Popov was providing information, the Soviets were correcting deficiencies in their delivery systems. Although they had about 1,000 TU-4s by 1953, they were ill-suited for an atomic attack on the U.S. The TU-4 had only enough range for a one-way attack. Being propeller-driven, it was vulnerable to jet interceptors.

In that same year, however, flight testing began on the Mya-4 jet bomber. The prototype made a flyover of the 1954 May Day parade in Moscow. With its speed increased over the TU-4, the Mya-4 would cut U.S. warning time in half.

The next year, ten Mya-4s were put on aerial display. Another new bomber was also shown—the TU-95. It was a four-engine turboprop design. A formation of nine was seen. It was apparent the Soviets were building bombers as modern as any in the U.S.[12] For the 1955 Air Force Day ceremonies, the Soviets put on an even more impressive show. First ten, then eighteen Mya-4 bombers flew over the heads of Western air attachés.

Estimates began to appear that projected a Soviet bomber force larger than that of the U.S. by the end of the 1950s. There were other disquieting developments; the pace of Soviet nuclear testing was picking up. The Russians' nuclear stockpile was growing. The U.S. was rapidly becoming vulnerable to a Soviet nuclear attack. In addition, Soviet activities were not limited to bombers. At a place called Kapustin Yar in the Ukraine, still another danger was taking shape. Since the late 1940s, the Soviets had been testing ballistic missiles—at first, captured German V-2s; then in the mid-1950s, Soviet designs. These intermediate range ballistic missiles (IRBMs) had ranges of about 1,000 miles (1,609 kilometers). This was enough to cover Western Europe from Soviet territory.

In trying to gain information on such developments, human agents, whether tourists, diplomats, or moles, faced a final, insurmountable difficulty. They could not see or hear more than a tiny part of the complete picture. An agent might know what was going on in his immediate vicinity but not what was happening down the road. "His" airfield might be quiet but that might not be true elsewhere. The tourist-spies could operate only in major cities and on main transport routes. Large areas were out-of-bounds. Moreover, a tourist or air attaché might photograph a new Soviet bomber, but that would not be proof it was in production. And although the mole's inside position made him the most valuable spy, he had to move cautiously and might have access to only certain information. With these limitations imposed on traditional

spies, more technical means of spying were sought. And so began the process which ultimately led to reconnaissance satellites.

Early Aerial Reconnaissance

Photoreconnaissance had proven invaluable during World War II. U.S. and British aircraft had roamed across occupied Europe, the Mediterranean, and into the heart of Germany itself. Photoreconnaissance played a large part in confirming agents' reports of German rocket development.

After the war, aircraft were used against the Soviet Union. The first penetration of Soviet airspace was made in the late 1940s. The flights were short-range, going only about 100 miles (161 kilometers) inside the border. Camera-carrying fighters and light reconnaissance bombers were used. They relied on speed to go in, take the pictures, then escape before defenses could react.[13]

Starting in 1950 and continuing through 1952, the U.S. used "featherweight" RB-36s. These aircraft were stripped of all unnecessary equipment, including all of the guns except the tail turret. Being so lightened, the RB-36s were able to reach altitudes in excess of 58,000 feet (17,678 meters)—beyond the effective reach of contemporary jet fighters. The jet fighters would use most of their fuel to climb to these altitudes. Once there, the jets, because of their higher wing loading, were barely controllable. A simple turn would tax the pilot's skill. Any mishandling of the unstable aircraft would cause it to go out of control or flame out its engine. The RB-36 had no such problems and could, despite its immense size, outmaneuver the interceptor. This gave the featherweight virtual immunity. On such missions, the RB-36 carried fourteen

RB-36s similar to this aircraft made overflights of Russia in the early 1950s. Stripped of all unnecessary equipment, they were able to fly above the reach of Soviet jet fighters. The B-36 was designed in the early 1940s as a long-range bomber, capable of hitting targets in occupied Europe from the U.S. mainland. When England did not fall, as was feared, the program was reduced in priority and the prototype B-36 did not fly until after the war. *U.S. Air Force Photo.*

cameras. Flights could last as long as 36 to 42 hours without refueling. This put considerable strain on the twenty-two- to twenty-eight-man crew.[14] The RB-36 replacement was the all-jet RB-47E. It was nearly twice as fast as the RB-36, but it was unable to reach the high altitudes attained by the old featherweight.

The RB-47E could make only limited penetrations into the Soviet Union. This meant most of Russia could not be photographed. The dangers of a deep penetration were underscored by the experience of an RAF Canberra reconnaissance aircraft in July 1953. The Canberra took off from a base in West Germany and flew across Russia at maximum altitude. It photographed the Soviet missile test center at Kapustin Yar. Before reaching Iran, the Canberra took several hits and barely made

it to safety. The flight was a joint CIA and British Secret Intelligence Service operation. The fundamental problem was that a conventional aircraft, like the Canberra, was simply too limited in altitude to make overflights with an acceptable degree of risk.[15]

Moby Dick

To solve the problem, the Air Force turned to man's oldest flying machine—the balloon. After World War II, thin, plastic material similar to that used for plastic sandwich bags was developed for use as the envelope of a balloon. Since the plastic was lighter than the rubberized fabric previously employed for balloons, the plastic balloons could reach altitudes above those of jet interceptors.

The U.S. reconnaissance balloon program originated from several sources. The first was a RAND Corporation study in 1946, which suggested that balloons be used to overfly the Soviet Union. Several years later, the idea was independently developed by Brig. Gen. George W. Goddard. While watching a high-altitude balloon flight, Goddard was surprised to see the balloon suddenly pick up speed and travel eastward. The balloon had been caught by the jet stream. General Goddard realized that a balloon in the jet stream could travel long distances and "take some mighty pretty pictures on the way." [16]

Development of high-altitude balloons was rather slow. The first Skyhook balloon had been launched from St. Cloud, Minnesota, on 25 September 1947. Although it was widely used to carry scientific packages to high altitudes, its reconnaissance applications

The RB-47 replaced the RB-36 for short-range penetrations of Soviet airspace in the mid-1950s. Nearly twice as fast as the RB-36, it was better able to cope with improved Soviet air defenses. The RB-47 was, however, vulnerable to supersonic jet fighters and so could only be risked in limited overflights. The B-47 first flew in December 1947; deliveries began in 1950 with over 2,000 being built. The last was retired in 1967. *U.S. Air Force Photo.*

were not investigated until 1951, in the form of a study of high-altitude wind currents. For the next 2 years, hundreds of balloons were launched from Tillamook, Oregon; Vernalis, California; and Edwards Air Force Base, California. The program was given the code name "Moby Dick." One of the participants later described the ballon by quoting Melville: "What a gleaming and dazzling thing, this splendid, great white whale!"

At the same time, under the designation WS-119L, work was beginning on the Skyhook balloon's camera, gondola, and recovery system at the Wright Air Development Center. There were four main problems. One was developing the necessary balloon. Another was ensuring the camera and electronics would not be damaged by more than a week of exposure to the low temperatures of the stratosphere. Still another was the parachute that would return the camera package to earth after the mission was over. The biggest problem was the midair recovery of the payload. C-119 aircraft were to be fitted with a trapezelike arrangement that would trail behind and below the aircraft. This would catch the descending parachute at 12,000 to 14,000 feet (3,657 to 4,267 meters), then reel it and the camera package aboard. Things did not go smoothly. By early 1953, Moby Dick looked like a complete disaster. One aircraft and crew nearly crashed in a recovery test. In the spring of 1953, both Moby Dick and WS-119L were transferred to the Atmospheric Devices Laboratory. The staff was able to pull the project together, and by August 1954 solutions to the technical problems were in hand and production could begin.[17]

On 15 April 1955, the 1st Air Division was activated at Offutt Air Force Base. The launch crews were from the Strategic Air Command (SAC). Their training began in May 1955. Several equipment problems—in particular balloon failures—cropped up during the training flights. Because of this, a smaller balloon was substituted, but this too caused a problem. The original, large balloon could reach an altitude of 75,000 to 80,000 feet (22,860 to 24,384 meters). With the smaller one, Moby Dick would cross Russia at only 45,000 to 60,000 feet (13,716 to 18,288 meters). This was within reach of Soviet fighters. Moby Dick's vulnerability was confirmed during the training program. The Air Defense Command (ADC) found that the balloon could be tracked both visually and on radar. At dawn and dusk, the sun's rays reflecting from the balloon's envelope turned it into a shimmering, multicolored rainbow—visible for miles. ADC also found jet interceptors could bring down the balloons.

By the New Year, the launch crews, balloons and camera packages were in West Germany, Turkey, and Norway. On 10 January 1956, the first eight Moby Dick balloons were launched on their journey across Russia.[18] The complete assembly consisted of the General Mills balloon, the suspension straps, a 24-foot (7.3-meter) in diameter recovery parachute, and the payload. This was two ballast containers (fine steel shot was used), a radio transmitter, and the camera gondola, which was the size of a home refrigerator. The total weight was 1,430 pounds (650 kilograms). The camera had two lenses and was pretimed to take 450 to 500 photos. The package carried a sign in Russian saying a reward would be paid for the safe return of the "meteorological package." This was the program's cover.

It would take 8 to 10 days for the balloon to cross Russia riding the prevailing winds.[19] The on-board system kept the balloon at a constant altitude—releasing ballast to compensate for the cooling of the gas at night,

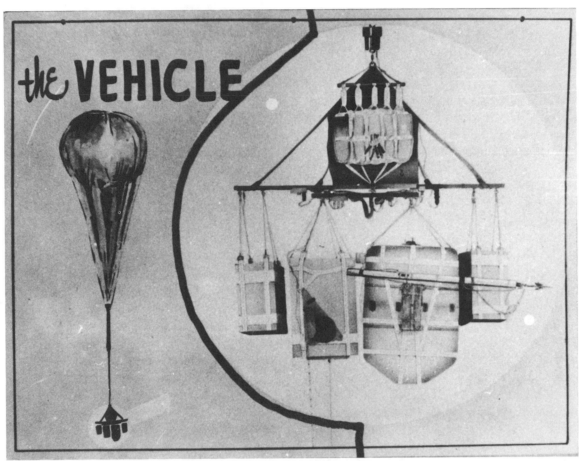

Soviet drawing of the Moby Dick/WS-119L payload. The small boxes on the ends contain iron-dust ballast, which was released through a magnetic valve to compensate for the altitude lost due to cooling and leakage. On the left center is the radio tracking beacon. It was turned on after a pre-set time (according to the launch site), which allowed ground stations to locate the balloon's position after it left Communist airspace. The large box is the camera gondola. The camera was protected from the cold by a thick layer of polystyrene foam. The camera had two 6-inch (15.2 cm) lenses and could photograph the ground 50 miles (80 km) in either direction. The exposure rate was pre-set according to predicted wind speed. The camera was turned on and off by a photo cell on the bottom of the gondola; this activated when there was sufficient light for photography. The gondola was slowly rotated by a small electric motor. After reaching friendly territory, the camera gondola was separated from the payload by a radio command and parachuted to a landing. The arrangement of the 1958 reconnaissance balloon payload was similar. *Photo via R. W. Koch.*

venting the gas as it heated up from the sun during the day. Based on the predicted wind speed, the Air Force could estimate when the balloon would reach Japan, Alaska, or the Bering Sea. Once it was spotted, a radio signal would cut the camera package free. If the C-119 missed the parachute, the gondola would float. A transmitter would signal its position for 24 hours.

After the first launches, the rate of launching was increased. During the first week, ten balloons were launched each day. Starting on 17 January, this was doubled to twenty a day. This was further increased on 25 January to thirty a day and on 28 January to forty. Both the small and the remaining large balloons were used. The initial launches apparently caught the Soviets by surprise. Soon, however, they learned how to deal with the ghostly intruders. The recovery rate began to drop. A temporary halt was ordered on 6 February 1956. The Russians protested the flights. On 9 February, the Soviets put fifty Moby Dick packages on display at a press conference. While portable generators provided light, Soviet army experts explained the equipment and showed photos—recovered from one of the balloons—of the Turkish/Russian border.

The situation started a row in Washington between Secretary of State John Foster Dulles,

A Navy Skyhook balloon being prepared for launch from the deck of an aircraft carrier. A similar approach was used for the 1958 series of reconnaissance balloon flights. The aircraft carrier was in the Bering Sea positioned so the summer jet stream would carry the balloons west across Russia, at an altitude of over 100,000 feet, to a midair recovery over Western Europe. The project was designated WS-461 L and launch was scheduled for July 4, 1958. During preparations, the cutdown timers were set for 400 hours; these would separate the gondolas. Gale-force winds forced a postponement. The three WS-461 L balloons were finally launched on July 7 but nobody had remembered to reset the timers, which had been running for some 80 hours. This caused the first and second balloons to come down inside Eastern Europe; the third balloon was found a year later in Iceland. President Eisenhower was extremely upset over the incident and cancelled all further balloon reconnaissance programs. *U.S. Navy Photo.*

who wanted the flights stopped, and the Air Force, who wanted to keep going. Dulles eventually won and the launches stopped on 1 March 1956. The U.S. continued to say the craft were only weather balloons. A nation, at that time, did not admit it spied.

The results of Moby Dick were mixed, partly for technical reasons. Once launched, the balloons were at the mercy of the winds. They might cross Russia at high speeds or could hover over one area for hours. Thus a specific target could not be photographed except by chance. The extreme conditions of high altitude also took their toll. At the low air temperatures the balloons encountered, the plastic became as brittle as thin glass. The shock from jettisoning ballast could shatter it. Subsequent analyses indicated that incorrect ballasting cut the flight time of some balloons by as much as two days, bringing them down inside Russia. The Soviet Air Force also brought some down. One wonders if any pilots became balloon ''aces.'' In addition to the fifty Moby Dick packages the Soviets put on display, others must have crashed into the trackless forests of Siberia and were never found.

In actual numbers, 516 balloons were launched. Of these, only 40 returned photos of communist territory. A total of 13,813 photos covered about 8 percent of the Russian and Chinese landmass.[20]

This total did not include one package that underwent a long odyssey. After crossing Russia, it landed in the Arctic Ocean. The package spent 2 years floating in the icy waters until it reached Iceland. There it was found by Dick Philbrick, who had worked on the project. He recognized the importance of the package and had it sent back to the U.S. The film had survived and indeed proved to provide some of the best photos of the program.

At about the same time the package was making a belated homecoming, Moby Dick was back in business. In early September 1958, the Soviets again protested the reconnaissance balloon flights. On 11 October, the wreckage of more balloons was put on display. Colonel Taransov said that the packages had been improved since the first series. The cameras could operate continuously for 18 to 20 hours. Each of the photos covered 300 to 460 square miles (800 to 1,200 square kilometers).[21] With this, the Moby Dick reconnaissance balloons passed from the scene; something better was available.

Open Skies

The major political event of this period was the July 1955 summit meeting in Geneva. It was the first such high-level gathering in 10 years. The meeting brought together President Dwight D. Eisenhower, English Prime Minister Sir Anthony Eden, French Premier Edgar Faure, and Soviet Prime Minister Nikolai Bulganin. The first 3 days went badly. In exchange for good relations, the Soviets demanded an end to the North Atlantic Treaty Organization (NATO). They also sought approval of a previously rejected plan to limit the size of each nation's armed forces and to completely ban nuclear weapons. The talks began to bog down. Then, on the fourth day, Eisenhower issued a surprise statement:

I should address myself for a moment principally to the delegates from the Soviet Union, because our two great countries admittedly possess new and terrible weapons in quantities which do give rise in other parts of the world, or reciprocally, to fears and dangers of surprise attack.

I propose, therefore, that we take a practical step, that we begin an arrangement, very quickly, as between ourselves—immediately. These steps would include:

To give each other a complete blueprint of our military establishments, from beginning to end, from one end of our countries to the other, lay out the establishments and provide the blueprints to each other.

Next, to provide within our countries facilities for aerial photography to the other country—we to provide you the facilities within our country, ample facilities for aerial reconnaissance, where you can take all the photos you choose and take them to your own country to study; you to provide exactly the same facilities for us and we to make these examinations and by this step to convince the world that we are providing as between ourselves against the possibility of great surprise attack, thus lessening danger and relaxing tension.

As Eisenhower's statement was being translated, there was a flash of lightning from the direction of Lake Geneva, and the lights went out. Even this could not compare to the drama of the moment. The Soviet delegation was stunned into silence. The Europeans were enthusiastic about "Open Skies." It was simple and direct, and it captured the imagination. Open Skies also had an element of practicality to it. To a weary and fearful Europe caught between the U.S. and Russia in a frigid cold war, Open Skies was a political and diplomatic triumph. U.S. opinion was also favorable. Here was something that could remove the great U.S. fear of the 1950s—that the Soviets, preparing in secret, could unleash a surprise attack. With U.S. aircraft roaming at will over Russia, the fear, and the chance

of a surprise attack, were substantially reduced.

The Open Skies concept had a long history. A 1946 plan for the international control of atomic energy had included the use of aerial surveillance to police the agreement. Later, the idea had been mentioned by a U.N. disarmament committee and in a plan by Secretary of State Dean Acheson for "international disclosure and verification" of the world's armed forces. Open Skies itself had been developed by Nelson Rockefeller and a team of experts on arms control and psychological warfare. In March 1955, several months before the summit, President Eisenhower had invited the men to the Quantico Marine Corp Base to come up with new ideas for a possible presentation. Open Skies was the result. Yet right up to the last minute of the Geneva Summit, Eisenhower had been undecided about presenting the concept. It was only the failure of the Soviets to offer a serious proposal that convinced him to speak. [22–24]

It was estimated by Sherman M. Fairchild, the chairman of the board of Fairchild Camera, that thirty-four RB-47s would be needed to cover all of Russia. Eighteen would be needed for communist China and one for Eastern Europe.

Eventually, the Soviets rejected Eisenhower's Open Skies proposal. Allowing U.S. aircraft to roam Soviet skies would mean Russians giving up their traditional secrecy. The Soviets considered this unacceptable. As they saw it, secrecy allowed them to have an added measure of security. An enemy could not bomb a military base if he didn't know its location or even of its existence. Open Skies gave the Soviets nothing they could not learn simply by reading American newspapers, magazines, and maps, since, as in a poker game, the West had virtually all of its cards face up. The Soviets, on the other hand, played their cards close to the vest. The Soviets would show only those cards such as bomber flyovers, military parades, or nuclear tests that enhanced the image of Soviet power—a mostly one-sided game. The concealment of Soviet strengths, and more importantly their weaknesses, was simply too critical for them to compromise. The time for Open Skies had not yet come.

In retrospect, the most interesting of the articles on Open Skies was written by Col. Richard S. Leghorn, USAF (Reserve), in *U.S. News & World Report* magazine. Leghorn argued that even if the Soviets turned down Open Skies, the U.S. could carry out overflights covertly without Soviet permission. Special reconnaissance aircraft could be designed to undertake such clandestine missions. Out of reach of Soviet defenses, the aircraft could fly with a minimal risk of detection or loss. The equipment and technology were available. The magazine in which the Open Skies article appeared was dated 5 August 1955. The day before, just such a "hand-tooled reconnaissance aircraft . . . superior to any air defense system" had made its first flight.[25]

NOTES Chapter 1

1. Foy H. Kohler, *Understanding the Russians,* (New York: Harper & Row, 1970) 99,100.
2. Harry Rositzke, *The CIA's Secret Operations,* (Pleasantville: Reader's Digest Press, 1977), 20,21,28.
3. *Khrushchev Remembers,* editor and translator, Strobe Talbott, (Boston: Little Brown and Co., 1974), 95.
4. John Barron, *KGB,* (New York: Bantam Books, 1974), 460,462.
5. Rositzke, *The CIA's Secret Operations,* Chapter 1.
6. Peer de Silva, *Sub Rosa—The CIA and the Uses of Intelligence,* (New York: Times Books, 1977), 55,61.
7. "Espionage: How the Deadly Game is Played," *Newsweek,* (11 December 1961): 43.
8. Rositzke, *The CIA's Secret Operations,* Chapter 4.
9. "Tale of a Tourist," *Newsweek,* (18 September 1961): 47.
10. William Hood, *Mole,* (New York: W. W. Norton, 1982).
11. Rositzke, *The CIA's Secret Operations,* 68,69.
12. Philip J. Klass, *Secret Sentries in Space,* (New York: Random House, 1971), 7,8.
13. William Green, *The World Guide to Combat Aircraft 2,* (Garden City: Doubleday, 1967).
14. Frederick A. Johnson, *Thundering Peacemaker,* (Tacoma: Bomber Books, 1978).
15. Stewart Alsop, *The Center,* (New York: Harper & Row, 1968), 216.
16. Brigadier General George W. Goddard and DeWitt S. Copp, *Overview,* (Garden City: Doubleday, 1969), 383,384.
17. W. W. Rostow, *Open Skies,* (Austin: University of Texas Press, 1982), 190,193.
18. Tom D. Crouch, *The Eagle Aloft—Two Centuries of the Balloon in America,* (Washington, D.C.: Smithsonian Institution Press, 1983), 644,646.
19. "How Far Up Is Home?", *Newsweek,* (20 February 1956): 22,23.
20. Crouch, *The Eagle Aloft,* 647,648.
21. Crouch, *The Eagle Aloft,* 648,649.
22. William Manchester, *The Glory and the Dream,* (New York: Bantam Books, 1974), 748,752.
23. "Trading Secrets," *Time,* (1 August 1955): 17
24. "The Periscope," *Newsweek* (1 August 1955): 11
25. Col. Richard S. Leghorn, USAF Res. "U.S. Can Photograph Russia from the Air Now," *U.S. News & World Report,* (5 August 1955), 70,73.

CHAPTER 2

U-2

His name is Clarence L. "Kelly" Johnson and he designed airplanes for Lockheed. Among that small fraternity, he had the reputation as an innovator—a designer willing to take on a challenge. He was also renowned for managerial skills that permitted development of his aircraft in a very short time and at a minimal cost.

In 1943, Johnson's team had taken only 143 days from receiving the contract for the XP-80 jet fighter until the aircraft was ready for its first flight. By the early 1950s, Kelly Johnson was director of Lockheed's Advanced Development Projects Group—better known as the "Skunk Works." At this time, while working on the F-104, Johnson became aware that the Air Force was looking into building a very high altitude reconnaissance aircraft. New jet engines matched with long-span wings offered the possibility of reaching unprecedented altitudes. The new design would correct deficiencies that caused the high loss rate of reconnaissance aircraft in Korea. The U.S. RF-80 photo plane could not outrun the Soviet MiG 15, and it was not armed to fight.[1]

The Air Force's goal was an aircraft able to fly above trouble. The specifications, issued in March 1953, were for an aircraft able to reach 70,000 feet (21,340 meters), with a range of 1,750 miles (2,800 kilometers), and able to carry a reconnaissance payload of 100 to 700 pounds (45 to 315 kilograms). Three companies—Fairchild Aircraft, Bell Aircraft, and Glenn L. Martin—were asked to make studies. The large aircraft companies, including Lockheed, were bypassed; it was felt that, in view of the limited number of planes to be built, a small company would give the project a higher priority. In early 1954, two aircraft designs were selected: a long-winged modification of the Martin B-57, to fill the short-term need; and the X-16 Bald Eagle, a large, twin-engine design by Bell. It would take several years, however, before the X-16 would be ready for test flights.[2]

Although Lockheed was not part of the study, Kelly Johnson submitted his own proposal in May 1954. The aircraft, named the CL-282, was an XF-104 with new, long wings. To save weight, a number of internal modifications were made. Instead of conventional landing gear, the plane used a dolly for ground handling and takeoff; the CL-282 would land on skids. The cockpit was not pressurized; the pilot would have to rely on a pressure suit.[3] Kelly Johnson's proposal was rejected. Two aircraft were already under development; there seemed to be no need for a third. Also, the CL-282 used a J-73 turbojet engine—deemed inferior for very high altitude

Clarence L. "Kelly" Johnson poses with a CIA-operated U-2C. Johnson was hired by Lockheed in 1933 and spent 44 years with the company. He assembled the "Skunk Works," a small group of talented engineers who operated independently of the main factory. Only 50 engineers were involved with the U-2 (some projects have larger purchasing departments); this procedure allowed rapid development at lower cost than conventional bureaucratic management. *Lockheed Photo.*

flight. Both the RB-57 and X-16 used modified J-57 engines, but the J-57 would not fit the XF-104 fuselage.

The story was not over, however. At this time, a presidential commission, headed by Dr. James Killian of the Massachusetts Institute of Technology, was looking at ways of preventing a surprise attack on the U.S. A subcommittee chaired by Edwin H. Land, of the Polaroid Corporation, dealt with intelligence matters. The subcommittee recommended that overflights of the USSR and Warsaw Pact countries be made for information on missiles, bombers, and nuclear weapons. Committee members looked at various aircraft proposals, including the rejected Fairchild and

Lockheed designs. Kelly Johnson told them that the CL-282's performance could be improved if the J-57 engine was used, and he assured them that the prototype would be ready only 8 months after the go-ahead—much faster than the X-16 Bald Eagle could be delivered. Johnson's promises impressed the subcommittee and, in November 1954, the CIA received President Eisenhower's approval to build the Lockheed aircraft.

The program was directed by Richard M. Bissell, Jr., of the CIA. The project's $22 million budget came from the CIA reserve fund. The Air Force supplied engines supposedly ordered for other aircraft projects. On 9 December 1954, Lockheed was given the order to begin work. The project was code named "Aquatone." Kelly Johnson and his engineers began putting in 100-hour weeks. The original CL-282 concept was modified somewhat. The fuselage was lengthened to carry more fuel and widened for the J-57 jet engine. The prospective altitude was increased by making the wingspan longer. The aircraft was also given a bicycle landing gear rather than skids. Outrigger "pogo" wheels were attached under the wings: on takeoff, they would drop off; at touchdown, the aircraft would balance on its main gear. Once it came to a stop, it would tip over onto one wing. The wing tips were turned down and reinforced to prevent damage. This arrangement saved a great deal of the weight inherent in a conventional tricycle landing gear. Another weight-saving effort was the elimination of an ejector seat. To bail out, the pilot had to open the canopy and go over the side. Unlike the original CL-282, the cockpit was pressurized. Everything about the aircraft was designed to be as light as possible. Each component was thinner and more delicate than its counterpart on a conventional aircraft. This single-minded concern with weight control meant the plane could reach altitudes never before achieved by an aircraft and maintain them for hours at a time. It also meant that in the event of a hit by hostile ground fire, the aircraft would come apart. There was concern that the aircraft would wear out if flown more than absolutely necessary.

Kelly Johnson's promises were kept, and the aircraft was ready for test flights by the summer of 1955. Because of the secrecy of the project, the prototype skipped the normal rollout ceremonies. Edwards Air Force Base was deemed too public for the program, so the aircraft was loaded aboard a C-124 cargo plane and flown to Groom Lake, Nevada, also

called Watertown Strip, a dry lake bed inside the Nevada nuclear test site. It was so remote, there was no danger of unexpected visitors discovering what was going on. The only access was by aircraft.

The pilot for the first flight was Tony LeVier, who had been Lockheed's chief test pilot since World War II. To cover his activities, LeVier used the name Anthony Evans (his mother's maiden name). After being reassembled, the aircraft was put through taxi and braking tests. During one braking test, the prototype lifted off and reached an altitude of 36 feet (11 meters). LeVier, unable to judge the aircraft's height above the ground, landed hard and blew both tires. Damage was minor, and new tires and larger brakes were soon fitted. Reference lines were painted on the lake bed to help the pilot determine his altitude.

On 4 August 1955, the prototype made its first flight. At the end of the flight, LeVier ran into a problem: he had been given all sorts of advice by the engineers on how to land the powered glider, but, when attempting to follow it, he found the aircraft simply refused to touch down. Due to its light wing loading, the plane floated just above the ground. LeVier kept trying to land, but he was unable to force the aircraft below a certain altitude. Finally, with fuel running low and the possibility looming of having to bail out, he made a final try. On this fifth attempt, he decided to handle the aircraft as a tail dragger—an airplane with a tail wheel. This meant stalling the aircraft just above the runway. He was successful. The aircraft thumped down safely.[4]

The landing attempt had left LeVier somewhat overwrought. As he was climbing out of the cockpit, he saw Kelly Johnson. LeVier made an obscene gesture and mumbled some-

Lockheed pilot Tony LeVier made the first test flight of the U-2 on August 4, 1955. His preparation for the U-2 tests included talking to B-47 pilots to familiarize himself with bicycle landing gear and becoming the oldest pilot (at 42) to pass Air Force pressure suit training. LeVier made 20 U-2 flights before going back to the F-104 program. *Lockheed Photo*.

thing about Kelly trying to kill him. Johnson responded with the same gesture and a loud "You Too," which the ground crew heard. Thus was born the U-2 label. Officially the "U" stands for utility, a category normally given to light aircraft in military service; but, in the U-2's case, the definition was a cover for its true purpose and the name's true origin.

In later test flights, the U-2 reached altitudes of more than 50,000 feet (15,240 meters) with ease. Kelly Johnson was on hand to greet Bob Matye when he set altitude records on three consecutive flights. Several problems appeared during the early tests. Fogging of the pilot's pressure faceplate was one.

A more serious problem was flaming out of the engine at high altitude. On his third high-altitude flight, Matye had such a flameout, which inadvertently proved the pressure-suit oxygen system and emergency-descent procedure. The flameout problem was cured by the use of a jet fuel (JP-TS), which did not evaporate at the low atmospheric pressure found at 80,000 feet (24,384 meters).

In late 1955 and early 1956, as the U-2 was undergoing flight tests, the CIA was recruiting the three groups of pilots who would make the overflights. Some of the criteria were obvious; others were more subtle. The pilots the CIA approached all had exceptional rat-

An aerial photo showing the sleek, uncluttered lines of the U-2. The U-2 was a demanding aircraft, intolerant of any errors by its pilot. At the same time, its altitude capability placed it above the reach of Soviet air defenses. *Lockheed Photo.*

ings and extensive flight time in single-engine, single-seat aircraft. They also had top-secret clearance. They were all reserve officers with indefinite enlistments, rather than regular Air Force officers, and they came from units that were being deactivated. The last two criteria were for security reasons. The resignation of a reserve officer would draw less attention than that of a regular officer. Amid the confusion of the reassignments, the disappearance of a few pilots would not be noticed. The CIA's first meeting with the individual pilot was very general—suggesting that an interest-

ing flying job was available if he wanted it. If the pilot expressed an interest, he was called to a second meeting, where the job was spelled out: he would be working for the CIA, he would be checked out in a new airplane that flew higher than any other, and he would overfly the USSR on spy missions. Those who volunteered began several months of briefings and went through a week-long physical at Lovelace Clinic in Albuquerque, New Mexico. Some of the tests the men underwent were later used on the Mercury astronauts.

After passing the physical, the pilots resigned from the Air Force, signed the 18-month contract, and were sent to Watertown Strip. Once at Watertown (which the pilots dubbed ''The Ranch''), an intensive, 3-month period of flight training and briefings on the reconnaissance equipment began.

The pilots were introduced to the special demands of flying the U-2. Although each pilot was experienced in jet aircraft, the U-2 was unlike anything they had ever flown. The takeoff roll was only a few hundred feet, even with a full load and on a hot day. As the aircraft lifted off, the pogos fell out and the U-2 began its spectacular ascent. Climbing at a 45 degree angle, it looked like a rocket. During the first few lift-offs, the pilots thought the U-2 would go right over into a loop. To prevent this, the trim was set nose down rather than nose up.

The climb of the U-2 is so steep that the aircraft can reach 25,000 feet (7,620 meters) and still be inside the boundaries of an air base. At 70,000 to 80,000 feet (21,336 to 24,384 meters), the U-2 levels off. Because of thin air, it has a very narrow speed range—the stall speed (the speed at which the wings are traveling too slowly to produce lift) is 404 knots (748 kilometers per hour). The Mach limit, which is the fastest the U-2 can fly, is about 412 knots (765 kilometers per hour). At this speed, the airflow over the wings is supersonic and generates shock waves that buffet the aircraft. If the U-2 drops below the minimum speed, it stalls, the engine flames out, and the aircraft must descend to 30,000 feet (9,144 meters) before it can restart. If the U-2 exceeds the maximum speed, the airframe will shake apart. The U-2 must be kept within the limits (called the ''coffin corner'') for 10 hours or more at a time. As fuel burns off and leaves the aircraft lighter, the U-2 will climb rather than fly faster.

Unlike conventional aircraft, the U-2 handles well at high altitudes. It was made to fly under these conditions. However, the narrow speed band in which it operates means that the aircraft must be handled with a light touch. All turns are gentle. If a tight turn is made, it is possible to have one wing tip stall and the other Mach buffet. The slightest difference in speed, because of the turning radius, is enough.

Because of the flight's precision and long duration, and the restrictions of the partial-pressure suit, the aircraft is flown by the autopilot. The pilot operates the reconnaissance systems and monitors the gauges and warning lights. On each of their training flights, the pilots broke the official world's altitude record of 65,889 feet (20,083 meters) set in a Canberra the year before. From their vantage point, the pilots could see the curvature of the earth. The sky was a deep blue-black and the sun, unfiltered by the atmosphere, was dazzling (part of the aircraft canopy was painted over as a sunshade). Below, the western U.S. spread out like a map.

Coming down from this lofty perch is a difficult operation. The pilot can't simply throttle back as on a normal aircraft. The engine would flame out. Instead, the engine bleed air valves and spoilers must be opened and the landing gear and flaps lowered to kill off lift but maintain speed during the long, slow descent. The U-2 must stay within the stall/Mach limits as long as it is at high altitudes. Once it is at lower altitudes and in thicker air, the speed band widens to normal proportions. At these lower altitudes, however, the pilot faces a new control problem. After spending several hours flying with a featherlike touch, the pilot must now manhandle the U-2, moving the control wheel through its full range of travel. Depending on its alti-

View of a U-2's instrument panel, taken in flight. In the center is the inverted periscope used to line up the photo target. The two knobs below convert it into a sextant for navigation. The glare on the instruments gives some idea of the brilliance of the sun at high altitude. The digital readout and navigation switches on the lower right were not on the original U-2. *Lockheed Photo.*

tude, the U-2 handles like two different aircraft.

During the U-2's final approach to landing, fuel is pumped from one wing to the other to balance them. Because of its long wingspan and light weight, the U-2 is greatly affected by crosswinds or turbulence. When a U-2 is 1 foot (0.3 meter) above the runway, the pilot stalls the aircraft, and it drops to a

landing. This was the procedure LeVier used to save the prototype. The last 3 seconds before landing are perhaps the most critical. If the U-2 is stalled too high above the runway, it can bounce back into the air, stall, and crash. The U-2 demands its pilot's complete attention at all times. A moment's inattention can destroy both plane and pilot. There is no margin for error.

Two of the CIA pilots were killed during training. In the first training class, a pilot was killed when a pogo wheel failed to drop off. He came back over the runway to shake it free; instead, the aircraft stalled and crashed. The other pilot, from the third training class, was apparently confused by lights at the end of the runway during a night takeoff. The U-2 flew into a telephone pole.[5]

The first CIA class completed its training in early April 1956. The 6 pilots, 4 U-2s, and 100 support personnel were designated the 1st Weather Observational Squadron (Provisional). They were sent to Lakenheath, England. The English were reluctant hosts, and the squadron moved to Wiesbaden, West Germany. The planes and pilots were ready. There remained only the political decision to set the project in motion.

Overflight

The act of flying a plane inside another nation's airspace, without permission, was provocative to say the least. Balancing this was one inescapable fact—the U-2 was the only way the U.S. could find out the things it needed to know about Soviet military strength. Agents, short-range penetrations, and the spy balloons all had failed.

The man who would have to make the final decision was President Dwight D. Eisenhower.

The CIA presented its plans for the overflight operation to the president on 31 May 1956. A month later, on 21 June, Eisenhower met with Killian and Bissell. The president agreed to the CIA's proposal. He stressed that he wanted the flights to cover the vital areas as quickly as possible. The men talked about the expected yield of intelligence, operating conditions, control and direction of the missions, and what would be done in the event of a crash. The president also stressed that an outside individual would have to be contacted before "deep operations" could begin. Presumably, this was the chancellor of West Germany, Konrad Adenauer. Bissell would contact the individual and report back.[6] This was done. On 2 July, Bissell, in his meeting with Col. Andrew Goodpaster (the president's staff assistant), said that the U-2 group was ready and eager to begin operations. In light of President Eisenhower's desire to cover high-priority areas, Colonel Goodpaster suggested a 10-day period of operations followed by a report. The next day, Colonel Goodpaster discussed his proposal with the president. Eisenhower agreed and gave the go-ahead to begin flights. Colonel Goodpaster informed Bissell of the confirmation at 2 P.M. on 3 July. Operations were to run through 14 July. Word of the go-ahead was relayed via code to Wiesbaden.[7]

From published accounts, it is possible to reconstruct the events that took place after the go-ahead reached Wiesbaden. Early the next morning, 4 July 1956, the pilot selected to make the first overflight was awakened, and he went to breakfast. Next he began the prebreathing and suiting-up process. The partial-pressure suit was very uncomfortable; under it was worn long underwear—turned inside out to prevent the suit seams from leaving welts. The suit itself was so tight that the pilot needed help to get into it. The fabric had little flexibility, which made movement difficult and even painful. In addition, the suit lacked ventilation, so the pilot became soaked with perspiration. After the suit was on, the pilot spent 2 hours breathing pure oxygen to remove the nitrogen dissolved in his bloodstream and prevent the bends in the event the cockpit depressurized. The oxygen was forced into the pilot's lungs under pressure, which reversed his normal breathing reflex; it required muscular effort to exhale. Normally, inhaling requires muscular action; exhaling is automatic. The reversed procedure was very tiring.

As the pilot prebreathed, he studied the maps. They were color coded. A blue line indicated the general flight path. Red lines indicated the target areas. These lines were to be followed as closely as possible. Also indicated on the map was when the photo or electronic equipment was to be switched on and off. Alternate targets were provided if cloud cover prevented photographing the primary target, so a last-minute change in the weather would not wipe out the flight. Brown lines showed routes back to alternate bases if Wiesbaden could not be reached. After prebreathing was completed, there was a last-minute weather briefing.

These procedures were familiar from the training flights, but for this mission, there was a difference: the intelligence officer asked the pilot if he wanted to carry a cyanide capsule. Another difference was apparent after the pilot was helped into the cockpit. On the instrument panel were two switches under safety covers. One was labeled "arm," the other "destruct." If it was necessary to bail out, the pilot was to throw the switches. The arm switch activated the self-destruct system. The second switch started a 70-second timer. At the completion of its cycle, 2.5 pounds (1.13 kilograms) of high explosives would detonate, destroying the cameras, film, and tape recorder—the proof of spying—carried only on overflights.

The pilot started the engine, checked the gauges, and, on signal, began his takeoff roll. The plane leaped into the air and was quickly lost from view. The pilot set his course east toward the unknown. After leveling off, he clicked the radio twice—the signal that everything was going as planned. Wiesbaden responded with a single click. The U-2 continued to follow the blue line on the map and crossed over into "denied territory." It is not hard to imagine the emotions the pilot felt at that moment. The act of flying the U-2, at least, kept him busy. Apprehension could be relieved by action—monitoring the engine RPMs and exhaust temperature, plotting fuel flow, and, most importantly, keeping the plane on the right side of the coffin corner.

Navigation was critical. As the plane headed deeper into Russia, the pilot had several methods of navigation. One, ironically, was provided by the Russians themselves. The U-2 would use Soviet civil radio stations for navigational fixes. The map showed each radio station's location and range. The pilot spent many hours listening to recordings of the stations to become familiar with them.

In an area free of clouds, the pilot used visual navigation. The U-2 had an inverted periscope. In the center of the instrument panel was the periscope's view screen. It allowed the pilot to watch for jet contrails from aircraft below. Although there had been intensive briefings on navigation and map study, much of the information was outdated. If the pilot saw a new city, airfield, or industrial complex, he would make a note of it on his map. During a photo run, the periscope was used as a drift sight to line up the aircraft on the proper course.[8]

The run-in began about 50 miles (81 kilometers) from a target. Corrections were made to place the U-2 directly over the target as it passed below. This was important, since the camera had only a limited field of view. The periscope was also used as a sextant for celestial navigation. It could be switched to look upward through a small bubble in front of the windshield. The U-2 pilot had an advantage often denied his lower-flying contemporaries. At 80,000 feet (24,384 meters), there were no clouds to obscure his view of the sun.

After about 3 hours of flight, the pilot caught sight of his main target—Moscow. When he completed his photo run over the city and suburbs, he turned north. The routine continued—checking the instruments, correcting the flight course, throwing the correct switches as indicated on the map, enduring the harsh glare of the sun and the discomfort of the cramped cockpit and pressure suit. One hour later, the U-2 passed over Leningrad, then turned west and flew along the Baltic coast, and from there headed back to Wiesbaden.[9]

The first flight had taken about 7 hours. The pilot was physically and emotionally exhausted. The tension that went with an overflight, the demands of flying the U-2, and the loss of perhaps a pound of body weight per hour from perspiration meant a mandatory 2 days grounding before the pilot was permitted to fly again.[10] His achievement represented a milestone. His flight marked the changeover from traditional to technical intelligence-gathering. In the end, the switch would change the world.

More overflights followed in short order. The Soviets detected the second or third mission on radar. It had been hoped the U-2 would fly too high to be picked up. By the end of the ten-day time period, five overflights had been made. The fourth and fifth overflights were made simultaneously. The Soviets publicly protested on 10 July. To avoid provoking them further, overflights stopped for several months. President Eisenhower received a full briefing. He was reportedly most impressed with pictures of the Kremlin. For several months, U-2 activities were limited to electronic surveillance flights along the border.

The second class of CIA pilots had completed their training at Watertown Strip in early August 1956. The 2nd Weather Observational Squadron (Provisional) was stationed at Incirlik Air Force Base in Adana, Turkey. One of the seven pilots was Francis Gary Powers. The unit began workup activities to become operational. The squadron's U-2s had to be reassembled and test flown. Each U-2 was handmade and had its own individual characteristics. One aircraft might be difficult to land; another might use fuel at a high rate. The pilots had to be familiar with the personality of each U-2. There were also briefings for the pilots, map study, and listening to the Soviet radio stations that were their unwitting guides.

By September 1956, electronic eavesdropping flights along the Soviet border were underway. Then in November, Eisenhower authorized the resumption of overflights. The 2nd Weather Observational Squadron's first overflight was made by Gary Powers. His aircraft crossed the border between the Caspian and Black Seas.[11] Several interesting targets were in range—Kiev, Stalingrad, naval installations on the Black Sea, and the Kasputin Yar missile test range.

By the end of the year, the first important result of the overflight program was in. The U-2 revealed that the Soviet bomber force was, in fact, a bluff. One U-2 photo showed nearly the entire Mya-4 force lined up at one airfield. The impressive flyovers had apparently been accomplished by putting every Mya-4 into the air and having them make a second pass.[12] By early 1957, the CIA had

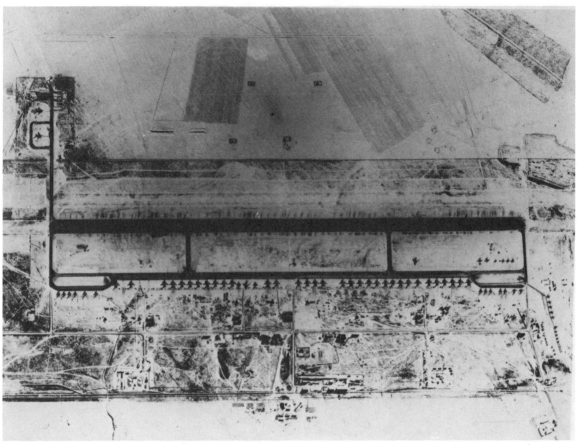

U-2 photo of a Soviet bomber base taken in 1956, which shows almost the entire production of Mya-4 jet bombers at that time. Photos such as this removed fears that the Soviets were building large numbers of bombers in the mid-1950s. The facilities at the base are rather spartan (no hangars, open air maintenance, a single runway and taxi strip). The facility at the upper left corner is probably a nuclear weapons loading/storage area. *CIA Photo.*

concluded the Soviets had only about fifty heavy bombers.[13] The U-2 had photographed enough airfields to prove there were no more. The disappearance of the "bomber gap" represented a milestone. For perhaps the first time, the U.S. had a real, solid number. Before, there had been only estimates—a range of possibilities based all too often on "five percent information and ninety-five percent construction."[14] A new confidence emerged —a glimpse of things to come.

Operations Continue

By this time, many of the program's unknowns had disappeared and the overflights achieved a certain routine. For one thing, it

U-2 photo of the SS-6 pad at Tyuratam in Soviet Central Asia, taken on a 1959 overflight. The launch pad is in the center with the blockhouse complex to the left at the end of a long ramp. The pile of dirt at the top left was removed in the digging of the flame pit. The first SS-6 test launch was made from this pad on May 15, 1957; it failed after 50 seconds. On October 4, 1957, *Sputnik 1* was launched from this pad, becoming the world's first artificial earth satellite. *CIA Photo.*

was clear that the Soviets lacked the means to shoot down the U-2s. Their two main jet fighters, the MiG 17 and MiG 19 could not come within 20,000 feet (6,096 meters) of the U-2's altitude. Moreover, the MiGs were armed only with cannons. Although large and hard-hitting weapons (a single hit would cause major damage to a jet fighter), the cannons' effective range was only about a mile. There was simply no way for the MiGs to get close enough to the U-2s to score a hit. On some early overflights, as many as thirty-five Russian fighters swarmed below the U-2 trying to reach it, so many, in fact, that Kelly Johnson would later call them "an aluminum cloud." [15] To lessen the possibility of a fighter making a visual sighting, the U-2, in 1957, was painted a sinister-appearing black to make them blend into the black sky.

The Soviets also had surface-to-air missiles (SAMs), but these too were ineffective against the U-2s. The SA-1 Guild was deployed in two rings around Moscow. [16] It was not, however, deployed in great numbers throughout the country. In any event, it lacked the performance and guidance needed to destroy a U-2.

The radar network that directed the various elements was spotty. Although Soviet radar could detect the U-2, it had difficulty plotting the aircraft's course or exact altitude.

The number of overflights was stepped up in 1957. Requests for coverage of specific targets were submitted to Bissell's Ad Hoc Requirements Committee, which would then plan the route the U-2 would take. Only occasionally were there return trips to specific targets. Most of the time, the targets were new. Known radar or SAM sites were avoided. The flight path was changed each time; a pattern was never allowed to develop. Weather conditions and the low, Arctic sun also affected mission timing. The overflights were organized, at least initially, into quarterly packages. They were submitted to the White House for approval. President Eisenhower would discuss the risks and advantages with the secretaries of state and defense. Each flight was personally approved by the president. [17]

The U-2 detachments were undergoing changes. In February 1957, the third group of pilots completed their training. The squadron was established at Atsugi near Yokohama, Japan. The first and second units were combined at Adana, Turkey. Also, use began of Operational Locations (OLs) such as Lahore and Peshawar, Pakistan. A detachment of more than twenty personnel was sent to the site. They would prepare the aircraft, make the overflight, and then pull out. Use of OLs had several advantages—the U-2s could cover a wider area than if they were tied to the main base. Also, the mountains along the southern Soviet border were less well defended.[18] There was less chance that a flight would be spotted there compared to one over the western part of Russia.

One spot in Russia was attracting an increasing amount of attention. In the barren desert just east of the Aral Sea, construction crews were building an intercontinental ballistic missile (ICBM) test center. By the end of 1956, the launch pad was completed and its checkout was underway.

On 15 May 1957 at 7 P.M. Moscow time, the first SS-6 Sapwood ICBM was launched. The Soviet rocket flew for 50 seconds before failing.[19] Several more launches were made during the spring and summer. Although they also failed, the Russians were mastering the large rocket.

The Soviet activities generated interest on the part of U.S. intelligence. The exact location of the Russian test site was unknown; pinning it down became a priority for the U-2s. The mission tracks were aligned with the main railroad lines. In the summer of 1957, a U-2 flying out of Pakistan came back with the first photos of the test site. They showed the launch pad standing at the edge of a huge flame pit. Along with analyzing the photos, it was necessary to name the facility. The practice at the time was to use the name of a nearby town. To find this, the CIA used maps of central Asia and Siberia prepared by Mil-Geo, the geographic division of the German Wehrmacht. (The fact that these maps were the best the CIA had shows just how little the West knew about Russia.) The analysts found the site of the launch pad on a 1939 map. It was at the end of a 15-mile-long (28-kilometer) rail spur leading into the desert, apparently to an open-pit mining area (probably worked by political prisoners). One of the quarries had been used as a flame pit. At the junction of the spur and the main Moscow-Tashkent rail line was the small station of Tyuratam.* Dino A. Brugioni, the CIA's chief information officer, thought this would be the best name. Others suggested Novokazalinsk or Dzhusaly. Brugioni successfully argued that these towns were too far away. So, Tyuratam it became.[20]

The missile tests had continued. On 3 August 1957, an SS-6 flew 3,500 nautical miles (6,482 kilometers). The missile's dummy warhead impacted on the Kamchatka Peninsula. On 26 August, the Soviets announced the flight. In contrast, only one U.S. Atlas ICBM had been fired, and it exploded soon after lift-off. On 7 September, a second successful SS-6 flight was made. At Tyuratam, observing the launch, was Nikita Khrushchev himself.[21]

Less than a month later, on 4 October

---

*Tjura-Tam on the map. This is the source of the hyphenated spelling one sometimes sees. The name means "arrow burial ground" in the local language.

1957, an SS-6 launched the first earth satellite. *Sputnik I* underlined the Soviets' newly acquired missile power in a way no announcement ever could. As Americans watched the satellite pass silently overhead, they felt a mixture of awe, surprise, admiration, and fear. *Sputnik*'s message was carried worldwide as it flew over the nations of the world. One month later, on 3 November, the Soviets launched *Sputnik II*. Much larger than its predecessor, *Sputnik II* carried a living passenger—a dog named Laika.

Late 1957 saw two events involving the U-2 program. At first, the overflights were viewed as a short-term effort. The pilots didn't think that the program would last the full 18 months of their contract. But by November 1957, their attitude had changed. The U-2 was proving a much more durable aircraft than originally thought. More importantly, the value of the data it was returning was apparent. It was decided to extend flight operations for another year.

The U-2 program's other development was ominous. Each November, the Soviets put on a military parade commemorating the Russian Revolution. The 1957 parade, coming as it did after the *Sputnik* launches, attracted more than the usual Western interest. The Soviets put on display a variety of new missiles—short-range artillery rockets, two battlefield support rockets, and a medium-range ballistic missile, the SS-3. One of the missiles was of particular interest to the U-2 pilots. Displayed for the first time was a new SAM— the SA-2 Guideline.

Slowdown

Soviet satellite launches and SS-6 tests continued into the new year. On 15 May 1958,

Sputnik III went into orbit. It was a 2,925-pound (1,327-kilogram) scientific satellite. The U.S. *Explorer I,* by contrast, weighed only 31 pounds (14 kilograms). Between August 1957 and late May 1958, the Soviets made six successful ICBM launches. Two additional attempts failed. The second of these failures occurred in July 1958. With this, the SS-6 tests stopped.[22]

At about the same time, the U-2 program was cut back from the high rate of 1957 and early 1958. From information available, the cutback appears to have been triggered by an overflight of the Soviet Far East on 2 March 1958. Four days later, the Soviets sent the U.S. a protest note about the flight. In a conversation with Secretary of State John Foster Dulles on 7 March, President Eisenhower indicated that he strongly believed the U-2 program should be stopped. He feared that the overflights, by miscalculation, could start a nuclear war. The Soviets could misinterpret a flight as preparations for a U.S. attack. Time and time again, Eisenhower would return to this theme—once referring to authorizing an overflight as the most "soul-searching" decision he had to make as president. In a February 1959 meeting, he commented that nothing would make him request authorization to declare war faster than violation of U.S. airspace by Soviet aircraft. Nor was this reluctance a new development. As early as 10 July 1956, during the first overflight series, Eisenhower said that he was very close to deciding not to continue the program.[23] He used a number of reasons to justify his reluctance. With development underway of reconnaissance satellites, hc felt that overflights could be kept to a minimum. Arguments that the reconnaissance satellites would not be ready until mid-1960 or early 1961 did not sway him. Eisenhower also argued that, with ICBMs, the traditional uses of intelligence (detecting preparations to attack and determining military strengths and weaknesses) were less important. ICBMs had only one method of attack—a sudden surprise knockout blow. There would be little or no warning of this type of attack.[24]

Eisenhower was in a difficult position. The U-2 was providing the U.S. with about 90 percent of its intelligence about the Soviet Union. The president did not want to make the overflights, yet the U.S. could not do without them. Because his approval was needed, Eisenhower could, however, control the number and timing of the flights. Many more were proposed than were actually flown.

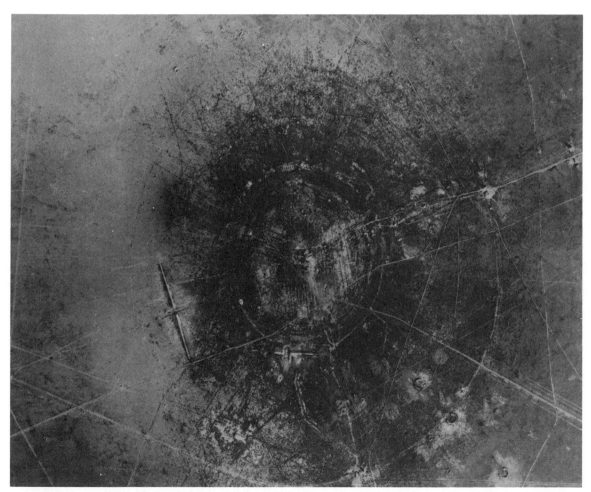

U-2 photo of a Soviet nuclear test site. Visible around Ground Zero are various foxholes, bunkers, revetments, and instrumentation. These are used to both monitor the explosion and to test the effects of the fireball, over-pressure, and radiation on military equipment. Such photos would indicate to the analyst the test goals of a series and the general sophistication of Soviet instrumentation. *CIA Photo.*

The result of the president's stance was that the U-2 overflights would remain sporadic for the remainder of the program. Eisenhower felt that, while one or two flights might be permissible, an extensive series was something he could not accept.

Despite the slowdown, the U-2s covered a number of important targets during this period. In 1958, an overflight was made of Sary Shagan—an area downrange from Tyuratam and near the western shore of Lake Balkhash. When the photos were examined, they showed that Sary Shagan was a test facility for developing antiballistic missiles. Construction on the guidance and intercept radars was already underway.[25]

Eisenhower's reluctance and concern about the overflights sometimes reached the point of indecisiveness, which was highlighted by the events of early 1959. At the

Overflight photo of a Soviet submarine base. The submarines are Whiskey- and Foxtrot-class boats meant to attack shipping rather than launch missiles. Also visible are the submarine tenders, barracks, headquarters building, and workshops. These would give the analyst an idea of the number of people assigned and the amount of maintenance and repair work the base could handle. *CIA Photo.*

start of the year, the president had approved an overflight of northern Russia. It was rated a number one priority, since the area covered by the flight was extremely important. By mid-February, however, unfavorable conditions had delayed the overflight until March, necessitating a new authorization.[26] On 4 March, Eisenhower said that he was disapproving any additional overflights.[27] Yet a month later, on 6 April 1959, he gave tentative approval for several reconnaissance flights; then, the next day, he reversed himself, having decided the political risks were too high. In fairness to Eisenhower, Secretary of State Dulles agreed with the president, saying that he would approve flights in eastern Russia, but not flights in northern or southern Russia. Eisenhower feared the terrible propaganda effect of a U-2 crash on the possibility of a summit meeting. At least there were straws in the wind that seemed to indicate that progress toward U.S.-Soviet cooperation could be made.[28]

Still, now and again, the need for information overcame Eisenhower's reluctance to approve the reconnaissance flights. The occasional U-2 overflights were being directed at suspected ICBM sites to try to determine the level of Soviet missile deployment. The question of whether to continue the overflights was becoming an increasingly controversial one as the decade of the 1960s neared.

The Missile Gap

The controversy known as the "missile gap" extended from 1957 to 1961. The gap refers to the difference between estimates of how many ICBMs the Soviets would deploy during the late 1950s and early 1960s versus how many the U.S. really would have. The question was a major one. The U.S., during this period, relied exclusively on bombers. (Both ICBMs and shorter range IRBMs were in the early test phase.) Although the U.S. bomber force was massive (214 B-52s, 1,650 B-47s, and 150 B-36s in 1957), it was concentrated at comparatively few bases, and it would take 2 hours for only 134 of the bombers to take off.[29] An ICBM would have a flight time of only 30 minutes, and the U.S. Strategic Air Command was simply not organized for such a short reaction time. If the Soviets built a massive ICBM force before the U.S. equivalent was ready, the Russians would be in a position to destroy most of the U.S. nuclear forces on the ground. The U.S. would be unable to launch a counterstrike. Its bombers would be charred rubble. Accordingly, the ability of the U.S. to deter the Soviets from making such an attack in the first place was in doubt.

The missile gap controversy pitted the U.S. Air Force against the Central Intelligence Agency (CIA). Army and Navy intelligence played more limited supporting roles. The controversy can be said to have started with the Gaither Report in late 1957. The Gaither committee (named for H. Rowan Gaither, its chairman) was originally organized in April 1957 to evaluate the feasibility of civil defense. Slowly, however, the committee broadened its examination into the vulnerability of U.S. nuclear forces. What it found was frightening. Although the CIA had predicted the Soviets would have an operational ICBM, it was the Gaither report that first brought to light how vulnerable the U.S. was to an ICBM attack—in particular, the lack of a fast reaction bomber force and the complete inability of the U.S. to detect an ICBM attack before the warheads exploded on their targets. The report itself and leaks about its conclusions generated a considerable amount of concern in Congress and with the public. The concern centered on the Soviets' production rate of the ICBMs and how soon the missiles would be operational.[30] The CIA estimated, in November 1957, that the Soviets might have perhaps 10 rockets on their pads by early 1959, based on the Soviets' pressing development at the maximum rate and there being no technical problems. The CIA further projected that the Soviets might have 100 ICBMs by early 1960 and perhaps 500 by early 1961. The Air Force estimates were for a large number of ICBMs operational by the end of 1959, 500 by mid-1960, and 1,000 by mid-1961, with this 500-per-year rate to continue into the 1960s.

The normal procedure for estimating a weapon's production rate is to identify the factories being used. Their floor space would be measured and estimates made about each shift's productivity using U.S. examples. The number of shifts (a normal 8-hour day versus a three-shift, 24-hour effort) could be determined from such factors as housing in the area, transportation, and stockpiled materials. For the missile gap controversy, this estimating procedure was not possible, since the U.S. did not know where the Soviet factories were. The Western world's lack of information did not end here. Also unknown was whether or not the Soviets had the capacity to produce the high-temperature alloys and precision electronics in the amount and on the same scale needed to support a massive effort.

Faced with a dangerous unknown, one tends to fear the worst. The fear was reinforced by the *Sputniks,* the surprise development of which showed a Western tendency to underestimate Soviet technology. Estimates became subjective and institutional demands and forces came to the forefront.

In July 1958, an added complicating factor appeared when the Soviets halted testing of the SS-6. Two interpretations emerged for the halt. The Air Force believed it meant the Soviets had solved the ICBM's technical problems and were ready to begin mass production. The estimates of 1,000 ICBMs by mid-1961 were, in their view, still valid. The CIA's estimate was different. They thought the test halt was a reflection of the technical difficulties of the SS-6. The CIA also pointed to the irregular intervals between firings as well as the variety of tests that had been made. Accordingly, the CIA pushed back their estimate of when the first Soviet ICBMs would be ready to late 1959. This was a year later than the Air Force estimate. The CIA also pushed back their projections of the total force to 100 missiles by 1961 and 500 by 1962.

The issue revolved around whether the number of Soviet launches was sufficient to discover all the bugs in the system. It had been estimated that from twenty to thirty launches would be needed to do this. If it were true, then the Soviets were only one-third of the way through the complete test cycle. In this case, the Air Force was wrong. If, however, the number of Soviet launches had been sufficient to discover all the bugs in the system, the first missiles would be ready by the end of the year.

The lack of tests made it impossible for the U.S. to resolve the question. No launches meant no new data. There was only the old information that had given rise to the question in the first place. This circular argument continued for nearly a year. The differing interpretations of the Soviet test halt were the first sign of how divergent the estimates would become.

It was the Soviets who finally resolved the question. In March 1959, SS-6 testing resumed, but with an important difference. Instead of the irregular research and development launches of 1957–1958, the new series followed a regular four-per-month pattern, indicating that the SS-6 was in the final stages of the test program. It would be only a matter of time before deployment began. By the end of 1959, U.S. estimates indicated that the Soviets might have ten missiles ready for launch. Accordingly, attention shifted in 1959–1960 to finding out the actual number of Soviet ICBMs rather than pursuing the more hypothetical estimates of future production.

The U-2 went hunting ICBMs. In addition to photographing suspected ICBM sites, they were also covering the Soviet railway system. The SS-6, with its four strap-on boosters, was simply too large to be moved by road (typically dirt and virtually impassable in bad weather). The stage sections would have to be brought to the launch site by rail. The U-2 photographs should show the stages in the process of moving from the factories to the launch sites. Even if neither was actually located, these missiles, caught in transit, would indicate the level of deployment. There was only one problem—the U-2s weren't finding any missiles: no ICBM sites were confirmed; no missiles were seen on the railroads.

Interpretation of this seeming lack of evidence spanned the complete range of possibilities. The Air Force yielded its position only slightly: they thought that the start of the Soviet missile buildup had been pushed back 6 months, but they stayed with their earlier high production rate estimate. The Air Force numbers were now 100 ICBMs in 1960, 500 by 1961, 1,000 by 1962, and a total of 1,500 by 1963. At the other end of the estimate spectrum was Army and Navy intelligence. They believed that no large-scale deployment was in fact underway or even planned. They

pointed to the numerous test failures of the SS-6. Of the fewer than twenty launches made between March 1959 and January 1960, less than half had made it to the target area. Also, Army and Navy intelligence noted the comparatively short range of the SS-6—only about 3,500 nautical miles (6,482 kilometers). Others argued that this was the maximum distance the Soviets could monitor the tests. In the Army and Navy's opinion, the lack of evidence meant a lack of missiles.

The CIA adopted a middle position. It offered two projections; the first was based on an ''orderly'' program. It assumed a production rate of 3 missiles per month, eventually working up to 15 per month, which would give the Soviets 36 ICBMs by the end of 1960 and about 400 by 1963. The CIA's other projection envisioned a crash program with production starting at 15 ICBMs per month. In this case, the Soviets would have 140 to 200 missiles by mid-1961 and 500 by 1963. (The variation is due to when production actually began.) In late 1959 to early 1960, the CIA concluded that the orderly program was the most probable, since a crash effort could not escape the notice of the U-2. The CIA acknowledged that the Army and Navy were correct about the SS-6's high failure rate. But the CIA noted that this was to be expected with the new technology. (Of the first thirty-five Atlases launched, seventeen were failures.) [31]

Although the Air Force's alarmist estimates were looking increasingly untenable as time passed, it was impossible to prove them wrong. The reason was the U-2's inherent limitations. The risk of overflights and Eisenhower's unwillingness to run them meant that few were made. Also, the aircraft's range was too short to cover some areas. Other areas, including Moscow, were photographed only once. Although large segments of rail lines had been photographed, it was not certain that they were statistically significant. Another problem was recognizing an ICBM site. The U.S. had photos of the Tyuratam test center. From them, the function of each building could be reconstructed and the general layout of the facilities mapped. The Army, Navy, and CIA all assumed that an ICBM base, once discovered, would follow the same pattern.

The Air Force argued that no particular arrangement should be assumed and that camouflage should be taken into account. ICBM sites might have been photographed, but the analysts had not recognized them as such. This is a common problem in photo interpretation. In practical terms, however, this argument meant that the Air Force would claim that any unusual construction was an ICBM site until proven otherwise. Objects in this category included ammunition storage sheds in the Urals, a Crimean War memorial, and a medieval tower. Analysts would later complain, "To the Air Force, any fly speck on a film was a missile."

The Political Dimension

The missile gap controversy was a bad situation. What made it even worse was the domestic and international political effects of the disagreement. Soviet Premier Nikita Khrushchev constantly referred to Soviet missile strength to support an aggressive foreign policy. In 1959, he claimed that mass production of ICBMs had begun. Later that same year, he said that he had visited a factory that turned out 250 missiles per year.

Space missions played a part in this psychological campaign. The *Sputniks* and *Luna* probes used the SS-6 ICBM as their booster.

Their achievements in space, greater than those of the U.S., were meant to be seen as indications of Soviet technical superiority. This effort even had a name—*Sputnik* diplomacy.[32]

Such Soviet claims of missile superiority became more frequent and strident as Khrushchev undertook his campaign to force the Western powers out of Berlin.

The missile gap controversy was also making itself felt in domestic U.S. politics. President Eisenhower steered a middle course, rejecting the alarmists' views. He stressed this cautious stance in speeches and at news conferences. Because of the emotional and political climate, however, his attitude generated a reaction opposite to the one he wished to convey. His leadership was questioned. The belief grew that he was not doing enough about defense, that, as Joseph and Stewart Alsop put it, ". . . the American government will flaccidly permit the Kremlin to gain an almost unchallenged superiority in the nuclear striking power that was once our speciality."[33] Even traditional Republican supporters, such as *Life* magazine, were becoming increasingly critical.

For the Democrats, the missile gap was a perfect symbol for their attacks on the Eisenhower administration as well as a vehicle to fulfill personal ambitions. Two individuals who made the missile gap "their" issue were Texas Senator Lyndon B. Johnson and Missouri Senator Stuart Symington (who also had been Air Force secretary in the Truman administration). Each was seeking the Democratic party's 1960 presidential nomination. Symington was very outspoken in the matter. From the evidence he had seen, he had concluded that the estimates on Soviet ICBM production were too low. When he complained to Eisenhower, he was invited to go to the CIA's

Board of National Estimates to see how they made the projections. Still not satisfied, Symington reiterated his opinion in August 1958. Three months later, the CIA issued revised estimates, based on the lack of U-2 photos of deployment. The lower projected numbers angered Symington. He felt the original estimates were too low. Now the CIA was presenting numbers that were still lower. At the January congressional hearings, Symington persistently questioned administration spokesmen about the numbers. Defense Secretary Neil McElroy was in a difficult position. He could not say why the estimate had been cut, since this would reveal the U-2 overflights.

Symington had not been told about the overflights despite his position on the Senate Armed Services Committee.[34] (He continued this position in 1960.) The estimates (again lower than the previous year), it will be remembered, came in the form of ranges for both an orderly and a crash ICBM program. To Symington, the estimates "have been juggled so the budget books may be balanced." The following week he said, " . . . if we compare the ready-to-launch missiles attributed to the Soviets on the new intelligence basis with the official readiness program for the U.S. ICBMs, the ratio for a considerable length of time will be more than 3 to 1." Senator Lyndon B. Johnson, at the same time, complained that the administration's "rosy and reassuring" picture "depended upon changing the yardstick for measuring the Soviet threat." [35] Even the approach of offering orderly and crash estimates was attacked, since the administration seemed to be saying that it knew what the Soviets intended to do.

This was all very confusing to the public. It must be remembered there were six estimates (Air Force, Army, Navy, CIA, Joint Chiefs of Staff, and State Department). All were leaked at one time or another, often without the source being labeled. There were, as well, contradictory newspaper reports. In January 1959, for instance, it was claimed that at least seven ICBM sites had been found in western Russia and five more in the Soviet Far East. Yet 2 months later, *The New York Times* reported that no ICBMs had been found.

Summits and SAMs

Another political feature of the time was a flurry of summit meetings. In September 1959, Khrushchev made a goodwill visit to the U.S. In return, he invited Eisenhower to visit Russia in the spring of 1960, after the four-power meeting in Paris. These meetings had a direct effect on the U-2 program. There had been no overflights since September 1959. Such layoffs were hard on the pilots and ground crews. With months passing between overflights, each new one was that much more uncertain. This was mixed with a feeling that time was running out.

As of early 1960, the U-2 program had been underway for nearly four years. In that time, Soviet air defenses had greatly improved.[36] MiGs were now carrying air-to-air missiles instead of cannons. The MiG 17 PFU and MiG 19 PM each carried four AA-1 Alkali missiles and associated radar. The Alkali missile was a beam rider: the fighter would hold its radar beam on the target, and the missile would be fired and would follow the beam to the target. With a range of 5 miles (8 kilometers), the Alkali posed a threat to the U-2.[37]

To counter it, the U-2 was equipped with a device called a "Granger." When the interceptor's radar locked onto the U-2, the Granger would transmit a false signal to break the lock.

Fighters ultimately proved not to be a threat. Being a first-generation effort, the AA-1 missile was limited. The interceptor's wings had to be kept level or the missile would fall out of the guidance beam.[38] Since the U-2 was far above the interceptor, such an ideal firing condition was not possible.

The Granger was a symptom of a problem with the U-2 since the first overflight. The black paint finish and added gear such as the Granger, ejector seat, and new reconnaissance equipment were increasing the U-2's weight and lowering the altitude it could reach. Since altitude meant safety for the U-2, increased weight was a serious problem. The Skunk Works made many studies of ways to increase the U-2's altitude and speed. They also examined ways of reducing radar reflectivity and infrared and radar jamming. Ultimately, only the altitude could be increased—by fitting a few U-2s with more powerful engines. A nastier problem was the SA-2 Guideline SAM. According to one account, the Soviet concentration on producing SAMs, at the expense of fighters, dated from the early probing missions in the late 1940s and early 1950s. Stalin had been advised that the best way to stop the flights would be with antiaircraft missiles.[39] The Guideline was a two-stage rocket—a booster and the missile itself. The high-explosive warhead weighed 286 pounds (130 kilograms). It was detonated by a proximity fuse when the missile passed within lethal radius of the target aircraft.[40] The SA-2 had the performance to reach 80,000 feet (24,384 meters). Moreover, the thin air at these altitudes did not slow down the high-velocity fragments, which meant that the warhead had a wide kill radius. The missile did not need to hit or even come very close to the U-2. An explosion within several hundred feet was sufficient to destroy the aircraft.[41]

Yet the Guideline had a flaw. Its ground guidance radar could not precisely control the missile, due to its high speed and the ineffectiveness of its aerodynamic control surfaces in the thin air.[42] As a result, the Guideline had only a 2 percent kill probability against a U-2 at high altitude.

As of February 1959, the Guideline was only a potential threat; no U-2 had been fired on during an overflight. The situation soon changed, however; some of the SAMs were coming dangerously close to the U-2's altitude. The pilots wondered how long it would be before the Soviets overcame their guidance problems.

President Eisenhower was also worried. In early 1960, as before, he was being pressed for more overflights. He had a new set of concerns revolving around the upcoming Paris Summit and his trip to Russia. In a meeting on 2 February 1960, Eisenhower said that his one, big advantage in a summit was his reputation for honesty. If a U-2 were to be shot down during the summit, the Soviets could display the wreckage in Moscow. The resulting publicity could ruin Eisenhower's effectiveness and cause the U.S. great embarrassment.[43] Given the president's long-standing reluctance to continue the overflights and his concerns over the Summit meeting, his next action puzzled historians and the public alike. He authorized two overflights for April 1960. The first was made on 9 April and was uneventful. The second overflight, scheduled for 28 April, was a departure from past practices. All previous overflights had taken off from and returned to the same base. This flight would go all the way across Russia. The U-2 would take off from Peshawar, Pakistan. Its route would take it over Stalinabad, the Tyuratam launch site, Chelyabinsk, and Sverdlovsk; across the Ural Mountains into European Russia; over Kirov, Archangel, Kandalaksha, and Murmansk; then out of Soviet airspace. It would land at Bodo, Norway. The planned flight path was 3,800 miles (6,114 kilometers), 2,900 miles (4,666 kilometers) of which was inside Russia. The overflight would take about 9 hours. One of the flight's prime targets was a suspected ICBM site at Plesetsk in northwest Russia. Earlier suspected sites had proven false, but Plesetsk seemed a stronger possibility.[44]

The U-2 pilot was Francis Gary Powers. He had been the backup pilot for the 9 April overflight. Powers was, by this time, one of the most experienced U-2 pilots. He and one other pilot were tied for total number of overflights. Because of the upcoming Paris Summit, a cutoff date of 2 May was applied to the flight.

Powers, his backup pilot, and his ground crew were deployed to Peshawar on 27 April. Bad weather over Russia, however, delayed the flight for 24 hours. This happened again on 29 and 30 April. Finally, on Sunday, 1 May 1960, the weather was acceptable and Powers took off. Cloud cover was a problem for the first hours of the flight. The only break was near the Aral Sea. The Tyuratam launch site was about 30 miles (48 kilometers) east. The pad area was covered by thunderheads, although the surrounding area was clear.

After Tyuratam, the clouds closed in again. They finally cleared about 50 miles (81 kilometers) south of Chelyabinsk. Then the U-2's autopilot malfunctioned. From here on, Powers would have to fly the aircraft manually—a much more difficult task. Despite the added workload, he decided to continue the flight. The weather had cleared and visibility

ahead looked excellent. The next target was Sverdlovsk. Powers made a 90 degree left turn 30 to 40 miles (48 to 64 kilometers) southeast of the town, and started his run over its southwest corner. He then noted an airfield not on the map and began writing the various instrument readings on his log. Suddenly, there was a brilliant, orange flash. It was the explosion of one or more of the fourteen Guideline SAMs the Soviets fired at the U-2. The aircraft was shoved forward and Powers felt a dull thump. The U-2 held steady for a moment; then the right stabilizer, damaged by the shock waves of the near miss, broke off. The U-2 became unstable and flipped over on its back. The wings were overstressed and broke off. The fuselage went into an inverted spin as it began the long fall towards earth. Powers, thrown around by the g forces, was unable to reach the destruct switches. He finally kicked free of the wreckage and parachuted to a landing and capture.[45–47]

The U-2 in Retrospect

In the nearly 4 years between the twin holidays of the Fourth of July 1956 and May Day 1960, about thirty overflights were made. A reasonable guess would be that two-thirds to three-fourths of them were made during the years of high activity (1956, 1957, and early 1958). Allen Dulles, CIA director during the program, would later write that the U-2 "could collect information with more speed and accuracy and dependability than could any agent on the ground. In a sense, it could be equalled only by the acquisition of technical documents directly from Soviet offices and laboratories."[48] The most important result of the U-2 program was its discovery that the

Soviets were not building large numbers of bombers. This knowledge alone had been called worth the cost of the entire program.

The importance of the overflights is not limited to this, however. Photos of the missile sites provided information about the training of launch crews for short- and intermediate-range missiles. The U-2 gave the U.S. information about the Soviet nuclear weapons program—development and testing, the location of stockpiles and storage sites, and the expansion of their facilities for producing fissionable material. The U-2 provided the U.S. with its first information on Soviet uranium mining and processing—the key element in Russia's ability to make nuclear weapons. The overflights also showed the status of Soviet air defenses, including the location of air bases and SAM sites and radars, their effectiveness, and the improvements being made. The U-2 also gave information on Russian submarine deployment, and provided a better understanding of the size and growth of the Soviet industrial base.[49]

The longest lasting result of the U-2 program was to show just how much could be learned from aerial photos. They could substitute for more traditional means of espionage. The impact of this realization extends to the present.

In the weeks and months following the 1 May overflight, however, what the U-2 had not provided was critical. It had not shown how many ICBMs the Soviets had. To settle this question, the U.S. would have to go into space.

NOTES Chapter 2

1. Larry Davis, *Air War Over Korea,* (Carrollton: Squadron/Signal Publications, 1982), 56.
2. John Sloop, *Liquid Hydrogen as a Propulsion Fuel, 1945–1959,* NASA SP-440.
3. Rene J. Francillon, *Lockheed Aircraft Since 1913,* (New York: Putnam, 1983), 473.
4. Society of Experimental Test Pilots, 1978 Report to the Aerospace Profession.
5. Francis Gary Powers and Curt Gentry, *Operation Overflight,* (New York: Holt Rinehart Winston, 1970).
6. Carrollton Press White House 333C 1979.
7. Carrollton Press White House 333D 1979.
8. Powers and Gentry, *Operation Overflight,* 30,31.
9. Lawrence Freedman, *U.S. Intelligence and the Soviet Strategic Threat,* (Boulder: Westview Press, 1977).
10. Orin Humphries, ''High Flight,'' *Airpower,* (July 1983), 10–27.
11. Powers and Gentry, *Operation Overflight,* 55.
12. Powers and Gentry, *Operation Overflight,* 57.
13. John Prados, *The Soviet Estimate,* (New York: The Dial Press, 1982), 46,47.
14. Prados, *The Soviet Estimate,* 19.
15. Clarence L. ''Kelly'' Johnson with Maggie Smith, *Kelly—More Than My Share of It All,* (Washington, D.C.: Smithsonian Institution Press, 1985), 127.
16. Andrew Brookes, *V-Force—A History of Britain's Airborne Deterrent,* (London: Jane's, 1982), 89.
17. Prados, *The Soviet Estimate,* 35.
18. Powers and Gentry, *Operation Overflight,* 61,62.
19. Ivan Borisenko and Alexander Romanov, *Where All Roads Into Space Begin,* (Moscow: Progress Publishers, 1982), 61.
20. Dino A. Brugioni, ''The Tyuratam Enigma,'' *Air Force Magazine,* (March 1984): 108–109.
21. Research Publications CIA 1544, 1984.
22. Research Publications CIA 2350 1982.
23. Carrollton Press White House 623A 1981.
24. Research Publications White House 1389 1982.
25. Prados, *The Soviet Estimate,* 152.
26. Carrollton Press White House 622A 1981.
27. Research Publications White House 2934 1982.
28. Research Publications White House 1389 1982.
29. Prados, *The Soviet Estimate,* 70,71.
30. Prados, *The Soviet Estimate,* 71,76.
31. Chronological Order of Atlas Flights (through 1976), General Dynamics Convair.
32. Philip J. Klass, *Secret Sentries in Space,* (New York: Random House, 1971), 36–44.
33. Prados, *The Soviet Estimate,* 80.
34. Klass, *Secret Sentries in Space,* 37,38.
35. Klass, *Secret Sentries in Space,* 46,47.
36. Powers and Gentry, *Operation Overflight,* 68,69.
37. Bill Gunston, *Modern Airborne Missiles,* (New York: Arco, 1983), 10,11.

38. Brookes, *V-Force,* 106.
39. Roy Medvedev, *Khrushchev,* (Garden City: Anchor Press/Doubleday, 1983), 151.
40. Bill Gunston, *Rockets & Missiles,* (New York: Salamander, 1979), 158,159.
41. Orin Humphries "High Flight," *Wings,* (June 1983): 10–31, 50–55.
42. Powers and Gentry, *Operation Overflight,* 67.
43. Carrollton Press White House 623A 1980.
44. Freedman, *U.S. Intelligence and the Soviet Strategic Threat,* 72.
45. Powers and Gentry, *Operation Overflight,* 74–84.
46. Johnson and Smith, *Kelly—More Than My Share of It All,* 128, 129.
47. Oleg Penkovskiy, *The Penkovskiy Papers,* (New York: Avon Books, 1965), 371,372.
48. Powers and Gentry, *Operation Overflight,* 58.
49. Statement by Mr. Allen W. Dulles, Director of Central Intelligence, to the Sentate Foreign Relations Committee, 31 May 1960.

CHAPTER 3

Into Space

The loss of Powers's U-2 was a devastating blow to U.S. intelligence activities. The U-2 had provided 90 percent of the information the West had on the Russians. Eisenhower knew it would be impossible to continue the overflights now that the Soviets could bring them down. The CIA U-2 units were brought home, and the number of pilots was cut from about twenty-five to seven.[1,2]

The tourist-spy operations were also hard hit. Up to this point, the KGB had turned a blind eye toward them. They had long known what the tourist-spies were doing, but they made little effort to stop them. If a tourist-spy went too far, he would be arrested and expelled. He would not, however, be imprisoned, nor would the incident be publicized. At times the treatment was gentle: a Western professor, observed by the police while photographing a military installation, was dealt with as a wayward child rather than as a spy.[3]

With the shooting down of the U-2, the situation changed. The Soviets used the incident to launch yet another antispy vigilance campaign. Travelers' luggage was checked more carefully at the border and surveillance was stepped up. Through the summer of 1960, the Soviets made almost daily attacks on the "provocative" activities of Western travelers. One American was expelled for photograph-

ing warships, another for giving a Russian a Bible.[4]

More seriously, one American was sentenced to 8 years in prison for photographing military installations in the Ukraine. Two Danes and four Germans were also convicted of spying. Their sentences of up to 15 years were more severe than that given to Powers (10 years). Their "confessions" had what *Newsweek* called "a Kafkaesque ring of self-derogation." One of the Germans called himself a "fool" and said that the Americans "are too cowardly to spy." Another was described as being caught taking pictures "with a trembling hand" and as "miserable and a nothing." The Soviets' intention was not only to convict and punish but to humiliate.[5,6]

The Soviet goal was to cut off any and all U.S. intelligence sources. The effort came at a very bad time—the missile gap controversy was becoming more heated as the projected time of maximum danger approached. The effort also came at the moment when the U.S. might actually be getting its first, hard information on missile deployment. Plesetsk seemed the best candidate so far for the first Soviet ICBM site. The loss of the U-2 meant that suspicions about the site could not be confirmed or denied. Yet the Soviets' efforts to cut off intelligence sources failed. Only 3 months after the U-2 was shot down, the CIA once more had reconnaissance photos of Russia. They had not come from the vulnerable U-2 but from an orbiting satellite.

Early Studies

The U.S. reconnaissance satellite program had its beginning in the years immediately following World War II. The V-2 had proven that a large rocket was feasible. A few far-sighted individuals saw in it the potential for an entirely new kind of reconnaissance—from space. The earliest, official U.S. study of reconnaissance satellites was made by the RAND Corporation. On 2 May 1946, the company published a 324-page report on earth satellites. In addition to scientific activities, such as the study of cosmic rays and the earth's magnetic field, the report found the satellite could perform military functions—in particular, reconnaissance, weather observations, and communications. Concerning reconnaissance, the report said that "the satellite offers an observation aircraft which cannot be brought down by an enemy who has not mastered similar techniques."

There were many problems in designing a reconnaissance satellite and the necessary boosters. The main technical one was electronics. Any system would have to rely on vacuum tubes. These were large and heavy and needed considerable electrical power. The sheer weight of the vacuum tubes and the batteries to power them would require a huge rocket with only a minimal payload.

In 1948, the situation changed. Bell Telephone Laboratories invented the transistor. Being smaller, lighter, and less power hungry than a vacuum tube, the transistor would en-

able the satellite's system weight to be reduced. In the years that followed, electronic components were further miniaturized, leading to, by 1960, the microcircuit, in which the equivalent of hundreds of transistors were fitted onto a chip less than an inch across.

Another outgrowth of solid-state electronics was the solar cell, which converted sunlight into electrical power, thus reducing the need for heavy batteries.[7]

The main problem with the reconnaissance satellite program was not technical but political. In those early, postwar years of austere defense funding, there was no money for anything as speculative as earth satellites. When, in 1948, a report by the secretary of defense briefly mentioned U.S. interest in satellites, it sparked an angry press and public outcry over the waste of taxpayers' money.[8]

Nevertheless, low-level studies continued at RAND. In April 1951, the firm's research scientists wrote a secret report entitled "Utility of a Satellite Vehicle for Reconnaissance." It was followed by a series of reports in 1952–1953 entitled "Project Feed Back." In March 1954, the reports were collected in a two-volume set.[9] To the uninitiated, all these classified reports sounded impressive. In fact, they were simply projections of what was possible. Without an active development program, the reports would remain just so much meaningless paper. And there seemed little likelihood of a program. Neither President Eisenhower nor Defense Secretary Charles E. Wilson was interested in the military uses of satellites. Indeed, within the military, "space" was a dirty word.

The real breakthrough that made reconnaissance satellites possible was in the area of nuclear weapons development. The U.S. had rejected the use of long-range missiles because atomic bombs were too large. All funding was dropped in 1947 and the money was shifted to unmanned cruise missiles. But by the early 1950s, the situation changed. The development of the hydrogen bomb meant that the ICBM could carry a warhead a thousand times more powerful than the early A-bomb. Further research indicated that the warhead could be greatly reduced in size, which made the technical problems of designing the ICBM easier.

By early 1954, various elements of the program had come together and the Atlas ICBM had received the highest national priority. It was soon followed by the Titan ICBM and Thor and Jupiter IRBMs. Although designed to deliver warheads over long distances, the missiles could also, with upper stages, orbit satellites. The power needed to boost reconnaissance satellites was at hand.

At this point, the Air Force (and the CIA) made the first serious move towards development of reconnaissance satellites. On 16 March 1955, a formal request was issued for contractor studies of a Strategic Satellite System under the designation WS-117L. Three contractors—the Martin Company, Lockheed, and RCA—made year-long studies. On 10 June 1956, Lockheed was selected as the winner. This was the beginning of Lockheed's 20-year dominance of the field. The Lockheed design used an Atlas with an upper stage propelled by a 15,000-pound-thrust (6,804-kilogram) engine. The engine was originally developed by Bell Aircraft to propel a flying bomb carried by the B-58 (this project was subsequently cancelled). Once in orbit, the upper stage would carry a payload of several hundred pounds of camera equipment. Work on the camera was conducted by Eastman Kodak and CBS Laboratories. The photos would

be developed on board, then radioed to earth. The program had the code name "Pied Piper." It was known within Lockheed as "Big Brother"—a more fitting name.

The same month Lockheed won the contract, RAND came up with another possibility for returning the photos to earth. Instead of radioing the photos, the film would be placed in a capsule with a heat shield, similar to those developed for ICBM warheads. The capsule would separate from the satellite, then retrofire and parachute to a landing.

Both methods had major technical uncertainties—in the first, designing the scanning equipment that would be able to transmit the photos quickly during a pass over a tracking station; in the second, the capsule would have to orient itself correctly, otherwise it could either reenter too steeply and burn up or be propelled into a higher orbit. Equally difficult was building a camera light enough to be carried and with sufficient resolution from the higher altitude to be useful. Another problem was stabilizing the satellite to make it an effective photo platform.[10]

Despite the need for information on Russia, work remained on a low-level, piecemeal basis. In mid-1957, Gen. Bernard A. Schriever, after considerable effort, was able to get $10 million for space activities. But, it was stressed, the funds were only for component development, not for complete systems.[11] Such a level of funding meant Pied Piper could not be launched before 1960.[12]

Eisenhower Approves

On the night of 4 October 1957, circumstances changed. With the launch of *Sputnik I,* the question became why the U.S. was not doing more in space. One beneficiary was Pied Piper. In late November 1957, funding for the project was quadrupled. In January 1958, President Eisenhower approved a major re-orientation of the program. Pied Piper could not fulfill the immediate need for a reconnaissance satellite, and the Atlas was still in the very early part of its test program. The new, near-term program, code named "Corona," would test the capsule-recovery method, which RAND felt had good prospects.[13] The booster was the Thor IRBM matched with the Lockheed upper stage, subsequently named Agena. The Thor was more advanced than the Atlas, having made its first successful full-range flight on 24 October 1957. Its drawback was that, being a smaller rocket, its payload was less. The reentry capsule was to be built by General Electric (GE).[14] The first ten to nineteen Corona satellites would test the complex launch and capsule-recovery sequence. They would not carry reconnaissance equipment.[15] The first launch was scheduled between July and October 1958, with operational status being achieved in the spring of 1959. The budget for the reconnaissance satellite program was $150 million.

As development began, work was also underway on the launch site. Cape Canaveral was poorly located for launches into the polar orbit. The launch site finally selected was on Point Arguello, 150 miles north of Los Angeles. A rocket could be fired to the south without passing over any land. A malfunctioning booster would fall into the vast Pacific. The site was a World War II army camp named Camp Cooke, subsequently renamed Vandenberg Air Force Base.

At the other end of the satellite's flight path, training exercises were also underway. Over Oahu, Hawaii, C-119J recovery aircraft flew daily practice missions. The aircraft from the 6593rd Test Squadron would catch the recovery capsule in midair as it descended under its parachute. The aircraft would fly in a circle, with a trapeze unit consisting of a pair of long, steel poles protruding from the open rear door. Strung between the poles was a nylon rope and hook. The C-119 would skim just above the capsule's parachute; then the hook would catch it and it would be reeled aboard.[16] The C-119 had about 10 minutes to catch the capsule before it drifted too low. If the capsule splashed down, recovery ships would pick it up. The trapeze unit was built by All American Engineering and was the same as that used for the midair recovery of the Moby Dick gondolas. On practice missions, the capsule would be dropped from high-flying U-2s from Edwards Air Force Base. Each U-2 carried two dummy capsules in place of the camera.[17]

Bad Streak

The first Thor Agena A was ready a year after Eisenhower's approval of the project. For the public launch activity, the project was renamed "Discoverer." To understand what happened next, one must remember that building rockets was still a very new art. The Air Force was trying to invent an entirely new technology. For the complete launch and recovery sequence to be successful, every one of several hundred thousand parts would have to work perfectly. Because they didn't, Discoverer endured the worst failures in any space program—U.S. or Russian.

The first attempt to launch *Discoverer 1* was made on 21 January 1959. During the countdown, a launch crewman inadvertently started the Agena's turbo pump. The resulting explosion damaged both the Thor and Agena

A.[18] The Agena was too badly damaged to use; the Thor was returned to Douglas Aircraft for repairs of its guidance system and the damage from the nitric acid spill from the Agena.

The next launch attempt came on 25 February; it was stopped a few seconds before liftoff.[19] The problems were corrected and, on 28 February at 1:49 P.M. Pacific standard time (PST), *Discoverer 1* was launched. Reporters watched from a press site on a dune 2 miles (3 kilometers) away. The Agena A separated and fired successfully to put the satellite into orbit. The Agena then began tumbling end over end, interfering with the radio transmissions.[20,21] For several hours after the launch, no signals were heard. Then ground stations began picking up sporadic signals, which lasted from 4 to 6 seconds at a time, during each 6-minute pass. Finally, after 4 days of uncertainty, the Air Force announced that *Discoverer 1* was in a 697- by 114-mile (1,122- by 183-kilometer) orbit. For this first attempt, no recovery capsule was carried.[22]

The Eastern bloc was less than enthusiastic. An East German radio broadcast attacked the U.S. for putting a military satellite into orbit ''without even talking to other states over whose territories the Discoverer is to perform espionage services.'' The broadcast contended that the U.S. was only hurting itself, as the Soviet Union would soon surpass the Western effort: ''United States Generals are thus, as in so many other fields, involving themselves in a race in which they should know from the start that they will be left far behind.'' [23] It was the first propaganda blast in the Soviets' verbal assault on U.S. reconnaissance satellites. The statements would become increasingly vicious over the next 4 years. (Why they stopped will be the subject of another chapter.)

Discoverer 2 was the first satellite to carry a recovery capsule. The capsule, 33 inches (84 centimeters) long and 27 inches (69 centimeters) in diameter, was covered by a heat shield and an after-body that contained the retro-rocket. These separated after reentry. The capsule's total weight was 195 pounds (88 kilograms). *Discoverer 2*'s weight in orbit was 1,600 pounds (726 kilograms), 300

The *Discoverer 2* capsule and adapter on its handling dolly. The Discoverer program was to test the feasibility of returning reconnaissance photos to earth by a reentry capsule. The program's importance increased with the missile gap and the loss of the U-2. *U.S. Air Force Photo.*

The *Discoverer 2* capsule and adapter being bolted onto the Agena upper stage. This was the first flight to carry a capsule. The adapter unit was designed to hold the camera, a modified version of the one carried by the Moby Dick reconnaissance balloon. *Discoverer 6* was believed to have been the first to carry a camera. It was not until *Discoverer 14* that photos were successfully recovered. *U.S. Air Force Photo.*

Blanket covers the *Discoverer 2* capsule and adapter in order to maintain the correct temperature before launch. It is released at liftoff. The fate of the *Discoverer 2* capsule is uncertain. Due to an error by a ground controller, it reentered over the Arctic near Spitsbergen Island. Although the parachute was allegedly sighted, the capsule was never found. It was suggested the Russians recovered it. Given the repeated failures of subsequent capsules, however, it is questionable whether it made a successful reentry landing. *U.S. Air Force Photo.*

pounds (136 kilograms) more than that of *Discoverer 1*. The Agena A was 19.6 feet (5.9 meters) long and 5 feet (1.52 meters) in diameter.

    After a 3-hour delay for fog, *Discoverer 2* was launched on 13 April 1959. The Thor fired for 2.5 minutes, then shut down. The Agena A separated and coasted briefly; then its engines ignited for 2.5 minutes. This placed *Discoverer 2* in a 225- by 152-mile (362- by 245-kilometer) orbit. The satellite then oriented itself into a tail-first position and maintained this position for the next seventeen orbits. Good signals were received by ground stations.

    In a press briefing beforehand, William H. Godel, director of planning for the Advanced Research Projects Agency, said that there was only a one-in-a-thousand chance for the capsule's successful recovery.[24,25] He was proven to be correct. Normally, the capsule was separated by a timer, which was updated by a ground signal. At a ground station at Kodiak, Alaska, a controller pushed the buttons to make the update. A monitor indicated that the command was incorrect; but, under the pressure of time, the controller forgot to press the reset button before transmitting a new command. This reset the timer for reentry over the Arctic.[26]

Tracking signals and sightings indicated that the capsule successfully separated and landed on or near the island of Spitsbergen north of Norway. Col. Charles G. "Moose" Mathison, director of the recovery effort, called Maj. Gen. Tufte Johnsen of the Norwegian Air Force Northern Command. The call set in motion a search by skiers, airplanes, and helicopters of the snow-covered hills and surrounding waters. After 6 days, the hunt was called off; nothing was found.[27,28] There was a suspicion that the Soviets, who operated coal mines on the island, may have retrieved the capsule.

The shortcomings of the Discoverer program so far were not that bad, given the primitive nature of space technology. The next flights began the bad streak that would plague the Discoverers for more than a year:

An early Discoverer booster being prepared in the pre-dawn darkness of Vandenberg AFB. The booster was a modified Thor IRBM with an Agena second stage. *U.S. Air Force Photo.*

- *Discoverer 3*   3 June 1959
  Agena failure; did not reach orbit.
- *Discoverer 4*   25 June
  Agena failure; did not reach orbit.[29]
- *Discoverer 5*   13 August
  Placed in a 454- by 135-mile (734- by 217-kilometer) orbit. Capsule separated next day, but no signals were received.*
- *Discoverer 6*   19 August
  Reached orbit successfully. Carried a new 300-pound (136-kilogram) capsule and possibly the first camera. Capsule separated the next day, but no signals were received.[31]

With five failures to recover a capsule out of five attempts, it was clear that additional work was needed. Flights were halted as the Air Force and GE began an effort to improve capsule reliability. The telemetry data indicated the capsules' batteries had become too cold, which prevented them from supplying power to the recovery beacon. The solution was to add painted patterns on the spacecraft, which would allow the capsule to absorb heat during the sunlit part of the flight, then retain it while the Discoverer was in the earth's shadow.[32]

Discoverer 7 was launched on 7 November, into a 519- by 99-mile (835- by 159-kilometer) orbit.[33] After three orbits, ground stations noted it had begun a slow tumble due to a failure in the electrical system. The failure also prevented the capsule from separating.[34]

Discoverer 8 followed on 20 November. During the launch, the Agena guidance system malfunctioned, putting the satellite into a 1,032- by 120-mile (1,661- by 193-kilometer) orbit. With the apogee twice as high as normal, the recovery plan had to be changed. Reentry on the seventeenth orbit would bring the capsule down too far west. The separation was changed to the fifteenth orbit, with a landing area only slightly southwest of the normal one. Despite the last-minute changes, separation and retrofire were normal. Telemetry, however, indicated the retro-rocket was separated late. The C-119 recovery aircraft picked up signals from the capsule beacon. They lasted only 2 minutes instead of the normal 20 or 30 minutes. The capsule, whose parachute apparently never opened, slammed into the ocean and sank.[35,36]

Modifications to the capsule continued to be made. A brilliant strobe light was fitted to make the capsule easier to see at night or in the water. Another addition was packages of chaff, which would be released during descent to provide radar targets.

In January 1960, the Air Force Satellite Control Center in Sunnyvale, California, was dedicated. From there, controllers could monitor data from the worldwide network of ground stations.[37] The Discoverer program

---

*The flight of *Discoverer 5* had an unusual postscript. On 1 February 1960, technicians at the Navy's SPASUR tracking network detected an unknown satellite. Subsequent analysis indicated it was the missing *Discoverer 5* capsule. From the orbital data, it appears the Agena A was incorrectly oriented. Normally, it traveled backwards and tilted down at 60 degrees. *Discoverer 5* orbited nose first, yawed 32 degrees to one side, and tilted up slightly. When the capsule retrofired, it was not slowed down but boosted into a higher orbit.[30]

was also extended to twenty-nine flights. This was subsequently increased to thirty-eight.

In February, flights resumed, but with no more success:

- *Discoverer 9* 4 February 1960
  Thor shut down early; Agena separated, but its burn was also short and it impacted 400 miles (644 kilometers) south of Vandenberg.
- *Discoverer 10* 19 February
  Thor went off course; it was destroyed 56 seconds after launch.[38,39]
- *Discoverer 11* 15 April
  First launch in 2 months. Capsule separated on seventeenth orbit; went into a high reentry trajectory, which prevented recovery.[40]

This latest failure sparked an intensive testing program of the recovery systems. The schedule of the next launch depended upon the study's outcome. An examination of the previous attempts indicated that when the capsule separated from the Agena, it would tumble or wobble due to the failure of the spin rockets to fire. To correct this, a system using compressed nitrogen was fitted. Drop tests of the new system were made at Holloman Air Force Base, New Mexico. A dummy capsule was carried to 100,000 feet (30,480 meters) by a Skyhook balloon. The capsule was then released and went through the retrofire; spin rocket operation; and finally, parachute deployment. An instrument package reported the results. The balloon flights began in early summer of 1960. Originally planned for nine tests, the program was extended.[41] In addition to the balloon flights, wind-tunnel tests of the parachute were made at the Arnold Engineering Center in Tennessee. Vacuum-chamber tests indicated that the retro-rocket performance became erratic due to the extreme cold of space.

Changes these tests suggested were incorporated into *Discoverer 12:* the capsule carried a special telemetry package, and the deployment of recovery forces was changed. A receiving station was set up on Christmas Island to receive signals if the capsule landed long. Added equipment was put in the Alaska and Hawaii stations. Five C-54 aircraft and three ships would provide added coverage during the capsule's return.[42] The need for Discoverer was greater now. Two weeks after *Discoverer 11*'s failure, the U-2 was shot down and the U.S. lost its prime source of intelligence.

On 29 June, *Discoverer 12* was launched. During the Agena's burn, interference from the telemetry caused a horizon scanner to fail, which made the Agena pitch down. The engine burn went as planned, but the incorrect angle caused the satellite to reenter the atmosphere and burn up.[43]

Success

The streak of launch failures and lost capsules made *Discoverer 13* sound like a bad joke. Superstition notwithstanding, the launch vehicle and worldwide network stood ready on 10 August 1960. Colonel Mathison monitored the launch from the Sunnyvale control center (known as "Moose's Mansion"). The launch was literally a last chance. Mathison already had received orders reassigning him to Washington, D.C.[44] *Discoverer 13* was launched at 1:38 P.M. Pacific daylight time (PDT) through a fog bank.[45] The satellite was placed into a 431- by 157-mile (694- by 252-kilometer) orbit, inclined 82.8 degrees. Next,

The *Discoverer 10* booster shortly before liftoff on February 19, 1960. A few seconds into the flight, the booster began to wobble, then head northwest towards the towns of Santa Maria and San Luis Obsipo. The destruct signal was transmitted 56 seconds after liftoff, and the rocket exploded into an orange fireball. Debris the size of cars rained down on Vandenberg AFB sending Air Force personnel running for cover. *U.S. Air Force Photo.*

the Agena A oriented itself into the tail-first attitude. Once the satellite's orbit was confirmed, Colonel Mathison boarded a C-130 transport for the flight to Hawaii. At Hickam Air Force Base, he monitored the deployment of the recovery fleet in preparation for the capsule's reentry the next day.

The capsule itself carried a special diagnostic package to observe its internal function, as well as the separation and recovery sequences. A five-channel telemetry unit transmitted its reading to ground stations. During the reentry blackout, the data would be recorded, then retransmitted after a 2-minute delay.

On 11 August, shortly before noon local time, the recovery force was assembled. It consisted of twenty aircraft, including the C-119 recovery aircraft, EC-121 radar aircraft, and two U-2s to track the capsule. Three ships also stood by. The recovery zone was 60 miles (97 kilometers) wide and 200 miles (322 kilometers) long.[46] Twenty-four hours and 37 minutes after the launch, *Discoverer 13*'s programmer automatically began the recovery sequence. The Agena pitched down, the explosive bolts fired, and springs pushed the capsule free. The Kodiak tracking station received signals indicating the capsule had spun up and the retro-rocket had fired. The capsule then stopped its spin and the retro was jettisoned. A few minutes later, the Hawaii station picked up signals indicating the heat shield had jettisoned and the parachute had opened. The recovery aircraft and ships picked up the capsule's beacon. A heavy layer of clouds prevented aircraft recovery. The capsule was spotted, however, as it splashed down.[47] Two C-119s circled the floating capsule as a call went out to the USNS *Haiti Victory* to send out a helicopter. Colonel Mathison flew out to the capsule in a seaplane,

but high waves prevented a landing. The helicopter arrived about 3 hours after splashdown. Its pilot, Lt. Albert C. Pospicil, flew over the strobe light and yellow-green dye marker. Bosun's Mate, Third Class Robert W. Carroll jumped into the swells, put the parachute into a bag, and attached a line to the capsule. It was winched aboard the helicopter and Carroll followed; then the helicopter flew back to the *Haiti Victory*.[48] A helicopter from Hawaii, carrying Colonel Mathison, flew out to the ship to pick up the capsule. With Colonel Mathison as escort, the capsule, its recovery beacon still beeping, was loaded aboard a C-130 for the flight back to the West Coast. The 16-month frustration of the "great capsule hunt" was over.[49]

The capsule, the first object recovered from orbit, was shown at a press conference at Andrews Air Force Base. It then went to the White House, where President Eisenhower was presented with a flag the capsule had carried. After several more stops, it was given to the Smithsonian Institute.[50]

While *Discoverer 13* was making the Washington social circuit, the Air Force made ready to launch *Discoverer 14,* which would carry a modified version of the camera used in the Moby Dick program.[51] The camera's resolution was 10 to 15 feet (3 to 5 meters)— sufficient to spot ICBM sites or airfields. The camera was fitted into the Agena just behind the capsule. The target was the suspected ICBM complex at Plesetsk. *Discoverer 14* was ready for launch on 18 August 1960. The countdown was delayed 15 minutes because the empty Agena stage from *Discoverer 13* was passing through the projected flight path.[52] It was feared its signals might confuse the tracking equipment. Launch came at 12:57 P.M. The Thor climbed from a low fog bank into a blue California sky.[53] *Discoverer 14*

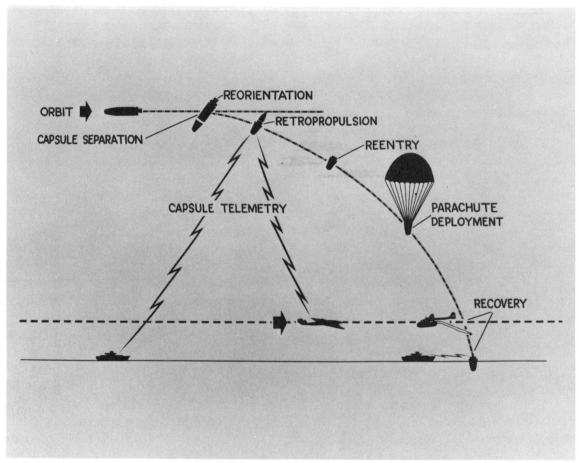

Recovery sequence for the Discoverer capsule. The process was started by either an onboard timer or a radio command from the Kodiak, Alaska, ground station. The Agena reoriented itself, and the capsule separated. The capsule was spun to stabilize it for the retro fire. Once the reentry was completed, the heat shield separated, the parachute opened, and the recovery beacon began transmitting. The nearest C-119 recovery aircraft then turned towards it. If the aircraft missed, ships made a surface recovery. Because of the number of steps and the primitive state of space technology, succcess was not achieved until *Discoverer 13. U.S. Air Force Photo.*

went into a 502- by 113-mile (808- by 182-kilometer) orbit. During the satellite's first orbit, the Kodiak station's data indicated it was in an incorrect attitude and using control fuel at a high rate. *Discoverer 14* soon stabilized and was able to make its photo run over Plesetsk. The early-afternoon launch time meant

tall objects would show a shadow. On the satellite's seventeenth orbit, the programmer began the recovery sequence. Several minutes later, one of the C-119s (call sign Pelican 9) picked up the descending capsule's radio signals. At 1:01 P.M., the aircraft turned towards it. As Capt. Harold E. Mitchell lined up on the parachute, the winch operator, T. Sgt. Louis F. Bannick, told him, "For God's sake, Captain, don't hurt it." The first pass missed by only 6 inches (15 centimeters). The aircraft did a fast turn and lined up for a second pass. It missed by 2 feet (0.6 meter). Time was running out; the cloud deck was at 8,000 feet (2,438 meters) and the capsule was only 500 feet (152 meters) above it. On the third try, the nylon cable caught the parachute. The entire process took only 13 minutes. The parachute and capsule were reeled aboard and placed into a container. In the aircraft, the mood was jubilant. Two years of practice had

The first object successfully recovered from orbit—the *Discoverer 13* capsule at Andrews AFB on August 13, 1960. The capsule was fitted inside a two-piece heat shield and retro-rocket which dropped off after reentry. At the top of the capsule is the parachute container. On *Discoverer 14* and subsequent missions that carried cameras, the film was threaded through the back of the capsule onto a take-up reel inside. When the photo run was completed, the film was cut and reeled into the capsule. The recovery sequence could then begin. *U.S. Air Force Photo*.

paid off. When the crew landed at Hickam Air Force Base, they were met by Gen. Emmett O'Donnell, Jr. and 200 Air Force personnel and their wives. Captain Mitchell was awarded a Distinguished Flying Cross; the other 9 crew members received air medals.[54] The capsule was sent immediately to the West Coast and ultimately was given to the Air Force Museum. In November 1963, the C-119 joined it in the collection.[55]

Discoverer 14's photos were sent to the CIA Photographic Intelligence Center in Washington, D.C. Although the photos were dark and of poor quality,[56] they showed rail lines that were not present on World War II maps. The question of how many ICBMs the Soviets had was still open, but, more importantly, the U.S. had resumed coverage of Russia.

The next launch was *Discoverer 15* on

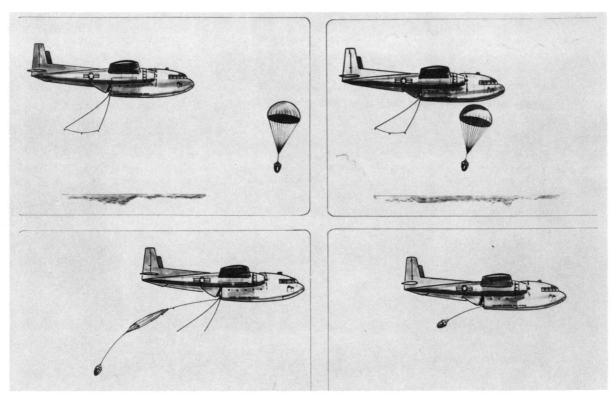

The C-119 recovery sequence. After the parachute opened, a recovery beacon aboard the Discoverer capsule began to transmit. The nearest C-119 then homed in on the descending capsule using its direction-finding equipment. Once the capsule was sighted, the C-119 turned towards it. It then skimmed just above the parachute, which was caught by lines stretched between two metal poles hanging below the open cargo door. The parachute collapsed, and the capsule was reeled aboard and placed in a container. This recovery system was originally developed for the Moby Dick reconnaissance balloon program. *U.S. Air Force Photo.*

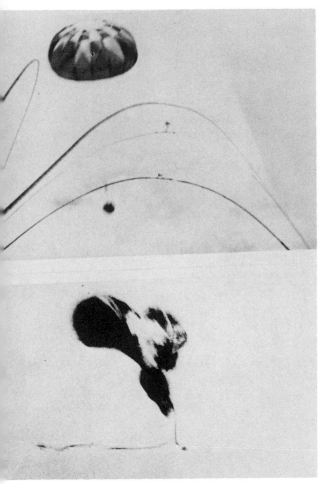
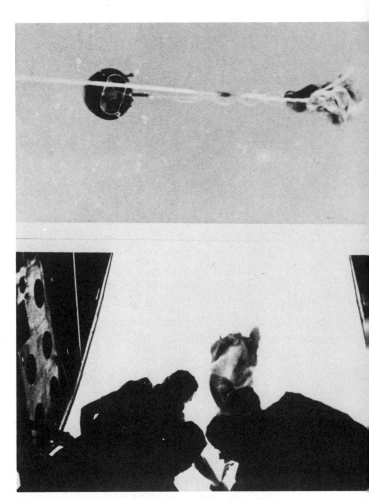

Discoverer 14: the first mid-air recovery of a film return capsule. The recovery line skims just above the parachute on the second attempt. On the third attempt the grappling hook catches the parachute, and the capsule is slowly reeled in. The crew is silhouetted in the open rear doors as capsule and parachute come aboard. The capsule contained photos of the Plesetsk area, site of a suspected Soviet ICBM base. *U.S. Air Force Photo.*

13 September 1960. The launch and orbit were normal. After the capsule was separated on the seventeenth orbit, it began to use up its control fuel at too high a rate. The capsule made a successful reentry and splashdown, but it was about 900 miles (1,448 kilometers) south of the recovery zone.[57] Search planes and a recovery ship picked up its radio signal as they headed towards it. That night, two recovery aircraft spotted the capsule's flashing

light; however, heavy rain squalls and high seas swept through the area and prevented recovery.[58]

Behind the scenes, there were important changes in the Discoverer program's management organization. The U-2 had been run by a small, informal group outside the normal bureaucratic channels, which was done in part to preserve the project's secrecy. For satellites, a more formal arrangement was needed. Originally, the authority was divided between the CIA and the Air Force; but this arrangement did not work out well and sparked several months of arguments between the White House, DOD, Air Force, and CIA.

On 25 August 1960, the National Reconnaissance Office (NRO) was formally established. The NRO was to be a national-level organization meant to eliminate uncertainty and use the satellite photos to the greatest extent possible. The NRO's first director was Under Secretary of the Air Force Dr. Joseph Charyk. The NRO was responsible for both routine control of the satellites (such as turning them on and off) and choosing the targets to be photographed. Because of the NRO's importance, its existence was secret.

The Air Force role in the Discoverer program was somewhat limited. It handled the launches, bases, and recovery of capsules. The NRO's orders were transmitted to the spacecraft through the Air Force's tracking stations. The satellite control center at Sunnyvale was the hub of the tracking, communications, and recovery network. The CIA handled research and development, as well as contracting and security.[59,60]

By the end of the summer, the U.S. had demonstrated the various elements necessary for reconnaissance satellites—launch, orbital stabilization, photography, capsule recovery, and the management organization to make use of the information. There still remained the demanding task of molding the still largely experimental satellites into a smoothly functioning, operational, intelligence-gathering system. This would take another year and still more difficulties.

NOTES Chapter 3

1. Carrollton Press White House 217A 1980.
2. Orin Humphries, ''High Flight,'' *Wings,* (June 1983): 10–31, 50–55.
3. Harry Rositzke, *The CIA's Secret Operations,* (Pleasantville: Reader's Digest Press, 1977), 60,63.
4. *New York Times,* 13 August 1960, 1.
5. ''Espionage: How the Deadly Game is Played,'' *Newsweek,* (11 December 1961): 43.
6. ''Tale of the Tourist,'' *Newsweek,* (18 September 1961): 47.
7. Philip J. Klass, *Secret Sentries in Space,* (New York: Random House, 1971), 72–76.
8. Lee D. Saegesser, ''U.S. Satellite Proposals, 1945–49,'' *Spaceflight,* (April 1977): 132–137.
9. Klass, *Secret Sentries in Space,* 76, 77.
10. Klass, *Secret Sentries in Space,* 14,15, 82–87.
11. Dr. Ernest G. Schwiebert, ''USAF and Space,'' *Air Force and Space Digest,* (May 1964): 161.
12. ''USAF Pushes Pied Piper Space Vehicle,'' *Aviation Week & Space Technology,* (14 October 1957): 26.
13. John Prados, *The Soviet Estimate,* (New York: The Dial Press, 1982), 106.
14. Klass, *Secret Sentries in Space,* 90,91.
15. Research Publications DoD 134, 1983.
16. *New York Times,* 25 January 1959, 1.
17. Jay Miller, *Lockheed U-2,* (Austin, Aerofax, 1983), 54,55.
18. ''Industry Observer,'' *Aviation Week & Space Technology,* (2 February 1959): 23.
19. ''Discoverer Aborted,'' *Aviation Week & Space Technology,* (2 March 1959): 27.
20. *New York Times,* 1 March 1959, 1.
21. *New York Times,* 2 March 1959, 1.
22. J. E. D. Davies, ''The Discoverer Programme,'' *Spaceflight,* (November 1969): 405–407.
23. *New York Times,* 6 March 1959, 10.
24. *New York Times,* 14 April 1959, 1,18.
25. *New York Times,* 15 April 1959, 1.
26. Russell Hawkes, ''USAF's Satellite Test Center Grows,'' *Aviation Week & Space Technology,* (May 30, 1960): 58.
27. *New York Times,* 18 April 1959, 1.
28. *New York Times,* 23 April 1959, 59.
29. William Roy Shelton, *American Space Exploration—The First Decade,* (Boston: Little Brown, 1967), 335.
30. Marshall Melin, ''The 'Unknown' Satellite,'' *Sky and Telescope,* (April 1960): 346.
31. Klass, *Secret Sentries in Space,* 94.
32. *New York Times,* 8 November 1959, 43.
33. Shelton, *American Space Exploration—the First Decade,* 335.
34. ''Discoverer Failure Caused by Inverter,'' *Aviation Week & Space Technology,* (16 November 1959): 33.
35. *New York Times,* 29 November 1959, 34.
36. Klass, *Secret Sentries in Space,* 95.
37. Julian Hartt, *The Mighty Thor,* (New York: Duell, Sloan and Pearce, 1961), 256.

38. Carrollton Press, DoD 288A 1977.
39. *New York Times,* 20 February 1960, 6.
40. Carrollton Press DoD 36C 1980.
41. Carrollton Press DoD 36D 1980.
42. Carrollton Press DoD 36C 1980.
43. Carrollton Press DoD 36D 1980.
44. *New York Times,* 13 August 1960, 1.
45. *New York Times,* 11 August 1960, 9.
46. *New York Times,* 12 August 1960, 1.
47. Carrollton Press DoD 36D 1980.
48. *New York Times,* 12 August 1960, 3.
49. *New York Times,* 13 August 1960, 1.
50. *New York Times,* 16 August 1960, 1.
51. W. W. Rostow, *Open Skies,* (Austin: University of Texas Press, 1982), 192, 193.
52. Carrollton Press DoD 36D 1980.
53. *New York Times,* 19 August 1960, 1.
54. *New York Times,* 20 August 1960, 1.
55. ''United States Air Force Museum and Its Aircraft,'' *Air Classics Quarterly Review,* (Winter 1975): 106.
56. Lawrence Freedman, *U.S. Intelligence and the Soviet Strategic Threat,* (Boulder: Westview Press, 1977), Chapter 4.
57. Carrollton Press DoD 36D 1980.
58. *New York Times,* 16 September 1960, 13.
59. *Washington Post,* 9 December 1973, A-1.
60. Jeffrey T. Richelson, *The U.S. Intelligence Community,* (Cambridge: Ballinger, 1985), 12–15.

CHAPTER 4

The End of the Missile Gap

With the loss of the *Discoverer 15* capsule, attention shifted to the second part of the U.S. reconnaissance satellite program—SAMOS. This was the descendant of the original Pied Piper of 1956–1957. After Eisenhower's approval of a step-up in early 1958, the Pied Piper's mission became the testing of a radio transmission reconnaissance satellite. In late 1958, the name was changed to Sentry. A year later, it was changed again to SAMOS (Satellite and Missile Observation System).[1]

The SAMOS satellite used an Agena Atlas booster. The total weight was 4,100 pounds (1,860 kilograms), about twice that of a Discoverer, with 300 to 400 pounds (136 to 181 kilograms) as the camera payload. The transmission to earth of reconnaissance photos was seen as technically more difficult than Discoverer's capsule return. Eisenhower was told, in the summer of 1959, that Sentry had not undergone a serious technical review to determine its operational feasibility, and that costs might increase.[2]

The technical details of the SAMOS camera are known because it was later used in the National Aeronautics and Space Administration's (NASA) Lunar Orbiter program.[3] It had two lenses—a short one for area surveying and a longer one for high-resolution photos. The camera used 70mm Kodak type SO-243 film, with fine grain and a resolution of 450 per millimeter. Being a slow film, it was not affected by radiation. The photos from both lenses were recorded on the same strip of film. The camera had a motion compensator to prevent blurring; it determined the spacecraft's motion over the ground and moved the film at the same angular speed. The film was developed on board using the Kodak "Bimat" process, in which a web, soaked in a developer/fixer solution, was pressed against the exposed film for 3.5 minutes. When developing was complete, the web and film were separated and the web was reeled up. The process is similar to that used in a Polaroid camera.

The Bimat process has several advantages for space use. It has only one step, so no storage tanks are needed. Each part of the film is developed by fresh chemicals and no deposits are left on the film.[4,5] After drying, the developed film is wound on a take-up reel. To be scanned, it must be rewound, which cannot be done until photographing is complete, since the Bimat web would stick to the dry, developed film. When scanning is to begin, the web is cut and wound up to clear the way. The photos are scanned in reverse order. The scanner, which uses a light beam only 5 microns in diameter, sweeps across the photo, and the light passes through the film. Its intensity varies according to the lightness or darkness of the negative. These variations are detected by a photomultiplier tube, which converts them into electrical signals. These are transmitted to earth and converted back into photos. If need be, the film can be seen several times.[6]

For use on the Lunar Orbiter, the SAMOS camera was modified to meet tighter size and weight restrictions; it weighed only 145 pounds (65 kilograms). It may have been fitted with a 40-inch (102-centimeter) focal length lens, which from 300 miles (483 kilometers) up, would have about a 20-foot (6.1-meter) resolution.[7] The SAMOS package also carried receivers to pick up Soviet radar transmissions.

SAMOS 1 was launched on 11 October 1960. The Atlas performed well but the Agena suffered a loss of control-gas pressure and did not reach orbit. The next SAMOS was scheduled for November.[8]

Discoverer Launches Resume

Discoverer 16 saw the introduction of the Agena B upper stage, which was 26.6 feet (8.1 meters) long, 7 feet (2.13 meters) longer than an Agena A. The added fuel increased the engine's burn time and payload. The Agena B's empty weight in orbit was 2,100 pounds (953 kilograms), compared to an Agena A's empty weight of only 1,700 pounds (771 kilograms). The Agena B's useful payload was nearly doubled to 1,000 pounds (454 kilograms).[9]

Discoverer 16 was launched on 26 October 1960 and was a failure. The Agena B did not separate from the Thor and both fell into the Pacific.[10]

Two weeks later was election day. By the next morning, John F. Kennedy had been

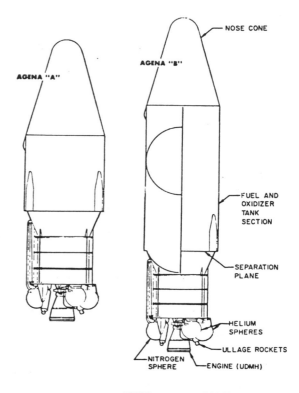

SECOND STAGE	AGENA "A"	AGENA "B"	
Weight—			
Inert	1,262	1,328	1,346
Payload equipment	497	887	915
Orbital	1,759	2,215	2,216
Impulse propellants	6,525	12,950	12,950
Other	378	511	511
TOTAL WEIGHT	8,662	15,676	15,722
Engine Model	YLR81-Ba-5	XLR81-Ba-7	XLR81-Ba-9
Thrust-lbs., vac.	15,600	15,600	16,000
Spec. Imp.-sec., vac.	277	277	290
Burn time-sec.	120	240	240

The Agena A and Agena B compared. Agena A was the first version, flown between February 1959 and January 1961. Agena B had longer tanks for a heavier payload. The first Agena B was *Discoverer 16* in October 1960, the last in 1966 (a NASA launch). Agena D was a B fitted with standard, modularized systems; Agena C was a study project that was never built. *DoD Drawing.*

elected president. One of his issues was the missile gap. At a VFW convention in August, he had said, "The missile lag looms larger and larger ahead." In September, he had urged a step-up in ICBM production.[11] Now, as president, Kennedy would have to deal with the missile gap.

In the next 2 months, as Kennedy was organizing his administration, two more Discoverers were launched. *Discoverer 17* went up on 12 November 1960. After it made thirteen orbits, the Air Force announced the flight would be extended a day.[12,13] On 14 November, while it passed over Alaska on its thirty-first orbit, the capsule separated. Ten minutes later, the crew of Pelican 2 spotted it descending by parachute. Pelican 2's pilot, Capt. Gene W. Jones, caught it on his second try at 9,500 feet (2,896 meters).[14]

Discoverer 17 was followed on 7 December by *Discoverer 18*. An improved Thor, with 10 percent more thrust, was used. This time the satellite remained in orbit nearly 3 days. Separation came on the forty-eighth orbit. The capsule was spotted at 25,000 feet (7,620 meters) and Pelican 3 turned toward it. The pilot, again Captain Jones, lined up on the parachute and caught it on the first try. Ironically, both times Captain Jones was a substitute pilot replacing those who were in school or ill.[15,16]

This brought U.S. reconnaissance satellite activity for the year to a close.* The *Discoverer 17* and *18* photos were disappointing; both sets were poor, apparently due to cloud cover.[17]

Kennedy and Reconnaissance Satellites

On 20 January 1961, John F. Kennedy was inaugurated as the thirty-fifth president of the United States. Immediately, he ordered major changes in the public affairs handling of the reconnaissance satellite program. Up to this point, there had been launch coverage and press conferences with the recovery aircrews. Now, there was to be a conscious effort to cut off all information about military space activities. The press plan for the *SAMOS 2* launch included only 1 day of advance notification rather than 5 days, no press briefing, and a severe cutback in information. This was only the beginning.[18]

The preparation for the *SAMOS 2* launch had, in the meantime, slipped into December and then into late January 1961. There had been two Atlas Agena failures, and the checkout of the satellite apparently took more time than expected.

On 31 January 1961, *SAMOS 2* stood ready. Technical and weather problems delayed it 2.5 hours. Lift-off came at 12:23 P.M. PST. The months of work paid off, and the satellite went into a 343- by 295-mile (552- by 475-kilometer) orbit, inclined 95 degrees.[19,20] The next several days would be taken up with photographing suspected ICBM sites and other areas of interest. Once photographing had been completed, the readout could begin, but it would take time. The Lunar Orbiter took 40 minutes to transmit one frame to earth.[21] *SAMOS 2* could probably transmit at a higher rate, but the number of photos

---

Discoverer 19 (20 December) tested infrared sensors in support of the Midas early-warning satellite rather than carrying a capsule.

Launch of the *SAMOS 2* from Vandenberg AFB. SAMOS was built to test the feasibility of radio transmission of reconnaissance photos to earth from orbiting satellites. The competing system was the recovery capsule used by Discoverer. The SAMOS program met with mixed results: of five launches, only two reached orbit. Additionally, *SAMOS 2*'s photos were poor quality. The radio transmission of photos would not be used again until the third-generation reconnaissance satellites. The large payload of SAMOS required use of the Atlas Agena booster, then the largest operational U.S. launch vehicle. *U.S. Air Force Photo*.

radioed down in a single, 10-minute pass was limited.

SAMOS 2 transmitted for nearly a month before it was shut down. Several months would be needed to fully analyze the pictures, which like previous ones, it is understood, were poor.[22] The photos could, however, be processed by computer to remove noise and other transmission errors, as well as improve contrast.

The next several months were eventful ones. The problem, as it had been since World War II, was Berlin. Khrushchev called it "a bone stuck in the throat." Like Stalin, he found the Western presence in Berlin unacceptable. Khrushchev meant to remove it. He had made his first attempt with Eisenhower. Backed by Soviet missile leadership, he demanded Berlin become a demilitarized "free city" and gave a 6-month deadline, which had been dropped when Khrushchev met Eisenhower in 1959. Now with Kennedy, he raised it again and with more vigor. In January 1961, he threatened to sign a separate peace treaty with East Germany. If the West tried to hold West Berlin, it would mean war.[23]

Soviet space achievements were to play a major role in the pressure campaign. In a meeting during late March, Khrushchev told East German leader Walter Ulbricht that Russia would soon put a man into space. This would place the Soviet Union in a position of unchallenged strength and prestige. The West then would be forced to agree to Soviet terms.[24] The first part of Khrushchev's plan was accomplished on 12 April, when Yuriy Gagarin made one orbit of the earth in *Vostok 1.* Five days later a brigade of Cuban exiles

landed at the Bay of Pigs. Seventy-two hours later, they were either dead or prisoners. Their defeat had far-reaching consequences: Khrushchev saw Kennedy's inaction as a sign of vacillation and weakness and began to think of ways both to advance Soviet interests and subject Kennedy to a test of strength.[25]

The test came at the Vienna Summit on 3 and 4 June 1961. In their private meetings, Khrushchev was rude and abusive and seemed almost ready to lunge at Kennedy. He gave Kennedy a virtual ultimatum. Russia would sign a unilateral peace treaty with Germany by 31 December 1961. On this, Khrushchev was adamant. The Berlin crisis had begun in earnest.[26]

After the dual successes at the end of the year, the Discoverer program went through another bad streak:

- *Discoverer 20*   17 February 1961
  Reached orbit, but stabilization failed.[27]
- *Discoverer 22** *   30 March
  Agena failed; did not reach orbit.[28]
- *Discoverer 23*   8 April
  Reached orbit, but began to tumble. Recovery attempted after 2 days but capsule went into higher orbit.[29,30]
- *Discoverer 24*   8 June
  Agena malfunctioned and impacted 1,000 miles (1,609 kilometers) downrange.[31]

The launch of *Discoverer 24* saw a singular act of courage. At T minus 30 seconds, a light in the blockhouse indicated one of six pins on the launch stand had not retracted, which had to occur before lift-off. A hold was called and two cars drove out to the pad.

---

*Discoverer 21 (18 February) carried infrared sensors for the Midas early-warning satellite program.

To the astonishment of watching newsmen, they stopped and five men got out. Two of them, Warren Dolezal of Douglas Aircraft and M. Sgt. Robert Gillham, walked to the launch stand and began removing the malfunctioning switch. Looming above them was the Thor booster and its load of highly explosive propellent—hissing and fuming. The men completed their task and left safely. Their heroism was in vain, however.

At the same time, U.S. intelligence was making another estimate of Soviet missile strength. In early June, just after Kennedy returned from Vienna, it was completed. The CIA believed the Soviets had 50 to 100 ICBMs; other estimates ranged from 10 or fewer (Navy) to 300 (Air Force). The CIA estimate was half of that it had made in 1960.[32] Although the satellite photos had not pinned down the number of missiles, they had, like those of the U-2, narrowed the range of possibilities. The Berlin crisis gave the differences greater importance. If the Air Force was correct, the Soviets had enough missiles to destroy the U.S. If the Navy or low end of the CIA estimates were right, the Soviets lacked the means to press their demands.

In the wake of Vienna and the Berlin ultimatum, Khrushchev and other Soviet officials made hard-line speeches. On 8 July, Khrushchev announced a one-third increase in Soviet defense spending. In response, Kennedy ordered a review of U.S. military strength and met with the Joint Chiefs of Staff. On 18 July, the U.S. formally warned the Soviets it would defend its rights in Berlin. On 25 July, Kennedy made a television address, saying, "If war breaks out, it will have started in Moscow and not in Berlin. Only the Soviet government can use the Berlin frontier as a pretext for war." He announced an expansion of U.S. military forces, a $3.5 billion budget increase, and a larger civil defense and bomb shelter effort. Kennedy combined this with a willingness to "see genuine understanding, not concession of our rights." [33,34]

As the crisis built, refugees poured out of East Germany into West Berlin—more than 30,000 in July and another 20,000 in the first 12 days of August. Many of them were doctors, technicians, and skilled workers.[35] On 6 August, the world received another demonstration of Soviet rocket power. Maj. Gherman Titov was launched in *Vostok 2*. He remained in space a full day before returning to earth. His flight was marked on 9 August by a Kremlin parade and reception. Here the relationship between Soviet space activities and Berlin were underlined. In his speech, Titov said that Russia had the power to "crush an aggressor should the enemies of peace launch another war." Khrushchev, at the reception, spoke of 100-megaton warheads and the rockets to carry them. A Paris newspaper headlined his speech, "Khrushchev—'I have enough missiles to crush the entire world!'." [36]

In the days ahead, Russia showed it could do more than threaten and bluster. A half hour after midnight on 13 August, East German troops began stringing barbed wire along the border between East and West Berlin. Next came brick, and 4 days later, the Berlin Wall was in place.[37] The wall both halted the flood of refugees and turned West Berlin into an isolated enclave under seige. The West seemed helpless in the face of this new challenge. No attempt was made to tear down the wall. The U.S. could only issue a "solemn warning," move 1,500 more troops into the city, and send Vice President Johnson to visit.[38]

East Germany and Russia were unimpressed and continued to tighten their grip.

Only one crossing point remained open to foreigners. West Berliners were ordered, by the East German authorities, to remain 330 feet (100 meters) away from the Wall.[39] The Soviets next demanded that the U.S., Britain, and France prevent West German leaders from flying into West Berlin. As Berlin's isolation increased, the Soviets were preparing another demonstration. On 30 August, they announced they were breaking the 3-year moratorium on nuclear testing. Bombs of 20 to 100 megatons would be exploded. The Soviets blamed the West's "threatening attitude" for the action. Khrushchev said he intended to shock the West into negotiating on Berlin and disarmament.[40] Within 24 hours, the first weapon, a 150-kiloton device, was exploded into the atmosphere. The U.S. and Britain issued a joint declaration, calling on the Soviets to halt their testing. The Soviets contemptuously rejected it and continued the series. Kennedy, in response, reluctantly approved underground testing. Explosions in the atmosphere were put off for the moment by the U.S.

On 8 and 9 September 1961, Khrushchev gave his terms for ending the Berlin crisis and nuclear testing—Western acceptance of a German peace treaty and complete disarmament. On 10 September, the Soviets increased the pressure by announcing they would begin testing long-range ICBMs into the Pacific. The first launch, which was made on 13 September, flew 7,500 miles (12,070 kilometers) downrange and impacted within half a mile of the target. The flights coincided with two nuclear tests. On 18 September, another successful launch was made. The same day, the Soviets detonated a 1-megaton weapon on the Arctic island of Novaya Zemlya.[41] Khrushchev was testing not only weapons but also the West's resolve. The activities were orchestrated to generate the maximum fear, to intimidate and break the will of the U.S. and Western Europe. Things seemed to be going Khrushchev's way. Gagarin, the Bay of Pigs, the Berlin Wall, and now nuclear tests were seen as Soviet victories. The West seemed on the run.[42]

As the 31 December deadline approached, the Soviet superiority in ICBMs offered the West few alternatives—the political humiliation of surrendering West Berlin and its inhabitants or the risk of nuclear war. That was the stark choice the West faced. In mid-September, Khrushchev seemed to be in the driver's seat.

The Missile Gap Closes

That summer, as war loomed over Berlin, the Discoverer program had been busy. The targets were suspected ICBM sites. *Discoverer 25* (16 June) broke the string of failures. It went into a 251- by 139-mile (404- by 224-kilometer) polar orbit and remained aloft for thirty-three orbits before the capsule was separated. The capsule reentered successfully but landed north of the recovery zone. One of the C-119s spotted it and radioed for a team of divers. Three frogmen parachuted into the ocean and secured the capsule on a twenty-man life raft. The three then waited out the night until they could be picked up by a destroyer.[43] The three divers, from the 76th Air Rescue Squadron, were T. Sgt. Leote M. Vigare, S. Sgt. William V. Vargas, and S. Sgt. Ray E. McClure; all received air medals.[44] The photos were of high quality but showed no missile sites.

Three weeks later, on 7 July, *Discoverer 26* was launched. It orbited thirty-two times before the capsule was commanded by the

A JC-130 aircraft with the recovery equipment deployed. JC-130 began replacing C-119 recovery aircraft toward the end of the Discoverer program. *Left:* The poles, lines, and winch are visible hanging out the open rear cargo door. The aircraft has cameras on each wing tip and under the tail to photograph the recovery sequence. *Right:* A capsule's eye view of a JC-130. The parachute of a descending capsule was caught by ropes stretched between two metal poles; the parachute was then collapsed and reeled aboard. *U.S. Air Force Photo.*

Kodiak ground station to reenter. It was caught on the first try by a C-119 flown by Capt. Jack R. Wilson.[45,46] Its photos also showed no missiles. While the satellite was in orbit, Khrushchev announced an increase in the Soviet military budget.

Exactly 2 weeks after *Discoverer 26*'s launch, *Discoverer 27* went up. Eighty seconds after lift-off, the booster malfunctioned and the range safety officer destroyed the missile.[47] The next attempt was equally unsuccessful. *Discoverer 28* (3 August) lifted off successfully, but the Agena failed and the satellite did not reach orbit.[48] The next mission, *Discoverer 29,* was launched on 30 August, the same day the Soviets announced they were breaking the nuclear test moratorium. *Discoverer 29* went into a 345- by 140-mile (555- by 225-kilometer) orbit.[49] As it passed over Russia, the camera photographed the Plesetsk area. *Discoverer 29* stayed aloft for thirty-three orbits. On 1 September, it reentered but landed outside the recovery zone. Three divers jumped into the Pacific and 40 minutes later had the capsule secured in a raft. They were picked up by the USS *Epperson.*[50,51]

The photos showed that Plesetsk was an ICBM facility. There were two above-ground launch pads, laid out the same as the pad at Tyuratam. Each pad had a SS-6 Sapwood ICBM plus a reload missile. Each carried a 5-megaton warhead.[52] After a 3-year search, the first Soviet ICBMs finally had been found.

Two Discoverers were launched in September. *Discoverer 30* (12 September) was successfully orbited; after thirty-three revolutions, it was brought down.[53] The recovery aircraft was a JC-130A flown by Capt. Warren C. Schenstad.[54] Five JC-130As had replaced the nine C-119Js previously used by the 6593rd Test Squadron. The first had been delivered in January with the conversion being completed in the spring.[55]

Discoverer 31 was launched on 17 September. It reached orbit, but the capsule did not separate and radio contact was soon lost.[56] On 9 September, an attempt to launch *SAMOS 3* was made. The Atlas Agena B exploded on the pad.[57]

By mid-September, the U.S. had recovered four capsules containing photos of suspected ICBM sites. Only those of *Discoverer 29* showed any missiles. Because the satellites had surveyed the whole of the Soviet land mass, the lack of missiles now could be proven. The four missiles at Plesetsk were the only ICBMs the Soviets had. The rest of the Soviet nuclear forces were made up of about 190 Mya-4 and TU-95 bombers and roughly 30 Zulu-, Gulf-, and Hotel-class ballistic missile submarines carrying about 90 missiles. The U.S., in contrast, had 78 Atlas and Titan ICBMs, 80 Polaris missiles aboard 5 submarines, and more than 1,500 B-52, B-47, and B-58 strategic bombers—in all about 2,260 nuclear weapons.[58,59]

As U.S. intelligence reconstructed the events later, the shortcomings of the SS-6 had caused Khrushchev to deploy only a token number in 1960. Instead, he decided to press development of the second-generation missiles—the SS-7 and SS-8. Flight tests had begun in early 1961. They were intensive, but initially had little success. Until they were ready, the Soviets cultivated a propaganda image of great missile strength, citing space activities to support their claim. The Soviets used this perceived superiority in missiles to back up an aggressive foreign policy.[60] In effect, Khrushchev was bluffing—relying on the inability of U.S. intelligence to discover just how few missiles the Soviets really had.

The satellite photos were confirmed by a mole—Col. Oleg Penkovskiy of the GRU, Soviet military intelligence. In late 1960, he made several attempts to contact Western intelligence. His approaches, coming at the height of the post–U-2 spy scare campaign, were believed to be provocations and were refused. Finally, on 10 March 1961, he was able to slip a letter and some papers to Greville Wynne, the head of a British trade delegation. The letter said he would be coming to London in a trade group. On 20 April, he met in a London hotel with two British and two American intelligence agents. Penkovskiy told them he feared that Khrushchev was intent on unleashing a nuclear war. He gave them details on Soviet missiles. At this and other meetings, he indicated the Soviets had only single missiles that were little more than prototypes, so it was not possible for them to launch a mass missile attack to destroy specific targets. He also said that a General Grigoryev was commander of a strategic missile brigade in the far North. The brigade had two pads that were able to fire one missile each per day.[61] (The facility, in retrospect, sounds like Plesetsk.) Taken together, the satellite photos and Penkovskiy's information both confirmed and explained the other's meaning.

The first public word that the missile gap was over came in a Joseph Alsop column in the 25 September 1961 *Washington Post*. He wrote that estimates had been cut from 200 to "well under 50 ICBMs." This was not nearly enough for the Soviets to contemplate a surprise attack. He noted that the lower estimates had been accepted by the U.S. Air Force air staff, signifying that the new numbers were "realistic." The Alsop column ended, "For the present, what is chiefly important is to make Khrushchev understand that if he pushes the Berlin crisis to the ultimate crunch, the great power that the U.S. possesses will not remain unused." Alsop had excellent defense sources and was close to President Kennedy.[62]

Khrushchev was quick to get the message. At the end of the month, he met with Belgium diplomat Paul-Henri Spaak. Khrushchev told him, "I realize that contrary to what I have hoped, the Western Powers will not sign the peace treaty. . . . I'm not trying to put you in an impossible situation; I know very well that you can't let yourself be stepped on." Finally, he said the key words, "I am not bound by any deadline." On 17 October, at the Twenty-second Party Congress, Khrushchev made it official: "The Western Powers are showing some understanding of the situation and are inclined to seek a solution. . . . If that is so, we shall not insist on signing a peace treaty absolutely before December 31, 1961."[63]

And so a new era dawned. At the Geneva Summit, Eisenhower had seen how important secrecy was to the Soviets. During an informal get-together, Khrushchev told Eisenhower he did not like the Open Skies proposal. He called it an espionage ploy against Russia. Eisenhower countered that it was to the advantage of the Soviets, who had repeatedly accused NATO of plotting aggressive war against Russia. Open Skies would allow the Soviets to keep watch on any NATO preparations. The West asked only reciprocal rights.

Khrushchev would have none of it.[64] Russia was to remain a closed society—its military, economic, and political activities concealed behind a wall of silence. This was Khrushchev's greatest weapon in the missile gap years. It gave him the ability to use bluff. His claims of power became substitutes for real power—deception akin to a poker player trying to win with a bad hand. The Discoverers had called Khrushchev's bluff. Once they had, Khrushchev learned, as all card players must, there was nothing left to do but fold. No longer would the U.S. have to operate in darkness, not knowing the strength of the enemy. With reconnaissance satellites, the U.S. could act with a measure of knowledge rather than from fear and ignorance. As a result, the U.S. was able to avoid both war over Berlin and an abject retreat. By providing an option other than the two extremes, reconnaissance satellites became a stabilizing influence.

Secrecy Descends

During the summer and fall of 1961, the Kennedy administration gradually cut back on the release of information on reconnaissance satellites. Soon after the end of the Berlin crisis, a curtain of secrecy descended. The November 1961 secrecy order banned the release of virtually all information. There would be no press coverage, no orbital characteristics, even a refusal to acknowledge there were any such things as reconnaissance satellites.[65] One reason for the order was to prevent the Russians from learning anything that might enable them to counter the satellites or lessen their effectiveness. The satellites were simply too important.[66]

Another reason for the secrecy was political. Constant reports of satellite achievements would underline how damaging they were to Russia, which might cause the Soviets to shoot them down.

There was no mistaking the rage the Soviets felt. In November, one Soviet newspaper accused the U.S. of "littering the cosmos," disregarding the needs of other countries, and intensifying the arms race. The accusations continued in a near-hysterical tone: "Actually, this is banditry on an international scale. . . . It should be dealt with as humanity has always dealt with this vice in all countries. Banditry should be outlawed." [67]

Because of the secrecy order, the name SAMOS became officially nonexistent. Subsequent launches would be unidentified. Discoverer would also be cast into the netherworld. A few more flights would be made before it too became unidentified.

Two SAMOS launches were made under the "unidentified" label. The first occurred on 22 November 1961. Twenty minutes after lift-off, the Air Force announced, "A satellite employing an Atlas Agena B booster combination was successfully launched today. The satellite is carrying a number of classified test components." [68] The SAMOS did not reach orbit, however.

The second unidentified launch, and last

SAMOS mission, was made exactly 1 month later. It went into a 467- by 145-mile (752- by 233-kilometer) orbit. The low point of its orbit was only half that of *SAMOS 2,* which would double the resolution of the camera. The larger Agena B would allow a heavier load of film and batteries. The satellite could make the first complete survey of the Soviet landmass since *SAMOS 2* the year before. To prolong its lifetime, the Agena engine was restarted to boost its orbit. The satellite remained in orbit until 14 August 1962, when it decayed.[69,70] This brought the SAMOS program to a close.[*]

The Discoverer program was also busy: *Discoverer 32* was launched on 13 October 1961. After eighteen orbits, the capsule reentered and was caught by a JC-130 flown by Captain Schenstad (who had recovered *Discoverer 30*'s capsule).[71,72]

The remainder of the first phase of the Discoverer program went as follows:

- *Discoverer 33* 23 October 1961
  Agena shut down early.[73]
- *Discoverer 34* 5 November
  Reached orbit, then Agena failed. No recovery attempted.[74,75]
- *Discoverer 35* 15 November
  Remained aloft for eighteen orbits. Capsule recovered in midair above Tern Island.[76]
- *Discoverer 36* 12 December
  Orbited for 4 days. Recovery on the sixty-fourth orbit. Capsule recovered from the ocean by the destroyer *Renshaw.*[77,78]
- *Discoverer 37* 13 January 1962
  Launch failure; Agena malfunction.[79]
- *Discoverer 38* 27 February
  Capsule reentered after a record sixty-five orbits. Caught in midair by a JC-130 flown by Captain Wilson.[80]

A Summing Up

The results of the first-generation reconnaissance satellites were mixed. The first third of the Discoverer program was a disaster. Yet once the bugs had been shaken out, it did not do all that badly. Of the twenty-six launches between *Discoverer 13* and the end of the first phase, seven failed to orbit. Of the remaining seventeen that carried capsules, twelve were successfully recovered (eight in midair, four from the ocean).[81]

SAMOS was less encouraging. Out of five tries, only two reached orbit. Additionally, *SAMOS 2*'s photos are known to have been poor.

Such accounting tells only part of the story, however. The satellites' purpose was to return information on Russia, for which they were tremendously successful. They exposed Khrushchev's missile bluff in time to defuse the Berlin crisis, which made all the effort, pain, and struggle worthwhile.

[*] In 1962, the policy changed slightly. The satellites remained unidentified, but the U.S. made regular reports of their launches to the United Nations. These included orbital characteristics and period. The only real change was to acknowledge launch failures.

NOTES Chapter 4

1. Philip J. Klass, *Secret Sentries in Space,* (New York: Random House, 1971), 91,101.
2. John Prados, *The Soviet Estimate,* (New York: The Dial Press, 1982), 108.
3. Klass, *Secret Sentries in Space,* 133,134.
4. *Catalog of Lunar Mission Data,* Goddard Spaceflight Center, (July 1977), 41,42.
5. Bruce K. Byers, *Destination Moon: A History of the Lunar Orbiter Program,* NASA TM X-3487, (April 1977), 64,67–69.

   As an aside, in 1963–64, Lockheed proposed using a complete Agena reconnaissance satellite as the Lunar Orbiter. The camera was ''a space proven package with the capability of performing high-resolution stereographic photography.''
6. Lunar Orbiter B Press Kit, NASA, 4 November 1966, 10–16.
7. Klass, *Secret Sentries in Space,* 104.
8. Carrollton Press DoD 36D 1980.
9. Klass, *Secret Sentries in Space,* 103.
10. *New York Times,* 27 October 1960, 17.
11. Klass, *Secret Sentries in Space,* 53,54.
12. *New York Times,* 13 November 1960, 1.
13. *New York Times,* 14 November 1960, 16.
14. Klass, *Secret Sentries in Space,* 103,104.
15. *New York Times,* 15 November 1960, 20.
16. *New York Times,* 11 December 1960, 64.
17. Lawrence Freedman, *U.S. Intelligence and the Soviet Strategic Threat,* (Boulder: Westview Press, 1977), Chapter 4.
18. Carrollton Press DoD 364B 1979.
19. Klass, *Secret Sentries in Space,* 104.
20. *New York Times,* 1 February 1961, 17.
21. Lunar Orbiter B Press Kit, 15.
22. Freedman, *U.S. Intelligence and the Soviet Strategic Threat,* Chapter 4.
23. William Manchester, *The Glory and the Dream,* (New York: Bantam Books, 1974), 908,909.
24. Curtis Cate, *The Ides of August,* (New York: M. Evans & Company, 1978), 128.
25. Arkady N. Shevchenko, *Breaking with Moscow,* (New York: Knopf, 1985), 110.
26. Manchester, *The Glory and the Dream,* 910.
27. *New York Times,* 18 February 1961, 1.
28. *New York Times,* 31 March 1961, 10.
29. *New York Times,* 9 April 1961, 31.
30. *New York Times,* 11 April 1961, 4.
31. *New York Times,* 9 June 1961, 14.
32. Prados, *The Soviet Estimate,* 89, 116,117.
33. Klass, *Secret Sentries in Space,* 61.
34. Manchester, *The Glory and the Dream,* 911.
35. Manchester, *The Glory and the Dream,* 912.

36. Cate, *The Ides of August,* 172.
37. Manchester, *The Glory and the Dream,* 912.
38. Klass, *Secret Sentries in Space,* 62.
39. Manchester, *The Glory and the Dream,* 913.
40. Glenn T. Seaborg, *Kennedy, Khrushchev and the Test Ban,* (Berkeley: University of California Press, 1981), 87.
41. Klass, *Secret Sentries in Space,* 64,65.
42. Seaborg, *Kennedy, Khrushchev and the Test Ban,* 87,88.
43. *New York Times,* 19 June 1961, 14.
44. *New York Times,* 21 June 1961, 3.
45. *New York Times,* 8 July 1961, 5.
46. *New York Times,* 10 July 1961, 1.
47. *New York Times,* 22 July 1961, 10.
48. *New York Times,* 4 August 1961, 3.
49. J. E. D. Davies, "The Discoverer Programme," *Spaceflight,* (November 1969): 405–407.
50. *New York Times,* 2 September 1961, 37.
51. Klass, *Secret Sentries in Space,* 106.
52. Freedman, *U.S. Intelligence and the Soviet Strategic Threat,* Chapter 4.
53. J. E. D. Davies, "The Discoverer Programme," *Spaceflight,* (November 1969): 405–407.
54. *New York Times,* 15 September 1961, 8.
55. "Industry Observer," *Aviation Week & Space Technology,* (5 December 1960): 23.
56. *New York Times,* 21 September 1961, 15.
57. Klass, *Secret Sentries in Space,* 106.
58. Prados, *The Soviet Estimate,* 119–121.
59. Robert Berman & Bill Gunston, *Rockets & Missiles of World War III,* (New York: Exeter Books, 1983), 88.
60. Carrollton Press CIA 56A 1975.
61. Oleg Penkovskiy, *The Penkovskiy Papers,* (New York: Avon, 1965), 63,164,165,339,342,343.
62. Klass, *Secret Sentries in Space,* 66,67.
63. Manchester, *The Glory and the Dream,* 913.
64. Dwight D. Eisenhower, *Mandate for Change, 1953–1956,* (Garden City: Doubleday, 1963), 521,522.
65. "Military News Curb," *Science,* 20 April 1962.
66. Klass, *Secret Sentries in Space,* 109,110.
67. *New York Times,* 7 November 1961, 5.
68. *New York Times,* 23 November 1961, 15.
69. Klass, *Secret Sentries in Space,* 111,112.
70. William Roy Shelton, *American Space Exploration—The First Decade,* (Boston: Little Brown, 1967), 339.
71. *New York Times,* 14 October 1961, 3.
72. *New York Times,* 15 October 1961, 57.
73. *New York Times,* 24 October 1961, 6.
74. *New York Times,* 6 November 1961, 8.
75. *New York Times,* 8 November 1961, 37.
76. *New York Times,* 17 November 1961, 16.

77. *New York Times,* 13 December 1961, 18.
78. *New York Times,* 18 December 1961, 30.
79. Shelton, *American Space Exploration—The First Decade,* 340.
80. *New York Times,* 4 March 1962, 39.
81. J. E. D. Davies, ''The Discoverer Programme,'' *Spaceflight,* (November 1969): 405–407.

CHAPTER 5

Second-Generation Reconnaissance Satellites

The tensions generated by Berlin were too great to simply vanish with the suspension of Khrushchev's ultimatum. The suspension was little more than a brief respite in the years of crisis. Berlin, Cuba, nuclear testing, and the new missile gap remained unresolved. Before they were settled, the world would be taken to the brink of nuclear war.

It was against this troubled background that the U.S. reconnaissance satellite program reached operational status. The pattern was to use two separate types of satellites. The first was the area surveillance satellite, which would carry low-resolution, wide-angle optics. The satellite would be used to photograph wide areas of Russia, either as a regular survey for new targets or to pin down activities, the exact location of which was not known. Because each photo covered a wide area, fewer photos would be needed.

Once the new targets were found, the second type of satellite, the high-resolution satellite, would make a detailed study. Using its photos, the analysts could both identify a new facility and assess its capabilities.

For area surveillance, the Air Force used the Discoverer satellite. In keeping with the secrecy order, the name was dropped, and it was referred to as "Program 162." [1] There were no major changes in operations from the earlier Discoverer program. The launch vehicle was a Thor Agena B. The satellite

would go into an orbit with a low point varying between 95 miles (153 kilometers) and 181 miles (292 kilometers). [2] After 3 to 4 days of operation, the capsule would separate, retrofire, and be caught in midair by a JC-130 aircraft. [3] The empty Agena would continue to orbit for 2 or 3 weeks before atmospheric drag caused it to reenter. The first of the unidentified Discoverers was launched on 21 February 1962. The Thor Agena B placed it into a 232- by 104-mile (374- by 167-kilometer) orbit.

The high-resolution mission required a new second-generation satellite. Development work began in late 1960, after the success of *Discoverer 13* and *14*. Lockheed won the contract for the Agena, Kodak provided the camera equipment, and GE provided the recovery capsule. [4] The larger camera system meant the spacecraft was heavier than Discoverer; in all, it weighed about 4,400 pounds (1,996 kilograms). Accordingly, an Atlas Agena B was the launch vehicle. To maximize photo resolution, the satellite went into an orbit as low as 92 miles (149 kilometers). Resolution probably would be in the 5- to 10-foot (1.5- to 3-meter) range. The satellite would orbit for 2 to 5 days. Once the photo run was completed, the recovery sequence would begin. Unlike the Discoverer, the high-resolution capsule did not have an on-board retro-rocket. Instead, the Agena's engine

would be restarted to begin the reentry. The capsule then would separate and be recovered in midair. The Agena would burn up.[5,6]

The first high-resolution satellite was launched on 7 March 1962. The Atlas Agena B placed it in a 420- by 156-mile (676- by 251-kilometer) orbit, and the satellite remained in space for 457 days. The high orbit and long lifetime were different from later flights; apparently, this one was either a failure or a test mission.

It was the next high-resolution mission, on 26 April, that established the pattern for subsequent flights. The satellite stayed aloft for 2 days before the Agena restarted and returned the capsule to earth.

During the spring and summer, the high-resolution satellites were launched at a one-per-month rate. The Discoverer area surveillance satellites were more active (two were launched in April, two in May, and four in June). The ninth launch, on 27 June, saw the introduction of the Agena D upper stage—the standardized, mass-production version. It used modular systems rather than the custom-built units of the Agena B. The Agena B would be phased out by the end of 1962.[7,8]

Operations

The first missions returned photos indicating the Soviets were deploying the SS-7 Sad-

An Atlas Agena launch. This booster was used to launch the second-generation high resolution satellite. It first flew in 1962 and continued in operation until 1967. The satellite would operate for several days before the Agena's engine would be fired to send the capsule back to earth. *U.S. Air Force Photo.*

dler ICBM, a large, two-stage rocket that used storable liquid propellents and carried a 3-megaton warhead. It had a range of 6,000 nautical miles (11,112 kilometers). The first rockets were deployed in complexes with two above-ground pads, arranged in clusters with eight or more missiles. The clusters made it easier to defend the complexes against bomber attacks and to reload the pads.

By the summer of 1962, analysts noted a change in the pads. Construction work had begun on "coffin"-type pads—above-ground concrete boxes in which the missiles were stored horizontally. A similar arrangement had been used for the early Atlas ICBMs. Three coffin sites made up each complex—an arrangement that gave the complex some protection from nuclear blasts.[9]

Defense Secretary McNamara gave little credence to these developments. The Soviets had only 75 ICBMs versus almost 300 for the U.S. Soon, Minuteman ICBMs would begin deployment, further increasing the U.S. lead. Moreover, the quality of U.S. missiles was superior. The Soviet SS-7 lacked the accuracy to destroy Minuteman missiles in their dispersed, hardened silos. In contrast, a single Minuteman warhead could destroy eight or more above-ground SS-7 missiles.[10]

To challenge the U.S. lead, in both numbers and quality, would require a huge investment the Soviet economy could ill afford. McNamara felt the Soviet's "rational" course would be to build only a "minimal deterrent," marginally expanding the Strategic Rocket Forces to between 300 and 600 missiles in all.[11] The U.S. would stress Polaris and Minuteman missiles. Protected by silos and the expanse of the ocean, the missiles could withstand a Soviet attack and still be able to strike back. Accordingly, McNamara requested an additional 200 missiles, for a total of 800 Min-

uteman ICBMs.[12] (McNamara rejected an Air Force request for 10,000 Minuteman ICBMs.[13,14]) At the same time, the number of bombers would decrease; the entire B-47 force would be phased out by the mid-1960s. The Atlas and Titan I missiles would also be retired. In all, more than 800 B-47s and 174 ICBMs would be eliminated.

Cuba

Despite their earlier failures, the Soviets remained determined to force the West out of Berlin. In January 1962, the Soviets began harassing air traffic into West Berlin. Airliners were buzzed by MiGs. By March, when the campaign was at its height, the intimidation included dropping radar-jamming chaff.[15]

In April, Secretary of State Dean Rusk warned the Russians that withdrawal of Allied troops from Berlin was nonnegotiable. The Soviets responded that the U.S. was taking an inflexible attitude.[16] At the same time, in response to the Soviet bomb-test series, the U.S. resumed atmospheric nuclear testing. The first bomb, code named "Adobe," was exploded on 25 April in the Pacific.[17]

As were U.S. officials, the Soviets were making decisions on strategic forces. Despite McNamara's view of their "rational" course, the Soviet military was determined to build a missile force superior to that of the U.S. It would take both vast sums of money and the rest of the decade. Khrushchev could not wait. In the spring of 1962, he found his "quick fix"—missiles in Cuba. They would close the new missile gap at low cost and within a few months.

Most attractively, the Cuban missiles could also be the key to force a solution to Berlin on Soviet terms. According to one rumor, the Soviets would sign a German peace treaty in November 1962. A total blockade of Berlin—road, rail, and air—would follow. If the West tried to airlift supplies, the planes would be shot down. Soviet conventional superiority in troops, tanks, and artillery would prevent a NATO rescue attempt. Khrushchev then would fly to the UN to make a deal. The trapped Western forces could leave in exchange for Berlin becoming a "free city."[18] Khrushchev believed that Kennedy was too weak-willed to take action before the trap was sprung. To prevent any internal opposition, Khrushchev limited discussion in the Presidium and restricted the number of people who knew of the plan.[19]

Khrushchev's Cuban adventure began with an increase in shipping—doubling, then

The "Missiles of October" on their way home—a pair of SS-4 MRBMs on the deck of the freighter *Bratsk,* November 5, 1962. The Cuban missile crisis was the closest the U.S. and Soviet Union have come to nuclear war in 40 years. Reconnaissance played a major role in the crisis. U-2 aircraft discovered the missiles and monitored their withdrawal; U.S. reconnaissance satellites indicated that Soviet nuclear forces were smaller than those of the U.S. Soviet reconnaissance satellites also monitored U.S. activities while the missiles were being prepared. *U.S. Air Force Photo.*

doubling again between July and September. The ships were bringing in a flood of military supplies to prepare the ground, both figuratively and literally, for the missiles.

As preparations in Cuba got underway, U.S. reconnaissance satellites continued to photograph Russian targets. Two Discoverer launches each were made in July, August, and September; and one high-resolution satellite was launched in August. Analyses of the photos continued to be assisted by the thousands of documents photographed by Oleg Penkovskiy. Although he was under KGB surveillance by this time, he nevertheless increased his output of information.[20,21]

Another aspect of the Soviet preparations was a step-up in the pressure campaign. In early August, the Soviets resumed nuclear testing. After several small tests, they exploded a 30-megaton weapon.[22] They also demonstrated their space prowess by launching *Vostok 3* and *4*—the first time that two manned spacecraft orbited together.

The first of the Soviet missiles arrived in Cuba on 8 September 1962, aboard the cargo ship *Omsk*. On 15 September, the ship *Poltava* brought in a second group.[23,24] In all, there were twenty-four SS-4 MRBMs at six sites and sixteen SS-5 IRBMs at four more locations. Each missile had a reload. The SS-4 carried a 2-megaton warhead and had a range of 1,100 nautical miles (2,037 kilometers). The SS-5 had a 4-to-6 megaton warhead and could strike targets 2,200 nautical miles (4,074 kilometers) away. The Soviet missiles covered the entire U.S., except for the far Northwest, and roughly doubled the megatonnage of Soviet strategic forces.[25,26]

As the Cuban missile sites were being prepared, Khrushchev met with the poet Robert Frost. Khrushchev told him that the West was too old and tired to stand up to a young dynamic nation like the Soviet Union.[27] Khrushchev's unguarded words provided an insight into the thinking that not only sent in the missiles but also motivated the years of crises.

Despite the Russians' security efforts, eyewitness reports of the missiles began to reach the CIA. By late September, the evidence pointed to suspicious activities in western Cuba, near the town of San Cristobal. The area was made a top priority for U-2 coverage, and a flight was made on 14 October 1962. The pilot was Maj. Richard S. Heyser. (Because of SAMs, Air Force pilots were used rather than CIA personnel.) After the U-2 landed, its film magazine was removed and flown to Washington, D.C. for development and analysis.[28] The next afternoon, 15 October, photo analysts examining the film could see missile transporters and launchers in a grove of trees. It was an SS-4 site. The Cuban missile crisis had begun.[29]

In charting their response, President Kennedy and his advisers could call on a wide range of information sources. Although no satellites were flown during the crisis, previous missions would give a clear understanding of Soviet strategic forces (the last mission, a Discoverer area surveillance flight, was launched on 9 October). Penkovskiy's data also played a role. According to some reports, he had provided the CIA with an SS-4 field manual. From it, analysts could determine just when the Cuban missiles would be ready.[*]

Ultimately, a blockade of Cuba was se-

---

[*]Oleg Penkovskiy was arrested on 22 October. His contact, Greville Wynne, was kidnapped by the KGB in Budapest on 2 November. Wynne was imprisoned and Penkovskiy was executed.[30]

lected, with an air strike as a last resort. Kennedy's television announcement on 22 October trapped Khrushchev, who had made no contingency plans in the event things fell apart. U.S. nuclear forces and naval superiority left him powerless to act. By transferring the Berlin crisis to the Caribbean, Khrushchev had moved to an area where the U.S. had all the advantages. He rushed to get out with a minimal loss of prestige.[31] Over the next several days, he sent several contradictory messages. In response, on 27 October, Attorney General Robert Kennedy proposed a deal to Soviet Ambassador Anatoly Dobrynin. The Soviets would remove the missiles in exchange for a U.S. promise not to invade Cuba. In addition, there was a secret understanding that the U.S. would remove its Thor and Jupiter IRBMs from Europe. The offer was coupled with an ultimatum—Robert Kennedy told Dobrynin that events were forcing the U.S. to take action. If the Russians did not accept the proposals within 24 hours, the U.S. would bomb the missiles. The next morning, 28 October, the Russians agreed to the proposals and the crisis was over.[32]

The Cuban missile crisis marked a turning point in U.S.-Soviet relations. It was the end of Khrushchev's aggressive attempts to browbeat the West. In many ways, it was Khrushchev's last, desperate gamble, which was reflected in the Soviets' general air of improvisation and lack of clear thought. Khrushchev's plan relied on speed and secrecy; yet the construction crews worked only in daylight and the sites were not camouflaged.

Although the cold war did not end with Cuba, there was a definite change in the atmosphere, evidenced by the hot line and limited test ban treaty. Another reason for the shift away from confrontation was that both U.S. and Soviet attention was moving away from Europe and towards Asia. The U.S. was becoming increasingly involved with Vietnam. Russia faced the ideological challenge of Mao and China.

Amid the celebration following the Soviet missiles' removal, there was a sour note. Soviet Deputy Foreign Minister Vasily Kuznetsov told John J. McCloy, "Never will we be caught like this again."[33] Within the Russian government, the humiliation hardened demands for a large missile force. Any second thoughts were answered by "remember what happened with Cuba?"[34]

U.S. reconnaissance satellite activities for 1962 were completed with a Discoverer launch on 14 December. In all, there were twenty Discoverer area surveillance flights and six high-resolution flights.[35]

New Boosters, New Sensors

A major change in 1963 was the use of a more powerful booster for the Discoverer satellites—the Thrust-Augmented Thor (TAT) Agena D. Three solid-fuel rockets were added to the Thor, each with a thrust of 50,000 pounds (22,680 kilograms) during their brief burn. With them, the TAT Agena D could put into orbit about 3,500 pounds (1,588 kilograms), a little more than 2,000 pounds (907 kilograms) of which was useful payload; the rest was Agena structure.[36,37] The added payload, compared to the Thor Agena, allowed the Discoverer to carry both a larger camera and film supply and small, piggyback, electronic intelligence satellites to pick up Soviet radio and radar signals.

The first TAT Agena–launched Discoverer (on 28 February 1963) ended in failure before reaching orbit. The same fate was met by the second attempt on 18 March. The first success was on the third launch attempt on 18 May. The Discoverer went into a 309-

A Thrust-Augmented Thor (TAT) Agena D stands ready for launch. This first-growth version of the original Thor Agena was used to launch some of the Program 162 reconnaissance satellites as well as the second-generation area surveillance satellites. The major change was the addition of three solid-fuel strap-on rockets, which increased the orbital payload. *U.S. Air Force Photo.*

by 95-mile (497- by 153-kilometer) orbit. Two more TAT Agena launches followed in June. The older Thor Agena–launched Discoverers continued to be flown between pairs of TAT Agenas. Eight of the new boosters were successfully launched along with five Thor Agenas (plus two Thor failures). The last Thor Agena was launched on 27 November 1963. With this, the booster was phased out.

The high-resolution satellites were much less active in 1963. The first was not launched until 12 July. Only three more Atlas Agena D launches were made during the rest of the year (in September, October, and December).[38,39]

When the photos were analyzed, they showed that the Soviets had ninety-one ICBMs by mid-1963. The year also saw the introduction of a new ICBM—the SS-8 Sasin—with two stages and a range of 6,500 nautical miles (12,038 kilometers). Because of the missile's huge size, Air Force intelligence believed the SS-8 had been designed to carry a high-yield warhead. During the 1961–1962 bomb-test series, the Soviets had tested 30- and 58-megaton weapons. Such warheads, exploded at high altitude, would cause an electromagnetic pulse (EMP), whose tremendous current would burn out transistors and other solid-state electronics in missile-guidance systems. A single warhead could thus disable several Minuteman ICBMs over a wide area without physically destroying their silos.[40]

Subsequently, it was determined that the SS-8 carried only a normal, 6-megaton warhead.* The SS-8 proved to be simply a backup in the event the SS-7 failed. It used nonstorable liquid propellent—technically less risky—but it also meant the SS-8 could not be fired on short notice. Once fueled, the liquid oxygen would quickly boil off. The missile, like the SS-7, was deployed in coffin sites.[41]

Deployment of the SS-7 also continued, with a new version carrying a 6-megaton warhead. Finally, in late 1963, the Soviets began test flights of a new, large ICBM—the SS-9 Scarp.

During budget hearings in early 1964, Defense Secretary McNamara gave his assessment of the Soviet developments. The U.S. estimates of Soviet ICBMs for mid-1967 were reduced from that of the previous year to only 325 to 525 ICBMs. To justify this reduction, McNamara noted that the estimates for mid-1963 in the 1959, 1960, and 1961 projections were all higher than the actual 1963 figure (91 missiles). By 1969, the U.S. estimated the Soviets would have 400 to 700 ICBMs. The high end of this projection was still less than the actual U.S. missile force (768 ICBMs). Use of liquid fuel meant the Soviet force was more costly to operate and had a lower readiness rate than that of the U.S. The use of above-ground and coffin sites made the Soviet missiles more vulnerable; and there were no hardened silos, although they were expected in the future.[43]

At the same time, work had begun in the

---

* The Soviets had attempted to build a heavy ICBM to carry large warheads. Designated the SS-XL, it was abandoned early on. The missile was eventually converted for space use as the Proton or D booster. As an ICBM, the Mod 1 version could carry a 35- to 45-megaton warhead. The Mod 2 was estimated to have a 45- to 55-megaton weapon.[42]

U.S. on new, more versatile sensors. One was a high-resolution camera—a modified Hycon LG-77 aircraft camera—which used a 66-inch (168-centimeter) focal length lens and a 4.5-inch square (11.4 centimeter square) film frame. It weighed about 400 pounds (181 kilograms). If the film had a resolution of 200 lines per millimeter (0.04 inch), the camera could spot objects as small as 2 feet (0.6 meter) from 100 miles (161 kilometers). The camera was tested in a high-altitude balloon flight from New Mexico during August 1963. It was fitted into a gondola similar to the nose of an Agena. The cold and the low air pressure would simulate space conditions.[44]

The improved resolution of the new camera would increase both the number of objects that could be detected and the amount of information in each photo. An ICBM site could be detected with a 10-foot (3-meter) resolution. At 5-foot (1.5-meter) resolution, it is possible to detect the type of ICBM. With 2-foot (0.6-meter) resolution, a photo interpreter can determine which particular version it is (i.e., SS-7 Mod 1 versus Mod 3). When the resolution is 1 foot (0.3 meter) or better, precise measurements can be made. A resolution for a given intelligence task varies according to the object. To even detect a supply dump requires 5-foot (1.5-meter) resolution. Spotting nuclear weapon components takes 8-foot (2.4-meter) resolution, while their precise identification requires 1-foot (0.3-meter) resolution. As a general rule, the larger the object, the less demanding the resolution (terrain, urban areas, and harbors are very easy). When technical intelligence is sought, the resolution must be measured in inches.[45]

Two other sensors were developed to counter Soviet camouflage and deception. The Soviets might try to build military facilities concealed by cloud cover and darkness. During clear weather and daylight, the facility would be covered by camouflage nets. Once completed, it would seem to be a clearing in the woods or some innocent-looking building. During World War II, camouflage on a large scale had become an art. West Coast aircraft plants had fake suburbs built on their roofs. They were complete with houses, trees, lawns, and cars. The one on the roof of the B-17 factory in Seattle used millions of square feet of chicken wire and lumber.[46]

Alternately, the Soviets might build fake silos to conceal their true weakness or to draw U.S. fire away from the real ones. They already had tried to build a missile force based on bluff. Why not one out of cardboard? There was historical precedent: In 1787, after Russia took the Crimea from Turkey, Catherine the Great sailed down the Volga to inspect the area. To impress her with its prosperity, Russian Prime Minister (and one of Catherine's many lovers) Grigori Potemkin had false-front buildings set up by the river bank. From the river, they looked like a real village. After Catherine passed, the buildings were taken down, then set up farther down the river. Such deception efforts have been given the name "Potemkin village."[47,48] In World War II, all sides used decoys and false targets—canvas tanks, fake factories and airfields, and a harbor.[49]

To detect camouflage, a technique called multispectral photography was developed. During World War II, it was discovered that vegetation and camouflage reflected infrared light differently. Although invisible to the eye, these differences are readily apparent on special infrared film. Rather than concealing the object, the camouflage netting would be like a sign advertising its presence. Multispectral photography extends the differences across the entire light spectrum. Each object reflects and absorbs specific wavelengths—not just infrared—differently, which gives each material a unique spectral fingerprint and allows an analyst to literally identify what an object is made of. The intelligence applications are obvious. Any camouflage would have to match the spectral fingerprints of actual vegetation. It was no longer enough to paint an object green.

For multispectral photography, several cameras on a common mounting were fitted with narrow-band filters. The specific wavelengths were selected to pick up areas of interest; for example, vegetation might strongly reflect a particular wavelength which would be absorbed by fabric netting. The difference between them would be maximized in the photos.

Multispectral photos were originally seen as a way of detecting secret underground nuclear tests. It was thought that the explosion might cause subtle changes in soil and vegetation, which the photos could detect. This would remove the need for vast numbers of drill holes to locate the explosion chamber. The first airborne tests were made in late 1962. A nine-camera unit built by Itek Corporation was fitted into a C-130. The results were encouraging and plans were made for spaceborne tests.

The infrared scanner, another new device, detected the heat given off by an object in the long infrared wavelength—beyond those picked up by infrared film. Because it detected differences in temperature, the scanner could function in complete darkness. Warm areas are "brighter" than cooler ones, producing a photolike image. On a satellite, an infrared scanner would be used to detect nighttime activities, such as heat sources, over wide areas. The heat given off by a camouflaged facility would stand out in infrared like a single light

in a dark forest. The scanner could also be used to determine the level of activity. In a normal photo of a power plant, it might be hard to tell if it is operating or not. In infrared, the hot boilers and exhaust plume would be readily apparent. Infrared could also be used to spot decoys. A decoy building might look like an actual one, but it would lack internal heating. There would be no warm room or heat leaking from the windows. Infrared would show the decoy to be cold and lifeless.[50]

The End of Discoverer

A number of milestones in the U.S. reconnaissance effort were reached in 1964. Most significant was the retirement of the Discoverer area surveillance satellite, and its replacement by a new, second-generation system. The decision itself was made in February 1964 between Under Secretary of the Air Force (and, as such, head of the National Reconnaissance Office) Brockway McMillan; Air Force Chief of Staff Gen. Curtis LeMay; and Gen. Bernard A. Schriever, chief of the Air Force Systems Command. They agreed Discoverer had reached the point of diminishing returns and the expense no longer justified the resources. The last Discoverer area surveillance satellite was launched on 27 April 1964. In all, the Discoverer program had lasted 5 years and a total of seventy-eight launches. From the rocky days of 1959 and 1960, it had developed into the first operational reconnaissance satellite.[51]

The design of the new, second-generation area surveillance reconnaissance satellite resembled that of Discoverer. The film was returned to earth in a capsule with its own retrorocket. After the capsule separated, the empty

Agena continued in orbit for 2 or 3 weeks.[52,53] The second-generation satellite probably had improved resolution, better reliability, and an increased operating lifetime over Discoverer's 3 to 4 days. The weight remained about 3,500 pounds (1,588 kilograms). The new satellite also used the TAT Agena D and went into orbits similar to those used by the Discoverers.[54] As a result, the debut of the new satellite was not noticed.

Introduction of the new satellite went smoothly with the first launch on 4 June. A second followed on 19 June. Subsequent launches were made one or two times per month. During 1964, two Discoverers were launched (plus one failure) along with eleven of the second-generation satellites. The launches were made between 1 and 3 P.M. to give long shadows in the photos, which would allow better estimates of the height of an object. The area surveillance satellites were also fitted with experimental infrared scanners during 1964–1965. The infrared energy the detector picked up was converted into electronic signals, which were recorded on tape, transmitted to Air Force ground stations, and converted back into images.

The Atlas Agena high-resolution satellite achieved year-round operation during the same period. The first was launched on 25 February 1964 and the last on 4 December. Previously, there were no launches during the first months of each year. In all, there were nine launches (with one launch failure) during 1964. Most satellites stayed aloft 4 to 5 days, the others 1, 2, 3, and 9 days. The launch time was between 11:30 A.M. and 2 P.M., providing direct illumination with the sun at the zenith, which was important for photography during the short days of fall and winter. Some high-resolution satellites were also fitted with experimental multispectral cameras built by

Itek. The flight tests would allow various filter combinations to be tried, as well as the bugs to be worked out. The film would be loaded into the capsule and returned in the normal way. More than doubling the number of high-resolution launches in 1964 over the previous year may have been related to the multispectral test equipment.[55,56]

The satellites' maturity was also reflected in the recovery of the capsule. From the hit or miss of the earlier Discoverer launches, the JC-130s were, by late 1963, catching 88 percent, which was close to the 90 percent recovery rate of practice missions.[57]

The satellites launched in 1964 spotted several developments. First, the Soviet ICBM force had doubled, going from 91 to 190 missiles. More importantly, work on the first Soviet silos had begun. More than forty holes were spotted. In budget hearings during early 1965, Defense Secretary McNamara continued to stress that estimates of future Soviet missile forces remained similar to those of previous years. The Air Force, traditionally a source of high estimates, seemed to agree. Its estimate overlapped part of McNamara's figures.[58] McNamara decided to freeze the Minuteman force at 1,000 missiles. He rejected an Air Force request for 200 more.[59]

McNamara's faith in a minimal Soviet ICBM force was repeated in an interview with *US News & World Report* magazine, in which he said, ". . . the Soviets have decided that they have lost the quantitative race, and they are not seeking to engage us in that contest. . . . There is no indication that the Soviets are seeking to develop a strategic nuclear force as large as ours." He justified this by saying, "The possibility of error is materially less than it has been at many times in the past because of the improvement of our intelligence collection methods."[60,61] The latter was a veiled reference to reconnaissance satellites.

As McNamara was making his confident prediction, the Soviets had begun testing a third generation of ICBMs. There were two distinct types. The heavy ICBMs were the SS-9 Scarp and SS-10 Scrag, both of which carried 20-megaton warheads and could be adapted for use as orbital nuclear weapons. As with the SS-7/SS-8 pair, the SS-10 was a low-risk, backup effort. The SS-10 used nonstorable propellents, while the SS-9 used storable ones. With improved accuracy and a larger warhead, both ICBMs could attack hard targets such as Minuteman control centers. Development of the SS-10 was abandoned in 1966 after 2 years of flight tests.

The light ICBMs included the SS-13 Savage, which was technically similar to the Minuteman (three solid-fuel stages, one small warhead, and even the same shape). The SS-13's lower cost made it possible to deploy large numbers. The other light missile, the SS-11 Sego, was designed to attack carrier task forces from Soviet territory. Once the carrier's position was found by Soviet aircraft, ships, or ocean surveillance satellites, the missiles would be fired. The 950-kiloton warhead could not be stopped by conventional air defenses. Because of the SS-11's antishipping mission, its range was only 3,000 nautical miles (5,556 kilometers)—half that of other Soviet ICBMs. All were to be deployed in hardened silos.

SS-7 and SS-8 deployment was also completed in 1965. In above-ground or coffin sites were 128 SS-7s. The last 69 were in silos. Only 23 SS-8s were deployed (14 in coffins and 9 in silos), which reflected the missile's inferiority due to nonstorable propellents.[62]

Because deployment was completed and the next generation of missiles was still being tested, the next 2 years were relatively quiet—thirty-five ICBMs were deployed in 1965 and sixty-eight in 1966. The latter were the first SS-9s and SS-11s.[63] Test flights of the SS-13 had revealed problems in its guidance system and last-stage motor, which derailed the plans for wide-scale deployment and sparked a search for a replacement. This was a modified SS-11 with extended range. Deployment began suddenly before flight tests had barely begun. The last-minute change caused few problems in Soviet planning, since aircraft carriers had diminished in importance. The nuclear strike role had been taken over by Polaris submarines. The SS-11, because of its small warhead and low accuracy, was targeted on such things as industrial centers.[64]

A stability in satellite operations marked 1965 and 1966. In 1965, thirteen area surveillance satellites (plus one abort) and eight high-resolution types (plus one failure) were launched—nearly the same as 1964. The effort was not stagnant, however; the launch time of the high-resolution satellites was changed. By 1965, they were being launched a few minutes before noon local time.[65,66] Both the multispectral cameras and the infrared scanner had proven themselves. It was decided to include both in the upcoming third generation of reconnaissance satellites.

A less visible change was in the National Reconnaissance Office (NRO). The CIA and the Air Force had continued to feud. The issue was the type of reconnaissance. The Air Force needed technical intelligence on the capabilities of Soviet weapons, which meant high resolution. The CIA put stress on simply counting, which required a lower level of resolution than technical intelligence. Since the CIA had control over research and development of new systems, the Air Force felt the CIA had too much power. The solution was to organize the National Reconnaissance Executive Committee to oversee the NRO's budget, structure, and research and development. The committee's chairman is the CIA director, but he reports to the secretary of defense, who has final authority. If there is a disagreement, an appeal can be made to the president.[67]

The major change in 1965, however, was the debut of the Air Force weather satellite program. Starting in 1960, the Air Force had used Tiros weather satellite photos to predict cloud cover over Russia, thereby preventing the waste of limited film. Use of the civil Tiros satellites ended in the mid-1960s because of a divergence in goals. The weather service needed wide-scale coverage to show patterns on a global basis. To do this, the weather satellites would be put into 800-mile-high (1,287-kilometer) orbits rather than the 500-mile (805-kilometer) ones previously used. The Air Force, on the other hand, needed to know if a specific small area would be clear. The new, higher orbit of the Tiros satellite meant resolution would be too low for this purpose. The Air Force had to undertake a separate effort, called the Defense Meteorological Satellite Program (DMSP).[68]

The DMSP's first operational launch was on 19 January 1965. The booster was a Thor Altar (a Thor with a small, solid-fuel upper stage). The early flights went into roughly 300-mile-high (483-kilometer) orbits, inclined 98.8 degrees. The satellite, because of this inclination, could photograph the same lati-

tude of the earth with the same sun angle day after day. For this reason, the orbit is called sun synchronous.[69,70]

Four DMSP satellites were launched in 1965. In September 1966, a new booster was introduced. The Thor Burner 2 was used to launch a heavier satellite. The satellites were placed into roughly circular 500-mile (805-kilometer) orbits.[71] Launches occurred at a two-per-year rate through the rest of the 1960s.

Activities evolved further in 1966. Specifically, the number of high-resolution launches increased while the total of area surveillance missions decreased. There were twelve Atlas Agena plus three third-generation Titan III B Agena missions. In contrast, there were only seven TAT Agena area surveillance missions, plus one third-generation satellite (one additional TAT Agena failed).

Previously, the ratio of high-resolution to area surveillance missions was the reverse, which may have been due to an increase in operational lifetime over the Discoverer satellites. More coverage per mission would mean fewer launches would be required for the same total coverage. It is known that the lifetime of the high-resolution satellites was increased to 6 or 7 days. The second-generation satellites were finally retired in early 1967 with three launches of each type.[72] With this, a third generation of satellites took up the watch on Russia.

NOTES Chapter 5

1. The Air Force in Space Fiscal Year 1964, Microfilm roll #30730, U.S. Air Force Historical Research Center, Maxwell AFB Alabama.
2. Anthony Kenden, "U.S. Reconnaissance Satellite Programmes," *Spaceflight,* (July 1978): 248–252.
3. Robert C. Filz and Ernest Holeman, "Time and Altitude Dependence of 55-Mev Trapped Protons, August 1961 to 1964," *Journal of Geophysical Research,* (1 December 1965): 5807–5821.
4. Anthony Kenden, "U.S. Reconnaissance Satellite Programmes," *Spaceflight,* (July 1978), 248–252.
5. Philip J. Klass, *Secret Sentries in Space,* (New York: Random House, 1971), 137–139.
6. Anthony Kenden, "U.S. Reconnaissance Satellite Programmes," *Spaceflight,* (July 1978), 248–252.
7. Andrew Wilson, "Agena—1959–1979", *Journal of the British Interplanetary Society,* (July 1981): 298–303.
8. Anthony Kenden, "U.S. Reconnaissance Satellite Programmes", *Spaceflight,* (July 1978): 248–252.
9. Robert Berman & Bill Gunston, *Rockets & Missiles of World War III,* (New York: Exeter, 1983), 87,91.
10. Klass, *Secret Sentries in Space,* 117,118.
11. John Prados, *The Soviet Estimate,* (New York: The Dial Press, 1982), 186–188.
12. Klass, *Secret Sentries in Space,* 112,113.
13. David A. Anderton, *Strategic Air Command,* (New York: Scribners, 1975), 126,254.
14. Berman & Gunston, *Rockets & Missiles of World War III,* 31,39.
15. Curtis Cate, *The Ides of August,* (New York: M. Evans Co., 1978), 492,493.
16. Klass, *Secret Sentries in Space,* 114.
17. Fact Sheet Dominic I, Defense Nuclear Agency.
18. David Detzer, *The Brink,* (New York: Crowell, 1979), 51,52.
19. Arkady N. Shevchenko, *Breaking with Moscow,* (New York: Knopf, 1985), 118.
20. Dr. Ray S. Cline, *The CIA Under Reagan, Bush and Casey,* (Washington, D.C.: Acropolis Books, 1981), 222.
21. Oleg Penkovskiy, *The Penkovskiy Papers,* (New York: Avon, 1965), 325–338.
22. Foreign Nuclear Detonations—Through 31 December 1981, Department of Energy.
23. Prados, *The Soviet Estimate,* 136,137.
24. Detzer, *The Brink,* 68–71.
25. Prados, *The Soviet Estimate,* 137.
26. Berman & Gunston, *Rockets & Missiles of World War III,* 72,85.
27. Cate, *The Ides of August,* 495.
28. Prados, *The Soviet Estimate,* 139.
29. James Daniel and John G. Hubbell, *Strike in the West,* (New York: Holt, Rinehart and Winston, 1963), 29.
30. Penkovskiy, *The Penkovskiy Papers,* 350.
31. Shevchenko, *Breaking with Moscow,* 118.
32. Detzer, *The Brink,* 247, 251–257.
33. Detzer, *The Brink,* 259,260.
34. Shevchenko, *Breaking with Moscow,* 118.
35. Anthony Kenden, "U.S. Reconnaissance Satellite Programmes", *Spaceflight,* (July 1978): 248–252.

36. Klass, *Secret Sentries in Space,* 131.
37. Philip J. Klass, "Military Satellites Gain Vital Data," *Aviation Week & Space Technology,* (15 September 1969): 55–61.
38. Klass, *Secret Sentries in Space,* 131.
39. Anthony Kenden, "U.S. Reconnaissance Satellite Programmes", *Spaceflight,* (July 1978): 248–252.
40. Lawrence Freedman, *U.S. Intelligence and the Soviet Threat,* (Boulder: Westview Press, 1977).
41. Berman & Gunston, *Rockets & Missiles of World War III,* 85,87.
42. Berman & Gunston, *Rockets & Missiles of World War III,* 89.
43. Prados, *The Soviet Estimate,* 188,189.
44. Klass, *Secret Sentries in Space,* 139.
45. Jeffrey T. Richelson, *The U.S. Intelligence Community,* (Cambridge: Ballinger, 1985), 110.
46. Edward Jablonski, *America in the Air War,* (New York: Time Life, 1982), 66.
47. Klass, *Secret Sentries in Space,* 142,143.
48. Nicholas V. Riasanovsky, *A History of Russia, 4th Edition,* (New York: Oxford, 1984), 266.
49. David Fisher, *The War Magician,* (New York: Berkely, 1983).
50. Klass, *Secret Sentries in Space,* 142–145.
51. The Air Force in Space Fiscal Year 1964, microfilm roll #30730.
52. Robert C. Filz and Ernest Holeman "Time and Altitude Dependence of 55-Mev Trapped Protons, August 1961 and June 1964," *Journal of Geophysical Research,* (1 December 1965): 5807–5821.
53. Robert C. Filz, "Comparison of the Low-Altitude Inner Zone 55-Mev Trapped Proton Fluxes Measured in 1965 and 1961–1962," *Journal of Geophysical Research,* (1 February 1967): 959–963.

These two issues have papers on measurements of cosmic radiation between August 1961 and June 1964 and during 1965. Since the measurements were made with equipment aboard the capsules, and the measurements continued after the Discoverer's retirement, the second generation satellites must also have used capsules.
54. Anthony Kenden, "U.S. Reconnaissance Satellite Programmes", *Spaceflight,* (July 1978): 248–252.
55. Klass, *Secret Sentries in Space,* 131–133,139,145,146.
56. Anthony Kenden, "U.S. Reconnaissance Satellite Programmes," *Spaceflight,* (July 1978): 248–252.
57. *New York Times,* 20 December 1959, 28.
58. Prados, *The Soviet Estimate,* 189,190.
59. Klass, *Secret Sentries in Space,* 132.
60. Freedman, *U.S. Intelligence and the Soviet Strategic Threat,* 104.
61. Prados, *The Soviet Estimate,* 190.
62. Berman & Gunston, *Rockets & Missiles of World War III,* Chapter 4.
63. Prados, *The Soviet Estimate,* 187.
64. Berman & Gunston, *Rockets & Missiles of World War III,* 73,85,87.
65. Anthony Kenden, "U.S. Reconnaissance Satellite Programmes", *Spaceflight,* (July 1978): 248–252.
66. Klass, *Secret Sentries in Space,* 139.
67. Richelson, *The U.S. Intelligence Community,* 284,298.
68. Klass, *Secret Sentries in Space,* 140,141.
69. William Roy Shelton, *American Space Exploration—The First Decade,* (Boston: Little Brown, 1967), 348.
70. The Soviet Space Threat Appendix I, Aerospace Defense Command, 12 December 1980.
71. Andrew Wilson, "Burner 2-Boeing's Small Upper Stage," *Spaceflight,* (May 1980): 210–213.
72. Anthony Kenden, "U.S. Reconnaissance Satellite Programmes", *Spaceflight,* (July 1978): 248–252.

CHAPTER 6

Satellite Operations Continue

A third generation of U.S. reconnaissance satellites made its debut in 1966. As before, two separate types were used. The high-resolution satellite used a Titan III B Agena D, which was a Titan III core with an Agena third stage. The two solid-fuel strap-ons and the transtage were not used. The booster could place 7,500 pounds (3,402 kilograms) into polar orbit. About 6,000 pounds (2,722 kilograms) was reconnaissance payload. This translated into a longer operational lifetime over the Atlas Agena satellite.

The Agena was fitted with one of three different camera payloads. One was a high-resolution unit, which could photograph objects as small as 2 feet (61 centimeters). Another was the multispectral package, the operational version of those tested on some second-generation satellites. Because of weight and volume restrictions, probably only four to six bands were covered. Use of a multispectral camera was indicated by a change in launch time. Some of the Titan III B Agenas were launched in the late morning rather than at noon. (The colors of objects varied according to sun angle and seasons.) The units were built by Itek Corporation.[1] The third system was a mapping camera. Its function—to precisely locate Soviet targets—was related to McNamara's nuclear strategy.

When McNamara came to Washington, D.C., he brought with him civilian strategists from RAND and the academic community. Rather than massive retaliation, they proposed a strategy of multiple options, which included limited nuclear strikes and avoiding attacking cities while hitting only Soviet military targets. In this way, the U.S. could respond to Soviet aggression with an appropriate level of force rather than an all-out spasm. Once the exchange took place, the U.S. could use its remaining forces as a bargaining chip to terminate the war on favorable terms by threatening further attacks.[2]

Attacking military installations required improved accuracy due to their smaller size compared to a city. This was further complicated when the Soviets began building hardened silos. A Minuteman's 1-megaton warhead could destroy an SS-7 cluster even if it exploded 5 miles (8 kilometers) away (assuming a 3.5-psi (246-grams/cm^2) overpressure). Soviet silos could withstand a 350-psi ($24,609$-grams/cm^2) overpressure. This required the warhead to land within 1,700 feet (518 meters) of the silo. (The crater that a 1-megaton warhead would dig is 1,000 feet (305 meters) across.)[3,4]

Conventional reconnaissance cameras could not locate a silo this precisely. Without knowing the exact location, the missile could not hit close enough to destroy it.

A Titan III B Agena lifting off from Vandenberg AFB. The booster was first flown in 1966 with the introduction of the third generation high-resolution satellites, which carried either a high-resolution camera, a multispectral package, or a mapping camera. With the introduction of the Big Bird satellite, the Titan III B was used to launch an improved high-resolution satellite. The Titan III B continued in service until the phaseout of the film-return satellites in the early 1980s. *U.S. Air Force Photo.*

The mapping camera used a low-resolution, wide-angle lens, which reduced the number of photos needed. The mapping camera would first photograph a landmark outside Russia whose position was known. Next, the exact distance and direction to a landmark inside Russia would be found, which would be used as a reference to precisely locate the target. Once a network of such reference points was established, the location of new targets could be found.

Because of the precision needed, the emphasis in the mapping camera was on "geometric fidelity" (i.e., lack of distortion and errors) rather than on high resolution. To achieve the fidelity, the film's temperature had to be controlled. A change of only a few ten-thousandths of an inch due to expansion or contraction would cause unacceptable errors. To correct for any small satellite attitude control errors, the camera would simultaneously photograph several stars. The results would determine the camera's exact vertical angle and provide the necessary geometric corrections. The photos were returned to earth by reentry capsule.

The first Titan III B Agena was launched on 29 July 1966. It went into a 155- by 98-mile (250- by 158-kilometer) orbit inclined 94.12 degrees. The satellite remained in orbit 7 days. The second Agena was launched on 28 September and remained aloft for 9 days. The third, on 14 December, also operated for 9 days.[5] The Soviets called this type "SAMOS M."[6]

The New Area Surveillance Satellite

The third-generation area surveillance satellite also used a new booster—the Long Tank Thrust-Augmented Thor (LTTAT) Agena D.

A Long Tank Thrust-Augmented Thor Agena D. This was the next of the growth versions of the Thor. The first stage was stretched in length and its forward end changed from a tapered to a cylindrical shape. This allowed a longer burn time for the first-stage engine, which meant a greater payload. The LTTAT launched the third-generation area surveillance satellites. Although the LTTAT launched its last reconnaissance satellite in 1972, the Thor (as the Delta) has continued to be upgraded with larger upper stages and nine solid-fuel strap-ons. *U.S. Air Force Photo*.

The Thor's first stage was increased in length by 14 feet (4.3 meters). Additional changes included new solid-fuel strap-ons and modifications to the first-stage engine. Development of the booster took only a year. To make launching more routine, a procedure was used that eliminated any checkout between the factory and the launchpad.[7]

The third-generation area surveillance satellite weighed about 4,200 pounds (1,905 kilograms), of which 2,600 pounds (1,179 kilograms) was reconnaissance equipment.[8] The 20 percent increase in payload weight was apparently used for a radio transmission camera system, similar to SAMOS, rather than a capsule. After the photos were converted into electronic signals, they were transmitted through a 5-foot (1.5-meter) in diameter dish antenna called the "Space-Ground Link System." The LTTAT Agena satellite also carried an infrared scanner for nighttime photography.

The infrared scanner was originally developed at the University of Michigan. The operational systems were built by HRB-Singer and Texas Instruments.[9,10] Some of the added payload was also used to carry an electronic intelligence subsatellite.

The photos and infrared images were received by ground stations and ships around the world. Ground stations were located in New Boston, New Hampshire; at Vandenberg Air Force Base; on Oahu; on Kodiak Island, Alaska; on Guam; and in the Seychelles islands in the Indian Ocean. A seventh station was located in an East African country that was not identified to avoid political embarrassment to the host government. Each ground station had one or more 60-foot (18.3-meter) in diameter dish antennas. To fill gaps in coverage, the Air Force used eight range instrumentation ships, which were converted Victory-class cargo ships—the USNS *Wheeling, Richfield, Range Tracker, Twin Falls, Watertown, Huntsville, Longview,* and *Sunnyvale.* They went to sea in the mid-1960s. Fitted with a 30-foot (9.14-meter) in diameter dish antenna, the ships were used for satellite tracking, launch monitoring, and surface recovery of capsules if the JC-130s missed.[11]

The stations record the signals on magnetic tape, but the signals are not converted back into photos. The tapes are flown to Washington, D.C. via special courier flights. A more complex approach is used for shipboard tapes. One method is for the courier aircraft to circle above the ships. The tapes are transmitted up to the aircraft and recorded on board; then the aircraft flies on to a nearby air base and from there to Washington, D.C.[12]

It is also possible to physically recover the tapes. They are placed in a canister, which is attached to a balloon with several hundred feet of line. The balloon is caught by a C-130, and the capsule is reeled aboard.[13] The tapes would be at the National Photographic Interpretation Center (N-PIC) within 2 or 3 days.

The first LTTAT Agena D launch took place on 9 August 1966. The satellite went into a 178- by 120-mile (287- by 194-kilometer) orbit inclined 100.12 degrees; it remained aloft for 32 days. Unlike the Titan III B high-resolution satellite, the Agena was not immediately phased into operation. Nine months would pass before the next LTTAT Agena launch, which could have been a reflection of technical problems with the satellite's systems. The next area surveillance launch was made on 9 May 1967. With this, the second-generation TAT Agenas were retired.

During 1967, six LTTAT Agenas were launched (along with the last three TAT Agenas). The launches took place between 10

A.M. and 2 P.M. local time. The satellites' lifetimes were between 15 and 64 days before they decayed. (It is not known how long they operated.) The improved lifetime that a radio transmission offered over a capsule system meant an increase in the coverage each satellite could provide. Six Titan III B Agena satellites (plus the final three Atlas Agena high-resolution satellites and a Titan III B abort) were also launched in 1967. The lifetime of the Titan-launched satellites ranged from 9 to 13 days before they were deorbited. This compared to 8 or 9 days for the Atlas Agena high-resolution satellites. The greater lifetime and added payload of the larger Titan III B and LTTAT satellites meant the U.S. could reduce the number of reconnaissance satellite missions, which lowered costs with no loss of coverage.[14]

Breakout

In 1967, the U.S. completed its missile-construction program. The U.S. had 1,000 Minuteman and 54 Titan II ICBMs in silos and 656 Polaris missiles aboard forty-one submarines. The number of missiles would remain frozen until the early 1980s.[15] At the same time, the photos returned by the third-generation satellites indicated that Soviet strategic forces were undergoing a major increase. In 1967, the Soviets deployed 278 ICBMs—a near doubling of the total—which included SS-9s, SS-11s, and the first few SS-13s. During late 1967, the first Soviet Yankee-class missile submarine began sea trials. Unlike the early, rather crude boats, the Yankee carried sixteen SS-N-6 Sawfly missiles (Gulf and Hotel submarines carried only three missiles each). The larger number of missiles per submarine along with the Sawfly's 1,300-nautical-mile (2,408-kilometer) range represented a major improvement. The first Yankee went to sea in early 1968 and several more were under construction.[16]

Unlike the U.S., the Soviets were also placing emphasis on defensive as well as offensive forces, specifically the deployment of an antiballistic missile (ABM) system. The Soviet ABM had a long history. The first attempt with the SA-5 Griffon began with deployment near Leningrad. The system was a failure and work stopped in 1963. A year later construction began outside Moscow on several ABM sites. They used the ABM-1 Galosh, a two-stage missile with a 250-mile (400-kilometer) range. During 1964 and 1965, work was sporadic, apparently reflecting design problems. By the second half of 1966, these were overcome and the project accelerated. Land was cleared for eight complexes arranged in a circle about 45 miles (72 kilometers) from Moscow. Most were old SA-1 SAM sites. Construction work went ahead at six sites, while the two in the south became dormant.

Satellite photographs showed, as McNamara described to Congress, "bulldozers clearing launching pads, excavating shovels digging trenches for cables and deep holes for launchers, concrete pourers laying out pads and access roads." Each complex had sixteen launchers and two radar "triads" (two sets of one large and two small radars). These tracked the incoming warheads and guided the interceptor missiles towards them. The command center tying them together was a "multi-level structure built entirely underground." Construction continued through 1967 at a modest pace on six of the eight sites.[17] These developments had made a mockery of both a Soviet minimal deterrent and McNamara's doctrine of flexible response. There remained the questions of how to deal with this. McNamara saw no point in increasing the number of U.S. missiles. The existing U.S. force could kill 50 percent of the Soviet population and destroy 80 percent of its industry. More warheads would not bring a significant increase in destruction potential.[18] He also had no faith in defensive measures, due as much to philosophical reasons as to technical shortcomings. McNamara's answer came in an 18 September 1967 speech. Immediate attention was given to the decision to equip U.S. missiles with multiple warheads and to build a "thin" ABM defense of U.S. cities. Multiple warheads would increase the number of surviving weapons after a Soviet attack, overwhelm the Soviet ABM, and maintain at least the appearance of U.S. superiority. McNamara was lukewarm about the ABM approval, saying no system could provide a defense for cities against a massive attack. The ABM was, rather, approved out of concern over Soviet intentions.[19,20] Far more important was McNamara's overall view of the cause of the arms race—the action-reaction phenomenon. The U.S., fearing the worst, made excessively high estimates of Soviet capabilities and so built more missiles than necessary. The Soviets, in turn, overreacted, which caused a further U.S. increase and so drove the arms race ever onward. The escape from this spiral was restraint: with no action, there would be no response and so U.S.-Soviet relations could be stabilized. This would be done under the doctrine of mutually assured destruction (MAD). Both the U.S. and the Soviets could destroy the other even if attacked first. With both sides equally vulnerable, a stable balance of terror would be established.[21] McNamara's observations soon were transformed by liberals into the "iron law of weapons innovation." Any doubts about the validity of action-reaction and MAD was heresy.[22]

The U.S. assumed that the Soviets would also see things this way. Intelligence estimates of the future Soviet ICBM forces ranged from 1,000 to 1,200. The low side was equal to the U.S., while the high side would allow them a modest numerical superiority for prestige.[23]

Year of Pain

Events in the wider world also had their impact on U.S. policy. During the 1960s, U.S. society had come under increasing pressure from various domestic issues—civil rights and its violent stepchild black power, Vietnam and the protest the war spawned, the generation gap and campus riots. In 1968, these stresses tore apart the fabric of American life. The key event was the Tet offensive in February. Vietcong forces briefly held several South Vietnamese cities and attacked the U.S. embassy in Saigon. Although beaten back with very heavy Vietcong losses, the Tet offensive broke the U.S. will to continue the war. More and more individuals in the government, press, and public came to oppose U.S. involvement. The vital spirit of national unity was gone. As the year went on, the setbacks seemed to follow one after another—the killing of Martin Luther King and Robert Kennedy, urban riots, student demonstrations, and the deadlock over the shape of the table at the Paris peace talks.[24]

The razor-thin victory of Richard M. Nixon in the November presidential election was a result of the yawning divisions in American society. Not since the Great Depression had they been so wide. They would get worse.

As events unfolded, the reconnaissance

satellites continued their rounds. In 1968, eight Titan III B Agena high-resolution satellites were launched along with eight LTTAT Agena area surveillance satellites—one less of each type than the previous year.[25] The trend towards fewer satellite launches per year had begun. The lifetime of the Titan III B satellites was extended somewhat, varying from 17.13 days (the first 1968 launch) to a low of 8 days (the last flight of the year). Most satellites orbited for 11 to 15 days. Operation of the LTTAT satellites was made more flexible by a new relay technique first used in late 1967 and early 1968. Rather than the

A dummy MiG-21 at an Eastern European airfield. The Soviets have made extensive use of dummy weapons and camouflage in their efforts to mislead U.S. reconnaissance satellites. This has included fake SAM sites, radar stations, aircraft, and submarines. Soviet ICBM silos have been disguised as oil storage depots and other innocuous facilities. *U.S. Air Force Photo.*

tapes being flown, they were transmitted via military communications satellites to ground stations near Washington, D.C. The photos could thus be in the hands of analysts within a few hours of their transmission from the reconnaissance satellites.

The photos indicated that the Soviets began thirty-six new SS-9 silos in the early months of 1968. After this sudden burst, there were no new silo starts through the summer and fall. Work continued on the sites started the previous year and, by mid-1968, the Soviet total had risen to 858 missiles—an increase of 288 ICBMs. By September 1968, the Soviets had 900 missiles, of which 156 were SS-9s. Another seventy-two silos were yet to be finished.[26]

Watching missile construction was made difficult by the Russians. In the early 1960s, Penkovskiy reported that the Soviet General Staff was deeply worried by U.S. reconnaissance satellites. Orders were issued to improve the poor camouflage of Soviet nuclear weapons and facilities.[27] In 1966, as the third-generation satellites were first entering service, the Soviets began to build dummy SAM sites and submarines.[28] According to one account, a ''Sub'' in Vladivostok harbor, photographed by a satellite, was bent at a right angle after a storm.[29] As part of this deception effort, the Soviets attempted to hide their silos. One set of satellite photos would show a silo under construction. A month later, new photos would show only a railroad or oil storage depot. The camouflage was brought in as the final step.[30]

Despite the increase in Soviet missiles, there were as yet no concerns that the Minuteman force was in danger. The SS-11 was too inaccurate, while the SS-9 force was still too small in number. This began to change with the testing of the SS-9 Mod 4. Carrying three 5-megaton warheads, it was the first Soviet ICBM to have multiple weapons. The first launch was made on 28 August 1968, with three more before the end of the year. The ''Triplet's'' warheads were each mounted on a track that slid down a rail. The weapon would be released by an explosive bolt or latch. Coinciding with the Triplet tests was a sudden resumption in silo construction. Six SS-9 silos were begun in a new missile field. Although few in number, they were the first silos started in winter (one can imagine the difficulty of digging a silo in ground frozen rock-hard).[31,32]

The outgoing Johnson administration believed that the Triplet was a multiple reentry vehicle (MRV) meant to destroy cities. The three warheads released in a ''shotgun'' pattern with the target in the center would cause more damage than a single 25-megaton weapon.

In the spring of 1969, some members of the Nixon administration came to a different conclusion. On 20 April 1969, the Soviets launched the first long-range Triplet test. Two more followed by mid-May. When analyzed, the warheads did not appear to be released simultaneously. Rather, they were varied. This meant each warhead's impact point could be adjusted. Thus the Triplet was a Multiple Independently Targeted Reentry Vehicle (MIRV). The spacing between the impact points seemed to match that between Minuteman silos, which varied between 10 and 15 miles (16 and 24 kilometers). With the Triplet, the Soviets could have the means to destroy the Minuteman force by the mid-1970s.[33]

These developments brought an added dimension to a fierce debate. On the surface, it was over whether to build a ''thin'' ABM defense, called ''Safeguard,'' for Minuteman silos. In fact, the debate went far deeper—to who was to blame for the cold war; the action-reaction phenomenon; and, ultimately, for all that was wrong with American society. The foes of ABM, primarily the "liberal wing" of the scientific community and former advisers to Democratic administrations, saw the enemy not in Russia but closer to home. The blame was with a powerful coalition of scientists and engineers whose research made the weapons possible, with the contractors who built them, with the congressmen whose districts benefited from defense spending, and with the military that used the weapons in such places as Vietnam. This group was given a name borrowed from an Eisenhower speech—the "military industrial complex." Because of them, the U.S., to quote one of the critics, was "fundamentally responsible for every major escalation of the arms race." Having placed the guilt on the U.S., the critics believed it was up to the U.S. to show restraint, even unilaterally, by not building the ABM.[34]

As the debate over the Safeguard ABM (and the other things) got underway in early 1969, various administration spokesmen gave details of Soviet activities derived from reconnaissance satellites. Defense Secretary Melvin Laird publicly announced that the Yankee-class missile submarines were in full-scale production at a large facility at Archangel-Severodvinsk. Eight or nine Yankee submarines had been launched and several were operational. During 1969, the Soviets also brought three ABM sites with forty-eight interceptor missiles to operational status. In congressional testimony in June, Defense Secretary Laird noted a change in the Soviet ABM effort. The radars at the ABM sites were originally pointed north towards the polar routes that U.S. missiles would take. By the summer of 1969, the configuration of the radars around the Moscow complexes was being changed—they were now pointing towards China.[35] The most fundamental change was the lack of any debate over numbers: unlike the missile gap years, the U.S. now knew how many ICBMs the Russians had, thanks to the reconnaissance satellites. Rather, the question was the meaning. Secretary Laird believed the buildup of large SS-9 missiles represented an attempt by the Soviets to acquire a first strike capability. The CIA, on the other hand, stated:

> *We believe that the Soviets recognize the enormous difficulties of any attempt to achieve strategic superiority. . . . Consequently, we consider it highly unlikely that they will attempt within the period of this estimate to achieve a first strike capability.*

The CIA believed that the cost would be too high and the task technically impossible, and that the Soviets realized that the U.S. would detect and surpass their effort.[36]

The source of all this information and debate was, of course, the reconnaissance satellites. Six Titan III B Agena high-resolution and six LTTAT Agena area surveillance satellites were launched in 1969. The drop in total numbers was continuing. The Titan III B Agena satellites remained in orbit between 11 and 16 days before they were commanded to reenter. This was up from the previous year. The LTTAT satellites stayed aloft for 3 to 4 weeks—a slight increase. The exception was the 19 March launch, which orbited for only 4.35 days.[37]

The satellites returned photos that kept the analysts busy and gave both sides much to think about. The growth of the Soviet missile force continued unabated—surpassing the U.S. total. There were 1,028 Soviet ICBMs in mid-1969, 1,060 by September, and 1,158 by the end of the year. Additionally, in the autumn, satellites also spotted the first deployment of SS-11s in medium- and intermediate-range ballistic missile sites in southwest Russia. These missiles were targeted on Western Europe. Accordingly, they were not included in the ICBM force. The end of Soviet deployment seemed nowhere in sight. Fifty-four new SS-9 silos were started in 1969. At this rate, the Soviets would have sufficient SS-9s to threaten the Minuteman force by mid-decade. Only the SS-13 program lagged. In September, there were only fifteen operational SS-13s and another twenty-five under construction.[38]

There was also a new factor in the equation. On 17 November 1969, the U.S. and Soviets began the Strategic Arms Limitation Talks (SALT) in Helsinki, Finland. After the talks began, the Soviets started only ten new silos for SS-11s and SS-13s in late 1969 or early 1970. No new SS-9 starts were made after August 1969, which was believed to be a signal of Soviet good faith in the SALT negotiations.[39]

On the other hand, construction on existing silos continued. By April 1970, 222 SS-9s were operational and another 60 were still being built. SS-11s continued to be deployed.[40] The Soviets also brought the fourth Moscow ABM site into operation, which raised the total to 64 Galosh interceptors.[41]

The Soviet moratorium on new silos held until May 1970. The Soviets then began twenty-four new SS-9 silos. Six were in an established field.[42] On 25 June, a Titan III B Agena high-resolution satellite was launched from Vandenberg Air Force Base. During its 11-day mission, it photographed Soviet missile complexes. On 6 July, its retros fired and the capsule was recovered. Within

48 hours, the film was flown to Washington, D.C., processed, and analyzed. On 8 July, Defense Secretary Laird announced the Soviets had broken the construction moratorium.[43] The watch on the new silos continued. A Titan III B Agena, launched on 23 October, rephotographed the sites. Examination of the photos from the 19-day mission indicated that the Soviets had dismantled the eighteen most recent silos and had slowed work at the other, more complete ones. Laird made the announcement of the slowdown in mid-December. Obscured by questions of whether the Soviets were actually observing a halt in new construction was the fact that deployment was continuing at the high level of previous years. By the end of 1970, 1,440 ICBMs had been deployed—an increase of 250 over the total at the start of the year. The situation was more ambiguous than simply numbers. Almost half of the ICBMs were SS-11s placed into medium- and intermediate-range ballistic missile fields aimed at Western Europe and China. Although this was their primary mission, they could still reach the U.S. Defense Secretary Laird began to include them in the Soviet ICBM total. Some felt this gave a picture of a force larger than it actually was. Yet leaving them out also gave an incorrect image.

Ambivalence also clouded the future of Safeguard. It was becoming clear that the system, particularly the radars, was vulnerable to a Soviet attack. The missiles and other components had been adapted from earlier ABM development programs for area and city defense, rather than being designed from the start for protecting silos. Moreover, the Triplet had failed to develop as had been earlier feared. By the end of 1970, it had been tested more than twenty times.[44] Its accuracy was still only 1 nautical mile (1.9 kilometers).[45] To destroy a Minuteman silo, the warhead would have to land within 0.25 nautical mile (0.46 kilometer). Safeguard was only a short-term solution to the vulnerability of land-based missiles. Long-term protection would be provided by "Hard Site," a combination of silo hardening and a specially designed ABM. This, however, generated little support in the Nixon administration. If Hard Site was deployed, it would be impossible to reach an ABM treaty with the Soviets. The U.S. could face the possibility of a Soviet ABM covering the entire Russian landmass. Ultimately, Nixon and National Security Adviser Henry Kissinger chose a treaty instead of an ABM. After April 1970, the U.S. sought limitations on ABM deployment. Minuteman would be defended by the SALT treaty.[46]

The foes of ABM, although unable to stop it, had not really lost. They had introduced into the political debate the idea that new American weapons only endanger the U.S. itself. Action-reaction had been elevated to dogma. In the future, every new weapons program would be opposed by vocal and influential elements.

Middle East Interlude

Satellite operations in 1970 continued the pattern of the previous few years. The number of launches decreased (five Titan III Bs and four LTTATs), while the time in orbit increased (11 to 21 days for the Titan satellites, 22 to 27.5 days for the LTTAT missions).[47]

It appears that one satellite mission reflected a major change in emphasis. Up to this point, reconnaissance satellites were used against Russia and China. It was not until the mid- or late 1960s that satellites began even general surveys of China. Coverage of specific, high-priority targets in China was still by aircraft—unmanned Ryan reconnaissance drones as well as manned aircraft such as U-2s and the Mach 3 A-12s and SR-71s. In other areas such as Cuba, Korea, Vietnam, or the Middle East, aircraft alone were used. Apparently, the first satellite used to cover an area outside Russia or China was a LTTAT area surveillance satellite launched on 22 July 1970. One difference was launch time—7:30 P.M. rather than the normal 10 A.M. to 2 P.M. The second was the inclination after launch. The booster made a dogleg maneuver that placed it into a 60-degree angle inclination orbit.[48] The normal inclination was 81 to 88 degrees. Only one other LTTAT satellite had been placed into a similar orbit (2 May 1969 launch; 64.97 degree inclination).[49] It was subsequently alleged that the satellite's target was not Russia but the Mideast.

The fragile cease-fire following the 1967 war had broken down in the spring of 1969. Egypt and Israel were exchanging artillery duels across the Suez Canal in late April. The Egyptians were trying to disrupt construction of Israeli fortifications on the banks of the canal. By July, the Israelis were launching air strikes deep into Egypt. The target was a belt of SAM sites west of the canal. The Egyptians retaliated with bombing missions into the Israeli-occupied Sinai. The war of attrition continued through the fall and winter of 1969 and into the new year. By January 1970, Israeli aircraft were bombing SAM sites near Cairo. In the spring, there were dangerous developments—Russian MiG pilots were flying combat missions. Several were lost in dogfights with Israelis. The battles had been going on for more than a year with no real gain. Both sides indicated a willingness to negotiate. As this was underway, the LTTAT was launched. Finally, Egypt and Israel agreed to a 90-day cease-fire. As part of the agreement, no new military equipment would be brought into an area 32 miles (50 kilometers) on either side of the Suez Canal after midnight, 8 August.[50,51]

The LTTAT reconnaissance satellite was in an ideal position to monitor events leading up to the cease-fire. It made its first pass over the Mideast at 7:30 P.M. local time. The sun would be low and objects such as SAM equipment would cast long shadows across the flat landscape. The other pass would be made at 4:30 A.M. The satellite's infrared scanner could pick up warm truck engines against the cool desert.[52,53] Within days, the Israelis were accusing the Egyptians of violating the agreement by moving in more SAMs. The satellite continued its unusual orbit until 18 August, when it reentered. On 3 September, *The New York Times* reported U.S. intelligence had confirmed the accusations through aircraft and satellite photos (U-2 coverage began 2 days after the cease-fire). About 200 SAMs had been added since 7 August. Fifteen batteries of six launchers had been deployed one or two nights before the cease-fire. Another fifteen were added between 15 and 27 August.[54] Although complete proof is lacking, evidence tends to indicate that the satellite played a role in discovering the violations.

Phaseout

The phaseout of the LTTAT, and apparently the first version of the Titan III B satellites, began in 1971–72. Before passing from the scene, they would be involved in a remarkable incident. It began with the launch of a LTTAT satellite on 18 November 1970 (the year's last mission). The satellite orbited for 22.8 days until 11 December. During December, analysts found ten new silos in SS-9 fields in central Russia.[55] The silos were much larger than regular SS-9 holes. The Soviet procedure is to put two or three fences around a 100-acre site. They would then dig a large hole about 100 feet (30 meters) across and 25 feet (7.6 meters) deep. This area housed workrooms, fuel pumps, generators, and other equipment. The silo itself was dug in the center of this large hole. The new silos used a funnel technique. At the top, the hole was slightly less than 30 feet wide (9.14 meters) and tapered as it went down, forming an inverted cone about 120 feet (37 meters) deep. The tapering was not apparent from the satellite, since the silo interior was in shadow. The new construction sparked fears that the Soviets were about to deploy a huge, new missile, reawakening the Minuteman vulnerability question.[56]

A Titan III B Agena high-resolution satellite was launched on 21 January 1971 and orbited 19 days until 9 February. On 17 February, a LTTAT launch was attempted, but it did not reach orbit (the first failure since 1966).[57] On 4 March 1971, the Senate Armed Services Committee was told of the new silos. Three days later, Senator Henry "Scoop" Jackson (D-Washington) revealed the new missile silos in a television interview, saying there was "forboding evidence" of "huge new missiles . . . big or bigger than the SS-9." Deployment, he estimated, would be at least sixty to seventy per year.[58]

The U.S. government privately expressed concern over the new silos. Their construction added urgency to the U.S. SALT efforts. The longer the talks continued, the greater the Soviet missile advantage. Just as the U.S. had its bargaining chip (ABM), the Soviet silo construction program was theirs. The Soviets reassured the U.S. that the new silos were not for new missiles (none had been tested) but just harder facilities. According to one report, the Soviets did more than give reassuring words—they put on a show. On 24 March, a LTTAT area surveillance satellite was launched. It orbited until 12 April. Its photos showed that the Soviets had lined up a complete set of silo liners by several of the new holes. They were in order of emplacement and faced upward to show their diameter. Also present was a missile canister, which indicated that the silo could not house a missile larger than the SS-9.[59] If the report is true, it indicates how reconnaissance satellites can be used to ease superpower concerns.

On 23 April, Defense Secretary Laird stated that the Soviets had 40 of the new silos at six different sites. The day before, a Titan III B high-resolution satellite had been launched. Its capsule was recovered on 13 May (21 days later). The photos showed 60 of the silos under construction and indicated there were two different types, with silo diameters differing by 4 feet (1.2 meters). That this slight difference could be detected gives some indication of the resolution the satellite could achieve. One-third of the new silos were at SS-9 fields, the rest at SS-11 sites. By August, analysts had counted about 80 of the new silos.[60] On 12 August, a Titan III B was launched, followed on 10 September by a LTTAT satellite.[61] The final count of the new silos was 91, 25 at SS-9 sites, the other 66 at SS-11 complexes. This brought the Soviet missile construction program to a close. There was some preliminary site work done in the autumn of 1971, but no actual construction was begun. The Soviets had a total of 1,618 ICBMs operational or under construction—209 SS-7s and -8s, 288 SS-9s, 970 SS-11s, 60 SS-13s—and the 91 new silos.[62]

During 1971, four Titan III B high-resolution satellites were launched along with two LTTAT area surveillance satellites (plus one failure). In 1972, only two LTTATs were launched before the type was retired. At the same time, it appears that the original Titan III B Agena high-resolution satellite was replaced by a new version with a longer lifetime. A new type of reconnaissance satellite also made its debut in 1971. It represented a departure from all that had come before. It was called "Big Bird."

NOTES Chapter 6

1. Philip J. Klass, *Secret Sentries in Space,* (New York: Random House, 1971), 162.
2. Lawrence Freedman, *The Evolution of Nuclear Strategy,* (New York: St. Martin Press, 1983), 228–239.
3. Robert Berman & Bill Gunston, *Rockets & Missiles of World War III,* (New York: Exeter Books, 1983).
4. *The Effects of Nuclear Weapons—1962,* Samuel Glasston, Editor, Atomic Energy Commission 1962), 134,135.
5. Klass, *Secret Sentries in Space,* 162–165.
6. Col. N. Gavrilov (Reserve), "Military Journal on U.S. Intelligence Satellites," *Zarubezhnoye Voyennoye Obozreniye,* (November 1984): 54–59.
7. DoD Weekly Report September 1966. Collection: CF DoD August 23–30, 1966, Box 119, Folder: DoD September 6–13, 1966, Lyndon Baines Johnson Library, Austin, Texas.
8. Philip J. Klass, "Military Satellites Gain Vital Data," *Aviation Week & Space Technology,* (15 September 1969), 55–61.
9. Klass, *Secret Sentries in Space,* 166,167.
10. Philip J. Klass, "Recon Satellite Assumes Dual Role," *Aviation Week & Space Technology,* (30 August 1971): 12,13.
11. *Jane's Fighting Ships 1975–76,* Captain John Moore (RN), Editor, (London: Jane's 1975), 521.
12. Klass, *Secret Sentries in Space,* 136,137.
13. "Aerial Data Snatch," *Aviation Week & Space Technology,* (10 February 1964): 26.
14. Anthony Kenden "U.S. Reconnaissance Satellite Programmes," *Spaceflight,* (July 1978): 248–252.
15. Berman & Gunston, *Rockets & Missiles of World War III,* 32–37.
16. Berman & Gunston, *Rockets & Missiles of World War III,* 93,94.
17. Lawrence Freedman, *U.S. Intelligence and the Soviet Strategic Threat,* (Boulder: Westview Press, 1977), 87–90.
18. Freedman, *The Evolution of Nuclear Strategy,* 247.
19. Freedman, *U.S. Intelligence and the Soviet Strategic Threat,* 119.
20. Klass, *Secret Sentries in Space,* 197.
21. Freedman, *The Evolution of Nuclear Strategy,* 254,256,335,336.
22. Freedman, *The Evolution of Nuclear Strategy,* 255,339,340.
23. Freedman, *U.S. Intelligence and the Soviet Strategic Threat,* 112,113.
24. William Manchester, *The Glory and the Dream,* (New York: Bantam Books, 1974), Chapter 33.
25. Anthony Kenden, "U.S. Reconnaissance Satellite Programmes," *Spaceflight,* (July 1978): 248–252.
26. Freedman, *U.S. Intelligence and the Soviet Strategic Threat,* 113,135.
27. Oleg Penkovskiy, *The Penkovskiy Papers,* (New York: Avon, 1965), 217,342.
28. John Prados, *The Soviet Estimate,* (New York: The Dial Press, 1982), 243.
29. Ralph Kenney Bennett, "U.S. Eyes Over Russia: How Much Can We See?," *Reader's Digest,* (October 1985): 142–147.
30. Freedman, *U.S. Intelligence and the Soviet Strategic Threat,* 101.
31. Prados, *The Soviet Estimate,* 208.
32. Freedman, *U.S. Intelligence and the Soviet Strategic Threat,* 135,141.
33. Freedman, *U.S. Intelligence and the Soviet Strategic Threat,* 138,139,143.

34. Freedman, *The Evolution of Nuclear Strategy,* 337–339,430,431.
35. Klass, *Secret Sentries in Space,* 197,198.
36. Prados, *The Soviet Estimate,* 217.
37. Anthony Kenden, "U.S. Reconnaissance Satellite Programmes," *Spaceflight,* (July 1978): 248–252.
38. Freedman, *U.S. Intelligence and the Soviet Strategic Threat,* 136,153,158,159.
39. Freedman, *U.S. Intelligence and the Soviet Strategic Threat,* 156,214.
40. Freedman, *U.S. Intelligence and the Soviet Strategic Threat,* 156,157.
41. Prados, *The Soviet Estimate,* 169.
42. Freedman, *U.S. Intelligence and the Soviet Strategic Threat,* 157.
43. Klass, *Secret Sentries in Space,* 200,201.
44. Freedman, *U.S. Intelligence and the Soviet Strategic Threat,* 143,144,148,158,159.
45. Berman & Gunston, *Rockets & Missiles of World War III,* 85.
46. Freedman, *U.S. Intelligence and the Soviet Strategic Threat,* 160,161.
47. Anthony Kenden, "U.S. Reconnaissance Satellite Programmes," *Spaceflight,* (July 1978): 248–252.
48. "Recon Satellite in Orbit Covering Mideast," *Aviation Week & Space Technology,* (31 August 1970): 13.
49. Anthony Kenden, "U.S. Reconnaissance Satellite Programmes," *Spaceflight,* (July 1978): 248–252.
50. *New York Times Index 1969,* New York Times Corporation, 1970.
51. *New York Times Index 1970,* New York Times Corporation, 1971.
52. "Recon Satellite in Orbit Covering Mideast," *Aviation Week & Space Technology,* (31 August 1970): 13.
53. Klass, *Secret Sentries in Space,* 201,202.
54. *New York Times,* 3 September 1970, 1.
55. Philip J. Klass, "Recon Satellite Assumes Dual Role," *Aviation Week & Space Technology,* (30 August 1971): 12, 13.
56. Freedman, *U.S. Intelligence and the Soviet Strategic Threat,* 164,165.
57. Anthony Kenden, "U.S. Reconnaissance Satellite Programmes," *Spaceflight,* (July 1978): 248–252.
58. Prados, *The Soviet Estimate,* 220,221.
59. Freedman, *U.S. Intelligence and the Soviet Strategic Threat,* 164–166.
60. Philip J. Klass, "Recon Satellite Assumes Dual Role," *Aviation Week & Space Technology,* (30 August 1971): 12, 13.
61. Anthony Kenden, "U.S. Reconnaissance Satellite Programmes," *Spaceflight,* (July 1978): 248–252.
62. Freedman, *U.S. Intelligence and the Soviet Strategic Threat,* 166.

CHAPTER 7

Big Bird and KH-11

Despite all the changes in generations, camera systems, and boosters, the pattern of U.S. reconnaissance satellite activities had remained constant. The satellites were either an Agena fitted with high-resolution optics or an Agena carrying a wide-angle camera for area surveillance. The fourth-generation satellite was a departure, in that both functions were combined on a single satellite. Having both high-resolution and area surveillance equipment in one package would speed up the gathering of intelligence. Previously, months would pass between the discovery of a new target by an area surveillance satellite and its examination by a specially launched, high-resolution satellite. With the fourth generation, this was cut to 2 weeks or so. The new satellite would also have a lifetime measured in months rather than days. Coverage would be increased, and only two or three launches per year would be needed. The emphasis in the U.S. program was shifting to the use of a few, very large, long-lived satellites.

The quest for extremely high resolution dates from the beginning of the U.S. reconnaissance effort. In late 1960, the Air Force studied a 40-foot (12-meter) focal length camera. A satellite equipped with this system could achieve 1-foot (0.3-meter) resolution from 250 to 300 nautical miles (463 to 556 kilometers) altitude.[1] Initial contractor design studies of the fourth-generation satellite began in 1965–66 and continued for a year or so. In 1966–67, Lockheed's Missile and Space Division once more was selected prime contractor. The satellite was given the designation "Program 612."

As Lockheed was starting development work, an extremely advanced study was being made. The June 1967 Mideast war indicated a need for satellite coverage of a fast-breaking crisis. The KH-X study envisioned a network of multiple reconnaissance satellites that would transmit their photos to earth in real time—as they were taken. The network could cover the entire world in a single day. One published account described 100 reconnaissance satellites and 4 or more manned space stations. The crews would maintain the satellites and "filter" the information relayed to the ground.[2] The KH-X plan had several problems. The cost would have been astronomical. Additionally, the number of photo interpreters and analysts would have to be increased, lest they be buried under the avalanche of photos.[3] The key importance of the KH-X study was to show the importance of real-time coverage.

By 1968, when the Program 612 code number was changed to 467, development work was well underway.[4] This included two space tests of the 20-foot (6-meter) in diameter antenna used to transmit the area surveillance photos. One of the tests lasted 20 days. Presumably, they were made on LTTAT flights.

The major technical challenge was the design goal of 1-foot (0.3-meter) or better resolution. This was approaching the ultimate limiting factor of the atmosphere. One problem was atmospheric dispersion—the scattering and absorbing of light as it travels through the air—which caused the contrast between an object and the background to fade. One can see this effect by looking at a range of hills extending to the horizon. The details on the nearest hills are clear—each tree and rock plainly visible. In contrast, the most distant hills are little more than an outline slightly darker than the sky. Ultimately, the scattering becomes so great that there is no difference in contrast between an object and the background, and it becomes invisible. Another factor in achieving high resolution is that the atmosphere also contains water vapor, dust, salt spray, clouds, and smog. These worsen the scattering and reduce the amount of light that reaches the earth's surface and is then reflected back into space. Finally, the resolution is lowered by atmospheric turbulence and inversion layers. The first can be seen in the twinkling of starlight; turbulence causes the image to be blurred. The second is a layer of cool air above lower warm air—an inversion which traps smog and other particles.[5,6]

During the early 1960s, some felt that even the best atmospheric conditions would be too poor to allow fine details to be seen. It turned out that atmospheric effects were less of a problem when viewed from above. These effects tended to be at the lower levels of the atmosphere.

Technical Details

The heart of the Program 467 satellite is the optical system. This is more than an enlarged aerial camera. It is believed the Program 467 satellite uses a 6- to 7-foot (2-meter) in diameter mirror.[7] A large mirror is needed to maximize angular separation. The wider the mirror, the closer two objects can be to each other and still be resolved. Rather than being a solid disk of glass, the mirror uses "egg-crate" construction—a front and back glass plate with glass ribs in between, which cuts the weight of a 72-inch (183-centimeter) mirror to only 1,100 pounds (499 kilograms).[8,9] The mirror is fitted into a Cassegrain telescope. Its focal length is reported

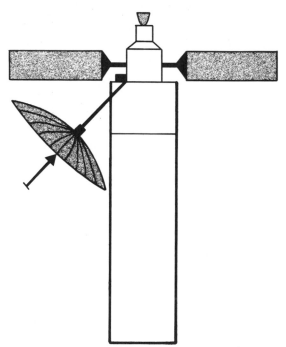

Drawing of a Big Bird reconnaissance satellite. It is based on limited information and is provisional.

to be 8 feet (2.4 meters), although it may be longer.[10] Behind the main mirror is the camera unit (shutter, motion compensator, and film magazine). Total weight of the telescope and camera unit is about 18,000 pounds (8,165 kilograms). It is built by Perkin-Elmer.[11] Exposed film is loaded into six reentry capsules, which are separated every few weeks and deorbited. Each has an on-board retro-rocket.

The Program 467 satellite also carries an Eastman Kodak area surveillance camera. Its photos are developed on board, scanned by a laser beam, then transmitted to earth via a 20-foot (6-meter) in diameter unfurlable dish antenna. The increase in size over the 5-foot (1.5-meter) dish used on the LTTAT area surveillance satellite allows a sixteen times higher transmission rate,[12] which would be particularly important if the camera had a multispectral capability.

Behind the telescope/camera unit is the Satellite Control Section (SCS), a ring-shaped structure 10 feet (3.05 meters) in diameter, 8 feet (2.4 meters) long, and weighing 3,125 pounds (1,418 kilograms) empty. The SCS has equipment bays fitted with electronic modules. If a problem appears during prelaunch checkout, the erratic module can be pulled out and a new one slipped in. Mounted in the center of the SCS is a 4,000-pound (1,814-kilogram) tank of hydrazine and a thruster (presumably based on the Agena). The thruster is used to correct for orbital decay, attitude changes, and to deorbit the satellite at the end of the mission.[13] At the rear of the SCS are two solar arrays to provide electrical power.[14]

The complete satellite is 50 feet (15.24 meters) long, 10 feet (3.05 meters) in diameter, and weighs about 29,260 pounds (13,300 kilograms). This is four to five times heavier than earlier reconnaissance satellites.[15] Because of its size, the Program 467 satellite acquired a nickname—Big Bird.

First Flight

The first Big Bird was originally scheduled for launch in late 1970, 3 to 4 years after work began. However, problems with the telescope/camera unit delayed the launch into the new year. By early summer 1971, the problems had been corrected and the first Big Bird stood ready on pad SLC 4 east at Vandenberg Air Force Base. The flight was also the first for the Titan III D booster, which was a Titan III C without the transtage and fitted with a radio guidance system rather than an inertial unit.[16]

Launch came at 11:41 A.M. on 15 June 1971. The Titan III D climbed into the California sky and headed downrange.[17] Several minutes later, the second stage shut down and the Big Bird separated. *Big Bird 1* was in a 186- by 114-mile (300- by 184-kilometer) polar orbit inclined 96.3 degrees. This inclination gave it a sun-synchronous orbit. The photos would have identical lighting conditions throughout the mission.[18] Being the first of its kind, *Big Bird 1* probably carried test instrumentation to report system operation, particularly the troublesome new camera. Day after day, the Big Bird continued its rounds, radioing down photos and releasing film-return capsules. The photos from the huge Perkin-Elmer telescope/camera had ''fantastic resolution''—twice that of the earlier Titan III B satellites. It was possible to spot people.[19] For the first time, technical intelligence could be acquired by satellite. *Big Bird 1* continued to orbit for 52 days. Finally, on

Launch photo of a Titan III D. This booster was used to launch Big Bird and KH-11 re-connaissance satellites until it was replaced by the West Coast version of the Titan 34 D. *U.S. Air Force Photo.*

6 August 1971, the on-board maneuvering engines fired for the last time.[20]

Even while *Big Bird 1* was aloft, the role of reconnaissance satellites was expanded. On 29 July 1971, the *New York Times* carried a report that, as part of President Nixon's upcoming visit to Peking, SR-71 and unmanned reconnaissance drone overflights were being suspended. Nixon was not going to risk another "U-2 incident." At this time, general surveys of China were made by LTTAT area surveillance satellites. If something of interest was discovered, an SR-71 overflight would be made. The Chinese, during the 1960s and early 1970s, had protested about 500 incursions.[21] The Nixon decision to suspend them meant satellites would now carry the full reconnaissance burden. Expansion of satellite coverage had been planned well before the China breakthrough. Work began in 1969 on a new satellite-tracking station at Nurrunger, Australia. It was completed in 1971. The Nurrunger station could receive the photos immediately after the satellite made its pass over China.[22]

SALT I

The second Big Bird was launched on 20 January 1972. It remained aloft for 40 days, somewhat less than the first mission. *Big Bird 2* also carried a P-11 electronic intelligence subsatellite. The launch was part of a step-up in U.S. reconnaissance activities. Between January and May 1972, three Titan III B high-resolution satellites and two LTTAT area surveillance satellites were launched. (Two of the Titan III Bs failed before reaching orbit.)[23] The reason for this increase was to make an up-to-date survey of Soviet strategic forces in preparation for the SALT I treaty,

which was signed on 26 May in Moscow. The treaty set limits on the number of "launchers" each side was allowed—1,710 for the U.S. (1,054 ICBMs and 656 aboard submarines) and 2,358 for the Soviets (1,618 ICBMs and 740 SLBMs). No more than 313 ICBMs could be "heavy" missiles, such as the SS-9. The Soviets could increase the number of submarine-launched missiles to 950. To do this, they would have to scrap one of the older SS-7 or -8 ICBMs for each new SLBM (i.e., 16 ICBMs for each new Y-class submarine).[24] There was no provision for on-site inspection. Rather, "national technical means of verification" were to be used, a euphemism for the reconnaissance satellites neither side was willing to admit existed.

With the signing of the SALT I treaty, the role of reconnaissance satellites was expanded. They were now to be policemen enforcing the treaty. Reconnaissance satellites were a central part of the arms control process. One cannot control what cannot be counted. Yet it went further. Given U.S. unwillingness to accept unpoliced agreements (dating from Khrushchev's violation of the nuclear test moratorium in 1961), the capabilities and limitations of the satellites would shape treaty provisions. Additionally, the treaty's ratification would depend upon the Senate believing the necessary verification was possible.

Operations

Big Bird 3 was launched on 7 July 1972 and remained in orbit 68 days. With this, the Big Bird became fully operational and the LTTAT area surveillance satellite was retired. (The last was launched on 25 May and orbited for 10 days.) *Big Bird 3* also established a pattern with the Titan III B high-resolution satellite launches. On 1 September (2 weeks before *Big Bird 3* reentered), a Titan III B was launched. It orbited for 29 days. The pattern of launching a Titan III B satellite during the final part of a Big Bird's mission would continue for the rest of the 1970s.[25] Clearly, this was part of the operational plan. Possibly the Titan III B satellite was to examine targets found by the Big Bird earlier in its mission, or to fill any gaps that might develop as the Big Bird ran low on film and fuel.[26]

From fragmentary evidence, it appears the Titan III B satellites working with the Big Birds were of a new type. The camera had the remarkable resolution of 6 inches (15 centimeters)—four times better than the camera flown on earlier Titan III B missions and twice as good as Big Bird's system. Film was returned via two reentry capsules.[27,28] Orbital characteristics and inclination remained the same (a low point of 84 miles (135 kilometers) and an inclination of 110.5 degrees). The only change was in lifetime. Titan III B satellites from the late 1960s until 1971–72 operated from 15 to 20 or more days. Because the increase was gradual, it is difficult to tell which was the first new Titan III B launch. The launches of 23 October 1971 (25 days), 17 March 1972 (25 days), 1 September (29 days), or 21 December 1972 (33 days) are all candidates.

The third and final Big Bird of 1972 was launched on 10 October. *Big Bird 4* remained in orbit 90 days, until 9 January 1973. Its companion Titan III B satellite was launched on 21 December 1972.[29,30]

New Soviet ICBMs

As the Big Bird became operational, a new generation of Soviet ICBMs made its de-

but in 1972–73. The first was the SS-16, a three-stage, solid-fuel missile with a range of 5,000 nautical miles (9,260 kilometers). It carried a single 650-kiloton nuclear warhead and was meant to destroy soft targets.[31] The first SS-16 test was made in March 1972—just before the SALT I treaty was signed. Tests continued sporadically "on a very modest scale," indicating technical problems.[32,33]

The next new Soviet missile, the SS-17, made its first test in September 1972. It was a two-stage, storable-liquid–fueled ICBM meant as a replacement for the SS-11. The SS-17 was a much larger rocket, however, with a payload four times greater than that of the SS-11. The increased payload was used to carry either four 750-kiloton MIRV warheads (Mod 1) or a single, 3.6-megaton weapon (Mod 2). The SS-17 also used a new "cold launch" technique. Compressed air or gas pushed the missile out of the silo before the engines ignited, which allowed the Soviets to remove shielding that protected the silo's interior from the rocket blast. This meant that a larger rocket could be fitted into an existing silo. The early test flights indicated some technical problems, although not as severe as those that beset the SS-16.[34,35]

The third of the new ICBMs was the huge SS-18. Larger than the SS-9 it was to replace, the SS-18 was a two-stage, liquid-fueled missile with a range of 6,000 nautical miles (11,112 kilometers). The SS-18 was designed to destroy hard targets such as silos. Like the SS-17, it used a cold launch system. SS-18 flights began in December 1972. Two versions of the missile were test flown—the Mod 1, with a single 24-megaton warhead; and the Mod 2, with either eight 900-kiloton or ten 550-kiloton warheads.[36,37]

The final new Soviet ICBM was the SS-19. It was also a two-stage, liquid-fueled replacement for the SS-11. Payload and accuracy were greatly increased. The SS-19 carried six 550-kiloton warheads. The missile's accuracy was estimated to be 0.19 nautical mile (0.35 kilometer)—better than any previous Soviet ICBM. Range was 5,200 nautical miles (9,630 kilometers). The SS-19 used a "hot launch"; the first stage was ignited in the silo. SS-19 flights began in April 1973, and the missile was soon achieving a 90 percent success rate.[38]

Policing SALT

The high rate of Big Bird activities continued into 1973—*Big Bird 5* (9 March), *Big Bird 6* (13 July), and *Big Bird 7* (10 November). As before, a Titan III B was launched towards the end of each Big Bird's flight (16 May and 27 September 1973 and 13 February 1974, respectively). An additional Titan III B failed on 26 June 1973, between the reentry of *Big Bird 5* and the launch of *Big Bird 6*. There was much to keep the satellites busy. The Soviet ICBM test program was underway in earnest. A key activity, however, was the policing of SALT.

The issue of SALT verification and Soviet compliance was a major one. The SALT talks were an ongoing process (SALT II discussions began in November 1972). The next treaty depended upon observation of the previous one.

To deal with suspected Soviet violations, a steering group on monitoring strategic arms limitations was set up on 1 July 1972. The chairman was CIA director Richard Helms. The group made quarterly reports to the National Security Council (NSC) (i.e., Kissinger). If it was felt that the evidence was sufficient, a protest would be made to the

Soviets at the Standing Consultative Commission.[39]

The first suspected violation was discovered in the spring of 1973, but it actually dated from the new silo construction of 1971. Satellite photos showed that there were 3 new types of silos under construction. The largest were called the III-X silos, one of which was being dug at each of several complexes. They were thought to be for either a new, special ICBM or a new command and control facility. At the signing of the SALT I treaty, 5 silos were under construction at heavy ICBM fields and 6 more at light ICBM sites. They were counted in the treaty as part of the 1,618 launcher ceiling. By the spring of 1973, it was discovered (possibly from *Big Bird 5*'s photos) that 150 III-X silos were being dug. If they were ICBM silos, they were a massive violation (representing about 10 percent of the allowable Soviet ICBM force). The U.S. protested in June 1973. The Soviets said they were command and control centers and this would become apparent as construction progressed. Their explanation was met with skepticism. The doors on the III-X silos were identical to those on normal silos: they could be blown clear moments before a missile was to be launched. Command and control centers would seem to have little use for this capability. The concern was that the III-X silos would later be converted into house missiles.[40]

As construction on the III-X silos continued, the Big Birds kept watch. *Big Bird 6* operated from July until October 1973. *Big Bird 7*, launched in November 1973, continued to orbit until early March 1974. It was apparently the first improved Big Bird. Previous flights had orbited for 2 to 3 months. *Big Bird 7*, on the other hand, remained aloft for 123 days. This increased lifetime was apparently one part of an in-house Lockheed improvement program that was underway by early 1972 (before Big Bird became operational). The new Big Bird also may have been able to transmit photos to earth in real time via relay satellite. This would bypass ground stations.[41] The longer lifetime put a greater demand on the fuel supply. The Big Bird's large size and low orbit meant regular engine burns to prevent reentry. A 3-day cycle was used. Once the satellite's orbit had decayed to around 152 by 100 miles (244 by 161 kilometers), a pair of burns raised it to 163 by 104 miles (262 by 167 kilometers). This also shifted the position of the low point of the orbit on the satellite's ground track. After about 3 days, air drag again lowered the orbit and the cycle began again.[42] The greater the number of cycles, the more fuel that would be used.

The watch on Russia and the III-X silos was continued by a Titan III B, launched on 13 February 1974 (shortly before Big Bird 7 reentered). It operated 32 days, until March. Big Bird 8 was launched on 10 April 1974. Two months later, the pattern of satellite operations was disrupted. On 6 June, midway through Big Bird 8's 109-day mission, another Titan III B was launched. The satellite orbited for 47 days, reentering shortly before Big Bird 8 did. This was a departure from the earlier pattern of a Titan III B launch at the end of the Big Bird's lifetime. As in 1973, the next launch was an out-of-sequence Titan III B. It was orbited on 14 August and operated for 46 days, finally reentering in late September. The final launch of 1974 was Big Bird 9 on 29 October. It operated for 141 days (a new record), until mid-March 1975. It was followed on 18 April 1975 (after Big Bird 9 reentered) by a Titan III B, which continued aloft for 48 days.[43,44]

The changes in the pattern of Titan III B launches and the 3-month delay between Big

Bird 8's reentry and the launch of Big Bird 9 (the normal interval was 1 or 2 months) implied that some development required a change in plan. The obvious candidate was the III-X silos; indeed, it was rumored that the 3-month delay was to allow modifications to Big Bird 9's sensors to penetrate Soviet concealment efforts.[45]

In any event, by early 1975 it was apparent the III-X silos were actually command and control centers. Accordingly, the original facilities were subtracted from the SALT totals. The Soviets were now allowed 308 heavy missiles rather than 313, and the number of light ICBMs was cut from 1,096 to 1,090.[46]

This did not end questions about Soviet compliance with SALT. During the eventful summer of 1974, the Russians were observed to place huge canvas covers over the Delta submarine construction yards at Severomorsk, near Murmansk in northern Russia. These covers prevented a count of the submarines under construction and observation of what missiles were being loaded aboard them. The concealment sparked a debate inside the State Department and intelligence community. The SALT I treaty prohibited concealment if it went beyond previous levels. The canvas covers had been put up intermittently before the treaty was signed. Defense Secretary James R. Schlesinger believed the covers were permitted.[47] A more important suspected violation was tests of the ''Square Pair'' radar used by the SA-5 Gammon high-altitude SAM to track ICBM warheads. Satellite photos showed several Square Pair radars at the Sary Shagan ABM test center. This was a violation, because the ABM treaty banned testing of SAM radars ''in an ABM mode.'' It was feared the testing would permit the Soviets to upgrade the 1,800 SA-5s to ABMs, which would give them a nationwide ABM defense, something also banned.[48] Each side was allowed only 200 launchers at two sites. (In 1974, this was cut to 100 ABMs at one site.) When the U.S. protested the radar tests, the Soviets claimed the Square Pair radars were range instrumentation used during tests. Soon after, the tests were stopped temporarily.[49]

Questions about Detente

The III-X silos and canvas covers and the ABM tests of the SA-5 radar were symptomatic of a more questioning attitude about detente than that which prevailed during the glory days of 1972. For one thing, the SALT I treaty had been oversold. Reality could never match Nixon's election-year rhetoric about SALT leading "the world out of the lowlands of constant war to the high plateau of peace."[50] But detente's shortcomings were not limited to rhetoric. Part of the skepticism was related to the new Soviet ICBMs. SALT was meant to protect the U.S. against the Soviets developing a first strike. To this end, the U.S. limited both the number of heavy missiles and changes in the size of silos to 15 percent. However, there was no definition of "heavy" ICBM in the treaty, only a unilateral U.S. declaration. The "insignificant" 15 percent increase (plus cold launch) allowed the Soviets to replace the SS-11 with much larger missiles.[51] By way of example, the payload of the "light" SS-19 (7,500 pounds/3,402 kilograms) is comparable to the "heavy" U.S. Titan II (8,000 pounds/3,629 kilograms).[52] It was projected that with the new missiles, the Soviets would have about 7,000 ICBM warheads to only about 2,000 on U.S. land-based missiles.[53] The test flights also indicated that the U.S. advantage in technology (guidance systems, small warheads), which once balanced Soviet superiority in numbers, was evaporating.

Not surprisingly, this shift in forces also caused a reexamination of U.S. assumptions about the arms race. In 1974, Albert Wohlstetter made several studies of now-declassified U.S. intelligence estimates. He found that throughout the 1960s, the U.S. had consistently *underestimated* future Soviet ICBM forces. This was the reverse of McNamara's claim, in the action-reaction phenomenon, of constant U.S. overestimates. Wohlstetter showed that from 1961 to 1964, the projected numbers for 1967 went down. Even when they began to go up, they still fell short of the actual 570 ICBM total. The breakout of 1967 onward was not predicted. Other estimates were equally false. The U.S. had expected the Soviets to phase out the SS-7s and 8s. They did not until SALT. The Soviets were also expected to stop building ICBMs when they equalled or marginally surpassed the U.S. The final total was half again as large as that of the U.S.[54]

One possible explanation for the underestimates of the 1960s was the overestimates of the 1950s. When the Soviets built only a minimal number of bombers and SS-6s, analysts came to believe this reflected a deliberate policy of having only a minimal deterrent (rather than technological shortcoming). This belief in a minimal Soviet deterrent lingered into the mid-1960s and the breakout.[55] In Wohlstetter's view, the arms race was not a simple mechanical process but a complex interaction of different goals, commitments, institutional factors, doctrines, and limitations.[56] Liberals, for their part, continued to hold action-reaction as the key principle. The new Soviet missiles were in response to the U.S. ABM. Once more the U.S. was to blame.

This was not the first time a president's foreign policy had been questioned. Events, however, were making it increasingly difficult for the Nixon administration to respond. In the spring of 1973, the Watergate scandal began. As the cover-up fell apart, foreign policy was pushed aside as Nixon tried to survive. The administration could not argue why detente should be supported. Nor was it possible to gain support by "getting tough" with the Russians. The Congress and the public were in no mood, after Vietnam, to support increased defense spending. An extensive U.S. ABM system was out of the question even before SALT. Strategic programs underway —the B-1 Bomber, Trident Submarine-Launched Ballistic Missile (SLBM), and Cruise Missile—all faced opposition. Many in Congress felt detente was incompatible with their development. The pain of the years from the death of Kennedy to the resignation of Nixon, the futility of Vietnam, and the inflation and recession caused by the oil embargo also brought a turning inward. People did not want to be bothered by foreign affairs. America had lost faith in its role in the world and in itself.

Despite these problems, the SALT process continued. In November 1974, President Gerald Ford and Leonid Brezhnev signed the Vladivostok agreement. This was a framework for the SALT II treaty. Each side was limited to 2,400 ICBMs, SLBMs, and bombers. Of these, only 1,320 could carry MIRVs. Increasingly, detente was seen not as a complex web of trade and technical agreements, scientific cooperation, and the like, but solely in terms of arms control.[57]

Big Bird Operations Continue

In late 1974, during the flight of *Big Bird 9,* the first SS-18s were deployed. By June 1975, when the Titan III B, launched on 19 April, reentered, analysts had counted ten SS-18s in silos. These were the single-warhead Mod 1 version, the MIRVed Mod 2 had technical problems. Deployment of the SS-17s and 19s also started in early 1975. By June, ten SS-17s and fifty SS-19s had been observed.[58,59]

On 8 June, *Big Bird 10* was launched. It orbited for 150 days (a new record). A Titan III B followed on 9 October 1975, shortly before *Big Bird 10* reentered. The Titan III B orbited for 52 days (another new record).

The major change on this flight, however, was the orbital inclination. Earlier launches had used inclinations of 110.45 degrees. The October launch used one of 96.41 degrees. This change had two advantages. It was sun-synchronous like Big Bird, which was important, given the fivefold increase in lifetime over the first Titan III B reconnaissance satellites. Also, the lower inclination would increase the payload the rocket could carry. The satellite's orbit was also slightly lower than earlier Titan III B launches (78 miles/125 kilometers versus 84 miles/135 kilometers). This altitude and inclination would become standard for future missions.[60]

The recovery forces were also upgraded. Starting in 1970, the tracking ships began to be retired. The last three—the USNS *Longview, Huntsville,* and *Sunnyvale*—left service in November and December 1974.[61] They had been used for ocean pickup of capsules. To replace them, the 6593rd Test Squadron used HH-53E helicopters. Aerial refueling, using HC-130Ps as tankers, extended the HH-53Es' time over the recovery area.[62] On a typical refueling mission, the transfer would be made 575 miles (925 kilometers) from base. The HC-130Ps could also be used for midair capsule recovery. The early model JC-130 recovery aircraft had been replaced

Aerial view of an HC-130 recovery aircraft. This is a more advanced version of the Hercules transport, which replaced the JC-130 in the late 1960s. *U.S. Air Force Photo.*

previously by HC-130Ns, fitted with advanced direction-finding equipment to home in on the descending capsules.[63]

The last Big Bird launch of 1975 came on 4 December. *Big Bird 11*'s photos showed another SALT violation. In December the Soviets began sea trials of four new Delta-class submarines. This required they dismantle fifty-one SS-7 and 8 launchers. The work had to be completed, according to a secret protocol to the SALT I treaty, 4 months after the sub-

marine first went to sea. Harsh weather conditions meant the construction crews were not able to finish work on forty of the coffin sites. President Ford was told of the violation and ordered a review before the issue was raised at the Standing Consultation Commission meeting set for 29 March 1976. Before the protest could be made, the Soviets officially told the U.S. of the problems and gave detailed information that confirmed U.S. intelligence. By the end of May 1976, the State

Department indicated the Soviets were rapidly completing the scrapping of the missiles. Others were less satisfied, noting the U.S. would have to take the Soviets' word, since no satellites were in orbit at the time.[64,65] (*Big Bird 11* reentered on 1 April 1976; the Titan III B, launched on 18 April, burned up soon after—on June 5. *Big Bird 12* was not launched until 8 July.) Even with the longer lifetimes of the satellites, gaps could develop at the worst times. Before the year was out, this would be corrected.

The violation was the last SALT activity in 1976. Negotiations were adjourned for the presidential election. President Ford was pressed by both the Democrats on the left and by the right wing of the Republican party. One reflection of this was the dropping of the word "detente" from Ford speeches. Ford survived a challenge by California Governor Ronald Reagan but lost the November election to Jimmy Carter, who campaigned on a platform of reduced defense spending and anti-Vietnam and Watergate emotion.

On 19 December 1976, as Carter was selecting his cabinet, a Titan III D lifted off from Vandenberg. It was not a Big Bird.

Bird of a Different Feather

The satellite went into an orbit with a low point of 153 miles (247 kilometers). Within 4 days this was raised to 212 miles (341 kilometers)—considerably higher than the typical Big Bird orbit of 101 miles (162 kilometers).[66] This was the first clue that the satellite was a new type. It carried the designation "Program 1010," although it would become better known by the CIA code name "Key Hole-11" (KH-11).

Program 1010/KH-11 developed from the KH-X study of 1967–68. By late 1970, the

Air Force and the CIA were ready to select a prime contractor. Lockheed, Hughes Aircraft, Philco-Ford, TRW, and possibly General Electric were expected to submit bids.[67] The contractor finally selected was not Lockheed. Rather, TRW would build the fifth-generation reconnaissance satellite. The Lockheed monopoly was over.

The KH-11 was to use a revolutionary new photo system. Until now, film supply had been the limiting factor on reconnaissance satellite operations. Once the film ran out, the satellite became useless—even if all other systems were working. The KH-11 avoided this by using a digital imaging system. Although specific details are secret, the basic design features of such systems are well known. The optical system focuses the light onto an array of charged coupled devices (CCDs)—light-sensitive capacitors about one- or two-thousandths of an inch on a side. The CCDs are assembled into an array with perhaps several hundred thousand elements. The light falling on each CCD generates an electrical current proportional to the amount of light received. (The principle is like that of a light meter.) The KH-11 optical system apparently scans across the ground in long, narrow strips. The current is read off and fed into an amplifier, which converts the current into a number that represents one of several hundred shades running from pure black to pure white. The picture is a string of such numbers—one from each CCD element. The numbers are then transmitted to earth, either directly to a ground station or via a Satellite Data System relay satellite.

Once on earth, the numbers are converted back into their corresponding shades of black, gray, or white. These are called "picture elements" or Pixels. The picture is built up Pixel by Pixel to form lines, which in turn form the strips. These are combined to make the complete picture. The process can be thought of as a high-tech version of paint-by-numbers.

The system has several advantages. First, nothing is consumed by the photographic process except electrical power, which is supplied by solar cells. The CCDs are not worn out or used up like film. The satellite can be operated not just for several months like Big Bird, but for several years. Limiting factors are now system reliability and fuel supply. The demands on the latter can be eased by use of a higher orbit, less affected by air drag.

The benefits also extend to analysis of the photos. Computer processing can remove electronic noise and transmission errors in the photos. Contrast can also be improved. The narrow range of tones can be stretched. A washed-out photo with little visible detail can be transformed into one that is finely detailed. The computer can also be used to compare old and new photos. Features that appear in both are cancelled out, leaving only the changes—a new building or an increase in the number and type of planes at an air base.

There are limitations. The KH-11's resolution is around 6 feet (2 meters).[68] Whereas the Big Bird can show people, the KH-11 is marginally able to spot a truck. The KH-11 was therefore limited to area surveillance. Its photos could show large-scale construction, such as communications facilities and airfields, and the presence of nuclear weapons components, aircraft, and missiles. They could not allow precise identification (i.e., they could distinguish between TU-95 and Mya-4 but not between TU-95A and F versions).[69] If more than day-to-day monitoring or spotting new targets was needed, the high-resolution cameras of the Big Bird or Titan III B satellites would have to be used.

Development began in mid- or late 1971, with the first launch set for 1976–1977.[70] Soon the bright promise of digital imaging began to fade. It was a new system pressing the state of the art. This translated into major cost overruns. By late 1974, the overruns were beginning to affect other satellite programs. The issue the NRO faced was whether to go ahead with a complex (and expensive) electronic intelligence satellite called Argus or use the money to keep the KH-11 on schedule.[71] Ultimately, Argus was dropped. The schedule was preserved but at the cost of a nearly $1 billion overrun. The cost problems were to have a continuing effect.

The KH-11, despite its new camera system, had a number of similarities with Big Bird. It was about the same size—50 by 10 feet (15 by 3 meters)—and had an on-board rocket for orbital correction. The KH-11 was lighter at 22,700 pounds (10,300 kilograms). Big Bird weighed 29,260 pounds (13,300 kilograms). The reason was the higher orbit. The Titan III D's performance dropped quickly as height increased.

When the first KH-11 launch was made on 19 December 1976, there was, of course, no announcement of the satellite's new nature. It was left to Western space analysts to sort out the clues. The first was the higher orbit. The second was lifetime. The satellite, carrying the international designation 1976-125A, was still in orbit at the end of 1977. It continued to operate for 770 days—far longer than a Big Bird. Confirmation of the new satellite would come in the fall of 1978. It was disastrous.

KH-11 Betrayed

William P. Kampiles was an operations desk clerk at CIA headquarters. He was also deeply frustrated and angry. It would be said later he held a "Hollywood" fantasy of the CIA—wanting to live the spy's life of derring-do. The reality he found was that very few

CIA employees matched wits with the KGB. Most, like Kampiles, just pushed paper. Kampiles decided to take action on his own. During late summer or early fall 1977, he removed copy 155 of the KH-11 system technical manual from a file cabinet. He folded it up, put it into an inside pocket of his coat, and walked out of the building. Kampiles resigned from the CIA in October 1977 under threat of dismissal. In February 1978, he went to Athens, Greece. On 23 February, he went to the Soviet embassy and told "an older, fat, balding security officer" he could provide information. The security officer demanded that Kampiles provide documents. Kampiles came back the next day with two or three pages from the manual, including a table of contents and an artist's conception. Kampiles turned them over to "Michael," actually Maj. Michael Zavali, a military attache and GRU agent. A week later, Kampiles gave him the rest of the manual. Michael handed him $3,000. For the price of a used car, the Soviets knew the satellite's characteristics, capabilities, and limitations; how its photographic system worked; and the quality of the photos and the process by which they were sent to users. The manual also told them the KH-11 geographic coverage, which would assist the Soviets in hiding their activities. Kampiles returned to the U.S. and told several friends about his contacts with the Russians (leaving out any mention of the manual). On 14 August, he met with several FBI agents, who found several holes in his story. During a further interrogation the next day, he confessed to selling the manual.[72]

Prosecuting the case raised major problems. The KH-11 manual would have to be submitted as evidence, which presented the risk of it becoming public. The trial judge agreed to restrict access to only the defense, prosecution, jury, and witnesses. Another problem came from inside the government. The Defense Department put "extreme pressure" on Attorney General Griffin Bell to claim the KH-11 satellite system had not been put into operation. Bell rejected this falsehood.[73]

The case went to trial in November 1978. After 8 days of testimony, the jury deliberated only 10 hours before returning a verdict of guilty on all charges. Kampiles was sentenced to 40 years.

Big Bird and KH-11 Operations, 1977–1980

Although the debut of the KH-11 brought the capability for year-round coverage with no interruptions, the satellite's dual role of monitoring Soviet activities and SALT enforcement continued. In January 1977, it was announced the Soviets had deployed 40 SS-17s, more than 50 SS-18s, and 140 SS-19s.[74] The ICBMs were placed into rebuilt silos. This involved a tremendous amount of concrete and moving the missile-support equipment from the lip of the silo, where it was vulnerable, to inside the silo. Once completed, the silo could withstand a 6,000-psi (4.22×10^5-grams/cm^2) overpressure. A Minuteman warhead would have to land within several hundred feet to destroy it. With the superhard silos, the Soviet missile force could ride out a U.S. attack.[75]

The second of the satellite's duties, SALT monitoring, was affected by the coming of the new Carter administration. It had very different ideas about SALT than did Nixon and Kissinger. Indeed, Carter rejected the heritage of the cold war. There would be no equivalent of Kissinger's domination of foreign policy. Rather, Carter sought a wide range of viewpoints. The administration included former Johnson administration officials like Secretary of State Cyrus Vance, hard-liners like NSC chairman Zbigniew Brzezinski, former anti-war activists; and "good ol' boys" from Georgia. Carter set the tone for SALT in his inaugural address when he spoke of his hope that "nuclear weapons would be rid from the face of the earth." As a start, Carter decided to go beyond the limits of the Vladivostok agreement. In March 1977, Secretary of State Vance went to Moscow with a proposal to cut the overall total of delivery systems from 2,400 to 1,800–2,000. Multiple-warhead systems would be similarly reduced from 1,320 to 1,100–1,200. Of this, only 550 could be MIRVed ICBMs. The Soviet heavy ICBM force would be cut to 150. The Soviets contemptuously rejected the proposal. The Carter administration's first attempt at "real arms control" ended in disaster.[76]

As the SALT II negotiations were resumed, an issue appeared that linked the monitoring of ICBM deployment, SALT enforcement, limitations of the satellites, and the politics of verification. The U.S. wanted the Soviets to accept several "counting rules"—specifically, that once an ICBM had been tested as a MIRV, *all* missiles of that type must be counted under the MIRV limit. An expansion of this was the "launcher-type counting rule." If a particular type of silo was used to launch a MIRVed ICBM, *all* such silos would be deemed to hold MIRVed missiles. Otherwise, the U.S. would have to keep watch on every silo all the time. The danger was the Soviets would secretly replace, under cover of darkness and clouds, single-warhead missiles with MIRVed ones. This requirement developed because of the missile fields near the small Ukrainian towns of Derazhnya and Pervomaisk. During the ICBM breakout of the late 1960s, the Soviets had deployed 60 silos near each town—120 in all—containing single-warhead SS-11s. Then in 1971, the Soviets began building another 30 silos of a new type at each field. Once completed, they were fitted with MIRVed SS-19s. The Soviets also rebuilt old SS-11 silos to the same configuration as the SS-19 holes. Thus, in the Derazhnya and Pervomaisk fields (or D and P) there were 120 single-warhead SS-11s and 60 MIRV SS-19s intermixed in identical silos. The only visible difference between the two was not in the silos but in their launch control centers. The SS-19 launch control centers had a domed antenna (called "the midget") not present at the SS-11 centers. Although the U.S. knew which silos held which missiles, there was concern this would not be true in the future. The U.S. wanted all the D and P silos counted as MIRVs under the launcher-type rule. The Soviets rejected this repeatedly during the summer of 1977, claiming the domed antenna was proof enough. Finally, the U.S. revealed it had observed the firing of a MIRVed missile from a test complex that lacked the domed antenna—it was irrelevant whether a silo held a MIRV. The D and P issue caused as much difficulty inside the Carter administration as it did with the Russians. Some felt the possibility of a secret substitution of missiles was a real danger. Others viewed it as a means of forcing the Soviets to accept the launcher-type counting rule. There was also a political dimension: a treaty giving the Russians the benefit of the doubt, on D and P or any issue, would not get through Congress. Some felt pressing the Soviets too hard on D and P was dangerous. It could force them to actually equip the silos with SS-19s—a similar situation that existed at Malmstrom Air Force Base (single-warhead Minuteman IIs and MIRVed Minuteman IIIs in similar silos). DOD rejected this, saying the Soviets intended all along to put in SS-19s and that Malmstrom was a phony issue. President Carter decided to bring up the D and P issue with the Russians. The limitations of the satellites, different outlooks, and the politics of verification transformed the D and P missile fields into a key test of Soviet flexibility and of whether SALT could continue.[77]

As the D and P issue became increasingly bitter, the satellite watch continued. The first KH-11 continued to operate. The first reconnaissance satellite launched during the Carter administration was a Titan III B, orbited on 13 March 1977. It operated for 74 days, reentering at the end of May. *Big Bird 13* followed on 27 June, remaining in orbit 179 days (a new record). Three months later, on 23 September, it was joined by a Titan III B, which operated for 76 days, finally reentering in December 1977. *Big Bird 13* was deorbited soon after. This was the first time all three types of U.S. reconnaissance satellites were in orbit simultaneously.

In September, President Carter met with Soviet Foreign Minister Andrei Gromyko. After several angry exchanges, Gromyko agreed to count all D and P silos as MIRVed. This was part of a deal that included a MIRVed ICBM ceiling, a lowering of total launchers, and the U.S. counting B-52 cruise missile carriers as MIRVed.[78]

Soon after the agreement was made, U.S. satellites observed SS-19s being loaded into the old SS-11 silos at D and P. This was part of the approximately 125 SS-17s, 18s, and 19s the Soviets were deploying each year. By early 1978, about 60 SS-17s were operational. The rate of SS-17 deployment was rather slow, probably to avoid having too many silos out of service at one time. There were also about 110 of the SS-18s (one-third

of the heavy ICBM force). Most numerous were the SS-19s, with more than 200 deployed and more silos being prepared.[79]

There was also a new Soviet missile that was becoming increasingly important in both U.S.-European relations and SALT—the SS-20 IRBM. It used the SS-16 first and second stages and carried three 600-kiloton warheads. The SS-20 was launched not from a silo but a large truck. It worried NATO because its mobility, multiple warheads, and accuracy were a greater threat than the SS-4s and 5s. The SS-20 was an IRBM with a range of 2,700 nautical miles (5,000 kilometers)—too short to be covered by SALT. It became an issue because it used the SS-16's first two stages. The fear was the Soviets would produce SS-16 third stages, fit them on the SS-20s, and turn them into ICBMs. The SS-16 had been tested from a launcher truck nearly identical to the one used by the SS-20. Because the missile was mobile, verification could be difficult. The SS-16/SS-20 would be fired from a presurveyed site—a road or forest clearing. For this reason, the U.S. pressed for a ban on SS-16 testing and production.[80] The U.S. also kept watch on the number deployed. Deployment began in 1977 with the first 20 launcher trucks. By 1978, this had grown to more than 100. They were deployed in both western Russia, to cover Europe, and in Siberia, targeted on China and Japan. Some were based at old SS-7 and 8 coffin sites.[81,82]

Nineteen seventy-eight was a quiet year for U.S. reconnaissance satellite launches. The first KH-11 was deorbited in January 1978 after 25 months in space. *Big Bird 14* was launched on 16 March and the second KH-11 followed on 14 June. There were no Titan III B launches during the year. *Big Bird 14* operated for 179 days, the same as its predecessor, reentering in September 1978. For the rest of the year, the second KH-11 continued coverage.[83] Their photos indicated that the deployment of new ICBMs was continuing unabated. By 1979, there were more than 100 SS-17s, more than 200 SS-18s, and at least 300 SS-19s.[84]

In October 1978, during the ceremonies at Cape Kennedy for NASA's twentieth anniversary, President Carter raised the curtain slightly. During his speech, he officially acknowledged that the U.S. used reconnaissance satellites. This ended the official blackout established by the Kennedy administration. Of course, by this time, use of reconnaissance satellites was common knowledge.

The Carter administration continued to wrestle with SALT issues such as finding a basing mode for the MX ICBM, cruise missile range, the Soviet TU-26 Backfire bomber, and limiting the number of warheads on Soviet ICBMs to four on the SS-17, ten on the SS-18, and six on the SS-19. This was important, since the "light" SS-19 was now seen as more dangerous, due to its greater numbers, than the "heavy" SS-18.

Progress was being made, but slowly, which increased the skepticism, a reflection of which was the direct challenge to the belief that the Soviets also accepted MAD. It was argued the Soviets believed they could fight and win a nuclear war. It would be the final apocalyptic struggle between two systems—socialism and capitalism. At its end, the Soviet goal was the U.S. and China destroyed, NATO occupied, Soviet domination of the Eurasian landmass, and the final triumph of socialism throughout the world.[85]

The question, going as it did to the heart of U.S. strategy and arms control policy, became a contentious issue. Both sides had impressive quotes from Soviet generals and officials to back up their views. From them, a dual picture emerges—that military superiority was necessary, and the outcome of any war (including nuclear) would be favorable to socialism if the necessary preparations were made. At the same time, the price of this political gain was a catastrophe far worse than the Soviet experience in World War II.[86,87]

Almost lost in the debate over war-fighting capabilities was something more important—the Soviet view that the "correlation of forces" was changing in their favor. This was the Soviet term for the sum of economic, political, moral, and military power of the U.S. and Russia. (As the Soviet economy was stagnant and the country was unable to feed itself, the part played by nuclear weapons was major.) From a U.S. nuclear monopoly at the end of World War II, Russia now had numerical superiority three decades later.

The Soviets believed this limited the West's room to maneuver and ability to resist—what they called "the realization by Western ruling circles of the new realities of our day and the corresponding correction of their political line." Some saw pressure or threats, rather than a Soviet first strike, as the real threat the West faced. This was called "Finlandization"—a still-independent nation is forced to submit to Soviet demands. The variation is "self-Finlandization" where the nation is dominated silently without overt threats.[88]

The Carter administration's ability to negotiate was also harmed by the cancellation of the B-1 bomber and the "neutron bomb." The latter put a strain on European relations and gave Carter a reputation for vacillation. Some began to speak of the "action-inaction phenomenon" to describe U.S. passivity in the face of a Soviet arms buildup.

Although most attention was on Big Bird, KH-11, and the Titan III B satellites, one could not overlook two satellites that supported them. The first was the Air Force weather satellite program. A new, more advanced version—the DMSP Block 5D—was introduced in the mid-1970s. The first launch was made on 11 September 1976. The Thor Burner 2 launch vehicle placed it in a 520-mile (835-kilometer) sun-synchronous orbit. Soon after, it began to tumble due to the failure of a solar panel to latch and a high-pressure nitrogen leak. The batteries were soon drained and the satellite seemed lost. However, on 5 October, the solar panels began to generate enough power for an attempt to be made to get it back under control. This effort became more urgent in February 1977, when a gyro problem appeared and became worse. By March, the software changes were radioed up to the satellite. On 24 March, one day before the gyro problem would have made it impossible, the weather satellite was stabilized. It was declared operational on 1 April 1977, and began returning cloud-cover photos over Eastern Europe, Russia, China, and other areas of interest. The second satellite followed on 4 June 1977. It also tumbled but the problem was corrected by 26 June. The third and fourth satellites followed on 30 April 1978 and 6 June 1979.[89]

The other support satellite was the Satellite Data System. Built by Hughes Aircraft Company, it was used both to communicate with bombers in polar regions and, more importantly, to relay KH-11 photos to earth. The first test satellite, SDS-A, was launched on 21 March 1971. It was followed by SDS-B on 21 August 1973. The first operational satellite, SDS-1, was launched on 10 March 1975. Two launches were made in the summer of 1976 (just before the first KH-11 was orbited). Subsequent launches followed at roughly

yearly intervals. SDS satellites are placed into an unusual orbit, with a high point around 24,545 miles (39,500 kilometers) above the Northern Hemisphere. Its low point was 194 to 256 miles (312 to 412 kilometers). The highly elliptical orbit meant it would spend several hours over Russia. From the ground, it would seem to move slowly across the sky. The KH-11 would transmit up to the SDS high above and the signals would be retransmitted back to earth. Contact could be maintained for perhaps 30 minutes at a time. Earth ground stations could receive signals for only 10 minutes. The photos would be in the hands of the analysts nearly "live." Because of its mission, secrecy envelopes the SDS. Unlike other military communications satellites, no photos of it have been released and the exact number of launches is unknown.[90]

As 1979 began, SALT was at the make-or-break point. It seemed that as one loose end was tied, another part would unravel. Outside issues, such as the establishment of formal diplomatic relations with China in December 1978, also complicated the talks. Looming above all was the difficulty of getting the treaty through the Senate. Finally, in early May, the last few issues were settled and the signing was scheduled.

At about the same time, the year's two reconnaissance satellite launches were made. *Big Bird 15* was first, on 16 March 1979. It continued to function for a record 190 days—more than 6 months. Next was a Titan III B on 28 May that also set a record—90 days, or ten times the lifetime of the first launches. The second KH-11, launched the year before, continued to operate.

The Soviets' missile modernization program was also nearing completion; more than

150 SS-17s, 308 SS-18s, and at least 300 SS-19s were operational. There were also about 160 SS-20s.[91]

The SALT II treaty and related documents, signed by Jimmy Carter and Leonid Brezhnev on 18 June 1979 in Vienna, were extremely complex. Each side was allowed 2,400 strategic nuclear delivery systems: 1,320 could be MIRVed ICBMs, SLBMs, and bombers with cruise missiles; no more than 1,200 could be MIRVed ICBMs and SLBMs; and MIRVed ICBMs were limited to 820. Of this subtotal, 308 were heavy missiles. The MIRV and launcher counting rules were included, as was a ban on SS-16 deployment. B-52s meant to carry cruise missiles and Mya-4s modified as tankers would have to be given "functionally related observable differences" to indicate their role. There was also a separate statement by Brezhnev promising that production of the TU-26 Backfire bomber would not exceed thirty per year and that it would not be given an aerial refueling capability or be based in areas that would enable it to reach the U.S. This was done to allay fears that the Soviets might turn it into a strategic bomber. In exchange, Backfires would not be counted under SALT.[92]

Many of the treaty's features reflected the role of reconnaissance satellites. The "functionally related observable differences" were, for example, to enable analysts to separate the specific types of bombers. The MIRV and launcher rules were to ease the job of keeping track of missiles. The SS-16 ban was to ease the job of verification. Other provisions of the treaty—the limit on total number of warheads on each missile, the limits of one new ICBM type, and changes in missile size—would be monitored by telemetry intercepts, radar tracking, and electronic intelligence sat-ellites. With its specific details, the SALT II treaty was a monument to the capabilities of modern intelligence. It was now up to the United States Senate to decide if it was enough. The congressional hearings and debate over SALT dominated the political news during the summer and fall. Then a totally unexpected event occurred that swept aside all else.

The Iran Hostage Crisis

It was a Sunday morning. On 4 November 1979, a mob of Iranian student militants climbed the fence surrounding the U.S. embassy and took the diplomats hostage. At first, the militants had planned to stay only a few hours to protest the admission of the Shah to the U.S. for medical treatment. The militants became overnight heroes and, with the support of the Ayatollah Ruhollah Khomeini, Iran's absolute theocratic ruler, no one dared challenge them. The militants-turned-kidnappers did not know what demands to make at first. They settled on the Shah's return for trial and execution, the return of the Shah's fortune, an apology for U.S. "crimes against the Iranian people," and payment of damages. Carter never gave them any serious consideration; instead he sought to negotiate the hostages' release through diplomacy. Contingency plans were also made. If the hostages were put on trial, Iran's seaports would be mined. If they were punished or killed, air strikes would be made. On 6 November, Carter also authorized planning for a rescue attempt. In the preparation of these plans, satellites played a large role. Carter would write later about ordering satellite photos to locate Iranian aircraft and other military forces. He also pored over photos of oil refineries and other strategic targets to be destroyed if the hostages were harmed.[93]

Intelligence in all its forms was especially critical for any rescue attempt. Delta Force, the U.S. antiterrorist unit, needed to know where the hostages were kept, how many guards there were, how they were armed, where reinforcements were and how quickly they could arrive—literally, volumes of information. The U.S. had no agents in Iran to supply such information.

In December, the Delta planners began receiving satellite photos of the embassy from the second KH-11. From the photos, analysts spotted antihelicopter poles the militants had placed in all likely landing zones. The analysts could also monitor day-to-day changes in activities. Any changes noted could illuminate another bit of information. Satellite photos were also used to select a refueling site for the helicopters; a hide site where the rescue team would wait out the daylight hours; and finally, an abandoned airfield where Delta and the hostages would transfer to jet transports to make their final escape. Satellite photos showed a likely refueling site about 200 miles south of Tehran—a flat area with a seldom-used dirt road that could serve as a landing strip. A hide site east of Tehran was also found, as was an abandoned airfield in a former bombing range west of the city.

On 7 February 1980, the third KH-11 was orbited. This was the first time two KH-11 satellites had been in orbit simultaneously.

By this time, the U.S. had developed other intelligence sources. One was the nightly TV news. The film showed how the gates were locked and what weapons the guards carried.[94] The U.S. also began sending in agents. They were Americans—professional intelligence operatives—who posed as foreigners and used business or media cover.

They were able to move freely. One agent was challenged because his (forged) German passport had the name Eric H. Schneider. A customs official questioned the lack of a middle name. The agent explained it stood for "Hitler." The customs officer was satisfied and allowed him to pass.[95] The agents assessed the situation on the ground, checking out whether a railroad near the hide site posed a problem, watching the embassy guards, and trying to learn where the hostages were being kept.[96]

By late March 1980, the diplomatic efforts seemed close to success. After an exchange of signals, control of the hostages would be transferred to the Iranian government as a first step towards their release. The signals were sent, but the Iranians could not carry through. The hostages were pawns in the internal struggle between the students, the street mobs, the Revolutionary Council, the "government," and Khomeini to control the revolution. Soon after, Iranian spokesmen announced some hostages would be tried as spies. Victims of Iranian kangeroo courts were shot within 12 hours of the rubber-stamp conviction.[97]

Diplomacy had failed, the Iranian government was powerless to act, Jimmy Carter was at the end of his patience—it was time.

The rescue plan was a complex one, due to distance and the embassy's location inside a major city. On the first night, Delta Force would enter Iran. Three MC-130s carrying troops and three EC-130s with fuel tanks would land at the refueling site (called Desert One). Soon after, RH-53D helicopters from the USS *Nimitz* would land. At least six of the eight helicopters would have to be operational to carry everybody and allow for failures later on. The RH-53Ds would be refueled and loaded and fly to the hide sites—one for the Delta troops, the other for the helicopters.

By sunrise, they would be under cover. The next night the Delta Force would ride to Tehran in trucks provided by U.S. agents. At the embassy, they would kill the guards, climb the wall, and storm the buildings. Once the hostages were safe, the RH-53Ds would land either in the compound, if the poles could be removed, or in a soccer stadium across the street. Air cover would be provided by two AC-130s; their 20mm cannons (able to fire 100 shells per second) would devastate any interference. Once the Delta troops and hostages were aboard, the helicopters would fly to the airstrip at Manzariyeh, which would be secured by a company of Rangers. Everybody would be loaded aboard C-141 transports and leave Iran under U.S. Navy fighter escort.

Satellite photos and agent reports indicated the guards had become complacent. President Carter approved the mission on 16 April.

The events of April 24 will be only briefly summarized. The C-130s, flying under the radar, reached Desert One successfully. The RH-53s ran into difficulty—one was forced to land, another turned back due to a dust storm, a third reached Desert One but was unable to continue. This left only five helicopters; the mission was aborted. As the helicopters were being refueled for the flight back to the USS *Nimitz,* one collided with an EC-130, killing eight crewmen.[98]

Like Johnson in Vietnam and Nixon with Watergate, Jimmy Carter saw his presidency being destroyed by the hostage crisis. The country was unified in a way it had not been since before Vietnam, but frustrated over the inability of the U.S. to gain the hostages' release. Some of the anger was directed at Carter. On 4 November 1980, the first anniversary of the embassy takeover, Carter was buried under a landslide vote for Ronald Reagan.

NOTES Chapter 7

1. "Industry Observer," *Aviation Week & Space Technology,* (12 December 1960), 23.
2. J. S. Butz, Jr., "New Vistas in Reconnaissance From Space," *Air Force and Space Digest,* (March 1968), 47,48.
3. Jeffrey T. Richelson, *The U.S. Intelligence Community,* (Cambridge: Ballinger, 1985), 300.
4. Curtis Peebles, "The Guardians," *Spaceflight,* (November 1978), 381.
5. James A. Fusca, "Improving Image Aids Reconnaissance," *Aviation Week & Space Technology,* (2 February 1959), 62–64.
6. James A. Fusca, "Satellite Optics Limited by Atmosphere," *Aviation Week & Space Technology,* (26 January 1959), 75,77–80,83,84.
   The scattering of sunlight by the atmosphere is also why the sky is blue.
7. David H. Smith, "King of the Space Telescopes," *Sky & Telescope,* (September 1985), 216.
8. R. J. Weymann and N. P. Carlton, "The Multiple-Mirror Telescope Project," *Sky & Telescope,* (September 1973), 162.
9. G. M. Sanger and R. R. Shannon, "Optical Fabrication Techniques for the MMT," *Sky & Telescope,* (November 1973), 280–282.
10. Philip J. Klass, "Recon Satellite Assumes Dual Role," *Aviation Week & Space Technology,* (30 August 1971): 12,13.
11. Philip J. Klass, *Secret Sentries in Space,* (New York: Random House, 1971).
    A minor irony: In the late 1970s, while an astronomy student at San Diego State University, I used a 16-inch telescope built by a division of Perkin-Elmer.
12. Philip J. Klass, "Recon Satellite Assumes Dual Role," *Aviation Week & Space Technology,* (30 August 1971): 12,13.
13. Dave Dooling, "Space Telescope Has Spies in Family Tree," *Space World,* (February 1983), 21,35.
14. "Industry Observer," *Aviation Week & Space Technology,* (7 June 1982), 11.
15. Philip J. Klass, "Recon Satellite Assumes Dual Role," *Aviation Week & Space Technology,* (30 August 1971): 12,13.
16. "Launch Vehicles," *Air Force Magazine,* (May 1976): 128.
17. "New Recon Satellite," *Aviation Week & Space Technology,* (21 June 1971): 15.
18. Philip J. Klass, "Recon Satellite Resumes Dual Role," *Aviation Week & Space Technology,* (30 August 1971): 12,13.
19. Philip J. Klass, "Big Bird Nears Full Operational Status," *Aviation Week & Space Technology,* (25 September 1972): 17.
20. Curtis Peebles, "The Guardians," *Spaceflight,* (November 1978), 381–385.
21. *New York Times,* 29 July 1971, 1.
22. Philip J. Klass, "Australian Pressure on U.S. Bases Eases," *Aviation Week & Space Technology,* (30 April 1973): 67,68.
23. Anthony Kenden, "U.S. Reconnaissance Satellite Programmes," *Spaceflight,* (July 1978): 254.
24. Lawrence Freedman, *U.S. Intelligence and the Soviet Strategic Threat,* (Boulder: Westview Press, 1977), 166,167.
25. Anthony Kenden, "U.S. Reconnaissance Satellite Programmes," *Spaceflight,* (July 1978): 254.
26. Philip J. Klass, "USAF Boosts Recon Satellite Lifetime," *Aviation Week & Space Technology,* (7 July 1975): 21,22.

27. "Space Reconnaissance Dwindles," *Aviation Week & Space Technology,* (6 October 1980): 18,19.

28. Craig Covault, "USAF, NASA Discuss Shuttle Use for Satellite Maintenance," *Aviation Week & Space Technology,* (17 December 1984): 14–18.

29. Anthony Kenden, "U.S. Reconnaissance Satellite Programmes," *Spaceflight,* (July 1978): 254.

30. Anthony Kenden, "Recent Developments in U.S. Reconnaissance Satellite Programmes," *Journal of the British Interplanetary Society,* (January 1982): 33–36.

31. Robert Berman & Bill Gunston, *Rockets & Missiles of World War III,* (New York: Exeter Books, 1983) 85,87.

32. "USSR Missile Comparisons, Development Pace," *Aviation Week & Space Technology,* (6 May 1974): 21.

33. Cecil Brownlow, "Soviets Developing New ICBMs," *Aviation Week & Space Technology,* (2 April 1973): 14.

34. "USSR Missile Comparisons, Development Pace," *Aviation Week & Space Technology,* (6 May 1974): 21.

35. Clarence A. Robinson, Jr., "Soviets Test Cold-Launch ICBM Firings," *Aviation Week & Space Technology,* (24 September 1973): 20.

36. Berman & Gunston, *Rockets & Missiles of World War III,* 85,88.

37. "USSR Missile Comparisons, Development Pace," *Aviation Week & Space Technology,* (6 May 1974): 21.

38. Berman & Gunston, *Rockets & Missiles of World War III,* 88.

39. John Prados, *The Soviet Estimate,* (New York: The Dial Press, 1982), 226–234.

40. Freedman, *U.S. Intelligence and the Soviet Strategic Threat,* 174,175.

41. "Industry Observer," *Aviation Week & Space Technology,* (8 May 1972): 9.

42. Anthony Kenden, "U.S. Military Satellites, 1983," *Journal of the British Interplanetary Society,* (February 1985): 62,63.

43. Anthony Kenden, "Recent Developments in U.S. Reconnaissance Satellite Programmes," *Journal of the British Interplanetary Society,* (January 1982): 33–36.

44. Anthony Kenden, "U.S. Reconnaissance Satellite Programmes," *Spaceflight,* (July 1978): 254.

45. Clarence A. Robinson, Jr., "Soviets Hiding Submarine Work," *Aviation Week & Space Technology,* (11 November 1974): 14–16.

46. Freedman, *U.S. Intelligence and the Soviet Strategic Threat,* 175.

47. Clarence A. Robinson, Jr., "Soviets Hiding Submarine Work," *Aviation Week & Space Technology,* (11 November 1974): 14–16.

48. Ralph Kenney Bennett, "U.S. Eyes Over Russia: How Much Can We See?" *Reader's Digest,* (October 1985), 142–147.

49. Prados, *The Soviet Estimate,* 237,238.

50. Adam B. Ulam, *Dangerous Relations,* (New York: Oxford University Press, 1983), 75.

51. Robert Hotz, "More SALT on the Table," *Aviation Week & Space Technology,* (29 April 1974): 11.

52. "Throw Weight Seen Critical in SALT 2," *Aviation Week & Space Technology,* (8 April 1974): 21.

53. Robert Hotz, "A Pinch of SALT," *Aviation Week & Space Technology,* (22 April 1974): 7.

54. Prados, *The Soviet Estimate,* 196–198.

55. Freedman, *U.S. Intelligence and the Soviet Strategic Threat,* 104–108,113,114.

56. Lawrence Freedman, *The Evolution of Nuclear Strategy,* (New York: St. Martin Press, 1983), 347.

57. Ulam, *Dangerous Relations,* Chapter 3.
58. Freedman, *U.S. Intelligence and the Soviet Strategic Threat,* 175.
59. P. H. Vigor et al, *The Soviet War Machine,* (London: Salamander Books, 1976), 212.
60. Anthony Kenden, "U.S. Reconnaissance Satellite Programmes," *Spaceflight,* (July 1978): 254.
61. *Jane's Fighting Ships 1975–76,* John W. Taylor, Editor, (London: Jane's 1976), 375.
62. "Industry Observer," *Aviation Week & Space Technology,* (13 January 1975): 11.
63. *Jane's All the World's Aircraft 1975–76,* John W. Taylor, Editor, (London: Jane's 1976), 375.
64. Prados, *The Soviet Estimate,* 241,242.
65. Clarence A. Robinson, Jr., "Another SALT Violation Spotted," *Aviation Week & Space Technology,* (31 May 1976): 12–14.
66. Anthony Kenden, "U.S. Reconnaissance Satellite Programmes," *Spaceflight,* (July 1978): 254.
67. Anthony Kenden, "Recent Developments in U.S. Reconnaissance Satellite Programmes," *Journal of the British Interplanetary Society,* (January 1982): 33–36.
68. Anthony Kenden, "Was 'Columbia' Photographed by a KH-11?," *Journal of the British Interplanetary Society,* (February 1983): 73–77.
69. Richelson, *The U.S. Intelligence Community,* 110.
70. "Industry Observer," *Aviation Week & Space Technology,* (8 May 1972): 9.
71. William Colby and Peter Forbath, *Honorable Men,* (New York: Simon and Schuster, 1978), 370.
72. James Ott, "Espionage Trial Highlights CIA Problems," *Aviation Week & Space Technology,* (27 November 1978): 21–23.
73. *Los Angeles Times,* 8 April 1982.
74. "Strategic Missiles," *Air Force Magazine,* (March 1978): 107.
75. Berman & Gunston, *Rockets & Missiles of World War III,* 91.
76. Strobe Talbott, *Endgame,* (New York: Harper Colophon Books, 1980), 39,60.
77. Talbott, *Endgame,* 109–119.
78. Talbott, *Endgame,* 128–130.
79. "Strategic Missiles," *Air Force Magazine,* (March 1978), 107.
80. Talbott, *Endgame,* 72.
81. "Strategic Missiles," *Air Force Magazine,* (March 1978), 107.
82. Berman & Gunston, *Rockets & Missiles of World War III,* 83,85,91.
83. Anthony Kenden, "Recent Developments in U.S. Reconnaissance Satellite Programmes," *Journal of the British Interplanetary Society,* (January 1982), 33–36.
84. "Strategic Missiles," *Air Force Magazine,* (March 1980): 132,133.
85. *Soviet Military Power 1985* (Department of Defense, 1985), 13.
86. Freedman, *The Evolution of Nuclear Strategy,* 365,366.
87. Rebecca V. Strode, "Soviet Strategic Style," *Comparative Strategy,* (Vol. 3 #4, 1982): 328–333.
88. Freedman, *The Evolution of Nuclear Strategy,* 366,367.
89. *The Aerospace Corporation It's Work: 1960–1980,* (Los Angeles: Times Mirror Press, 1980) Chapter 9.
90. Jeffrey T. Richelson, "The Satellite Data System," *Journal of the British Interplanetary Society,* (May 1984): 226–228.
91. "Strategic Missiles," *Air Force Magazine,* (March 1980): 132,133.
92. Talbott, *Endgame,* 295–326.
93. Jimmy Carter, "Keeping Faith," *Time,* (18 October 1982): 52,53.

94. Col. Charlie A. Beckwith USA (Retired) and Donald Knox, *Delta Force,* (San Diego: Harcourt, Brace, Jovanovich, 1983), 192,196.
95. Jimmy Carter, "Keeping Faith," *Time,* (18 October 1982): 57.
96. Beckwith and Knox, *Delta Force,* 239.
97. Benjamin F. Schemmer, "Presidential Courage—and the April 1980 Iranian Rescue Mission," *Armed Forces Journal,* (May 1981): 60–62.
98. Beckwith and Knox, *Delta Force,* 253–257.

CHAPTER 8

Reconnaissance Satellites in the 1980s

The second and last reconnaissance satellite launch of 1980 came after the failure of the Iran hostage rescue effort. *Big Bird 16* was orbited on 18 June 1980. The satellite's lifetime was 261 days—nearly 9 months. It was not deorbited until 6 March 1981.

The pattern of reconnaissance satellite operations had changed from that of the mid-1970s (a Titan III B launch near the end of the Big Bird's lifetime). One reason was the multiyear coverage the KH-11 could provide. The number of launches could be cut to two per year. The other was the KH-11's high cost. The overruns caused the Carter administration to halt production of the Big Bird and Titan III B film-return, high-resolution satellites. In 1980, there were enough Big Birds to last through 1983. The four remaining Titan III B satellites were held in reserve for high-priority coverage.

The decision was controversial. The KH-11 had greater flexibility than the shorter-lived, film-return satellites. It could provide repeated coverage of targets, which could not be done by film-return satellites that were limited by film supply and reentry capsules.[1] Photo interpreters need such regular coverage to determine the flow and rhythm of a place. In this way, changes can be spotted. Emphasis on the KH-11 meant the U.S. could better monitor Soviet activity. However, the dimin-

ished number of film satellites meant fewer high-resolution photos, which lessened the amount of technical intelligence that could be gathered (the KH-11's resolution was too low). This came at a time when the need for such information was growing.

Some argued the problem was much broader. Because of the satellite's long lifetime, high reliability, and cost, only two launches per year were made. Although this could provide the necessary coverage, it was "technically fragile" in that there was no margin for failure. The program had great capabilities but lacked depth.

To replace the high-resolution satellites, a modified KH-11 was under development. The KH-12 Ikon satellite also used digital imaging but with a resolution like that of film-return satellites, which could be as good as 6 inches (15.2 centimeters). The KH-12 Ikon would perform both high-resolution and area surveillance; photos would be relayed to earth in real time. The first launch would be in 1984, with the booster being either a Titan III D or the space shuttle. A four-satellite network would ultimately be established. The KH-12 Ikon could also be serviced in orbit by shuttle astronauts. This preventative maintenance, repair, and refueling would greatly extend the satellite's lifetime.[2–4]

On 20 January 1981, Ronald Reagan became the fortieth president of the United States. The new administration faced several problems inherited from the last year of the Carter administration. The problems came from the Soviet invasion of Afghanistan in December 1979. In response, Carter had withdrawn the SALT II treaty, embargoed grain and technology sales, and boycotted the Moscow Olympics. The Western alliance was less than united over the sanctions. European leaders, although saying Soviet troops must leave Afghanistan, stressed the need for continuing East-West dialogue. Europe was unwilling to take actions against Russia so long as its vital interests were not directly threatened. The Soviets exploited this split. They depicted the new cold war as being due not to Soviet actions but to the U.S. response. Europe itself was also divided. In December 1979, NATO announced it would deploy cruise missiles and Pershing II IRBMs to counter the Soviet SS-20. The decision was not unanimous. Belgium and the Netherlands were very reluctant; West Germany strongly supported it.[5] The move sparked a growth in the ban-the-bomb and unilateral disarmament groups.

The SALT II treaty was in limbo. Reagan, during his presidential campaign, had called it "fatally flawed." After the election, he stated the U.S. would observe it even though it had not been approved by the Congress. The emphasis was not on arms control but on a buildup. The MX ICBM, Trident SLBM, and B-1 and stealth bombers were all to go ahead. Conventional forces, such as the Navy, were also to be improved to correct a decade of neglect.

Many in the Reagan administration viewed arms control with suspicion if not outright hostility. On the other side, the U.S. peace movement was gaining strength. In late 1981, this coalesced around a nuclear freeze on de-velopment, testing, and deployment of nuclear weapons. City councils began passing resolutions supporting a freeze. At the same time, President Reagan made his first arms control proposal. On 18 November, he called for the Soviets to dismantle their SS-4, 5, and 20 missiles in exchange for NATO's abandonment of its modernization plans. The "zero option" was rejected, but talks on medium-range missiles in Europe began on 30 November.[6]

Two U.S. reconnaissance satellites were launched during 1981—a Titan III B on 28 February and the fourth KH-11 on 3 September. This was a replacement for the second KH-11 which was deorbited on 23 August 1981 after 38 months of operation (*Big Bird 15* was deorbited on 6 March). The third KH-11 continued to operate through the year. Thus, except for 11 days, there were two KH-11s in operation simultaneously.[7,8]

The satellites' photos indicated the Soviet ICBM modernization program was completed. The final force was 150 SS-17s, 308 SS-18s, and 360 SS-19s, all deployed in the superhard silos. The remainder were 520 SS-11s and 60 SS-13s. The SS-17 and 18 silos can be reloaded for follow-on strikes. Plans had been made for the survival of personnel and equipment needed to decontaminate and reload them.[9]

At the other end of the military spectrum, reconnaissance satellites were also used to monitor Soviet military activities in Afghanistan. According to one account, on several occasions Soviet troops were caught in their own chemical weapons. Satellite photos were claimed to show Soviet troops undergoing chemical decontamination at front-line locations.[10]

Reports about an unsuspected capability of the KH-11 surfaced in 1981. On 12 April space shuttle *Columbia* was launched on its first flight. About 2 hours after lift-off, television transmissions showed several damaged heat-shield tiles. Although not in a critical area, they raised concerns that other tiles were missing from the shuttle's underside. This could cause its loss during reentry. NASA requested DOD use ground-based telescopes to photograph the shuttle. Telescopes had been used since the mid-1960s to observe Russian satellites. Their resolution was several feet for a low-orbit satellite. These attempts were unsuccessful, but according to some reports, "other DOD resources" were being used. The fears proved groundless; no other tiles had been lost and *Columbia* reentered and landed successfully. Six months later, *Aviation Week & Space Technology* magazine indicated a KH-11 had been used to photograph the shuttle. Later, Anthony Kenden, a writer specializing in U.S. military space activities, analyzed the claims. Using tracking data from the Goddard Spaceflight Center, he found the second KH-11 made a close approach 28 hours after the shuttle's launch. The third KH-11, also in orbit at the time, never came closer than 3,554 miles (5,720 kilometers). The encounter came when the shuttle's and KH-11's orbits crossed over northern Missouri (neither made any special maneuvers); the KH-11 was at an altitude of 196 miles (316 kilometers) while the shuttle was 169 miles (272 kilometers) high. Closest approach came at 28 hours, 42 minutes, 12 seconds after the shuttle's launch; minimum separation between the two was 51 miles (82.5 kilometers). At this distance, the KH-11's resolution (allowing for the lack of atmospheric distortion) would be about 15.75 inches (40 centimeters). The encounter was fleeting; the orbits crossed at a right angle and the shuttle was in photographic range for only 13 sec-

onds. The angular motion between the two was 7.54 degrees per second. The rate for ground photography was between 0.86 and 1.79 degrees per second. Also, even at closest approach, resolution was marginal (a black tile is 8 inches (20 centimeters) square). All this led Kenden to question the KH-11's ability to return any meaningful data.[11] Although claims have continued, the KH-11's satellite-inspector role is as yet not proven.

The peace movement continued to grow in strength during 1982—demonstrations in the U.S. and Europe grew larger, antinuclear books crowded the shelves, speakers and classes increased in colleges. The movement also scored political gains. Freeze resolutions were introduced in the Senate and House. Although both failed, the margin in the House was 204 to 202. Freeze resolutions were also on the ballot in nine states, the District of Columbia, and twenty-nine cities and counties. The resolution passed in most elections. In Europe, the impact was stronger. The British Labour Party endorsed unilateral disarmament. In West Germany the "Greens"— a left-wing, environmentalist, antinuclear group—were influencing main-line parties.

Often the peace movement seemed less a political effort than a moral crusade, which was reflected in church involvement. In May, the World Council of Churches said, "The time has come when the churches must unequivocally declare that the production and deployment, as well as the use, of nuclear weapons are a crime against humanity." The same month, the Russian Orthodox church (i.e., the Soviet government) sponsored the World Conference of Religious Workers for Saving the Sacred Gift of Life from Nuclear Catastrophe.[12]

Russian support for the peace movement was not limited to conferences. In November

1981, the Soviets announced a halt in construction of SS-20 bases in European Russia. In mid-March 1982, Brezhnev offered to halt deployment if NATO gave up its modernization plans. This was followed in May by a unilateral freeze on base construction.

U.S. reconnaissance satellites told a different story. In late January 1982, the Soviets deployed new SS-20s at a new base, bringing the total force to 280 at thirty-two bases (200 missiles and twenty-two bases in Europe, the rest in Asia). Several more bases were under construction. By September, three new bases were operational and the number of SS-20s had increased to 324.[13,14]

There were three reconnaissance satellite launches in 1982. The first, a Titan III B launch on 21 January, was a special mission. The satellite, with the international designation 1982-6A, was placed in a sun-synchronous, 330- by 88-mile (531- by 141-kilometer) orbit. It seemed to be an ordinary high-resolution reconnaissance satellite.[15] Anthony Kenden found otherwise. Normally, a Titan III B satellite follows a daily cycle of orbital corrections because of its low altitude. Kenden found, in contrast, 1982-6A raised its orbit the day after launch to 401 by 344 miles (645 by 553 kilometers), far above those used by Titan III B reconnaissance satellites. On 29 January, the low point was raised to 362 miles (582 kilometers). By 9 March, it had maneuvered six more times, ending up in a 407- by 387-mile (655- by 622-kilometer) orbit, where it remained for a week. It then began lowering its orbit. By 22 April, it was in a 381- by 373-mile (613- by 601-kilometer) orbit. Five days later, it made a major maneuver, raising its orbit to 401 by 393 miles (645 by 633 kilometers). It had maneuvered sixteen times in 96 days. Finally, on 23 May it was

deorbited after 122 days of mystifying behavior. Kenden examined various military space missions and found 1982-6A was most probably to test systems for the KH-12 Ikon.[16]

The other two satellites were more conventional. First was *Big Bird 17,* launched on 11 May 1982. About a month later, the third KH-11 malfunctioned. Kenden found, from Goddard tracking data, a change in orbital behavior at this time.[17] The satellite was apparently over the Soviet Union when it malfunctioned. There was concern it had fallen victim to a Soviet ground-based laser. It was doubted it had been attacked, but the satellite was out of radio contact when the malfunction occurred.[18] The third KH-11 continued to orbit until November, when it reentered. It had been aloft for thirty-three months.

The fifth KH-11 was launched on 17 November 1982. A new pattern had appeared—a pair of KH-11s would conduct routine reconnaissance. Each year, a single film-return satellite would be launched for high-resolution coverage. The Titan III B and Big Bird would alternate—the Titan III B in odd years, Big Bird in even ones.

A critical test of the peace movement's power would come in 1983. It was an election year in Europe, strategic weapons negotiations (called START for STrategic Arms Reduction Talks) had resumed, and the Pershing II and cruise missile would be deployed at the end of the year. Results were mixed. The U.S. House passed a freeze resolution while the Senate rejected it. The Roman Catholic church issued a pastoral letter attacking U.S. defense policy. The English elections in June were never in doubt. Conservative Prime Minister Margaret Thatcher was riding a wave of popularity from the Falklands victory the year before. Less than a quarter of Britain's population supported unilateral nuclear disarmament (a key point in the Labour Party's campaign). Far closer were the West German elections in March. Soviet interference with the campaign was blatant; in January, Soviet President Yuriy Andropov openly endorsed the center left Social Democrat party. The election results were a resounding victory (48.8 percent) for Chancellor Helmut Kohl's Christian Democrats, who had supported the NATO missiles. The antinuclear "Greens" picked up 5.6 percent, which translated into twenty-seven Bundestag seats.

Despite the lack of political success, the demonstrations grew larger as missile deployment approached. The climax was reached in late October with an "Action Week." On 22 October, two million people demonstrated in Europe. London, Rome, Bonn, Hamburg, West Berlin, and Stuttgart all saw several hundred thousand protestors. The first cruise missile arrived at Greenham Common in England on 14 November. On 22 November, the Bundestag voted overwhelmingly to approve deployment of the Pershing IIs. The next day, the first arrived. In retaliation, the Soviets walked out of the arms reduction talks.[19]

U.S. reconnaissance satellite activities remained at a low level but with a change in the pattern. As before, two reconnaissance satellites were launched, but in 1983, both were film-return satellites. A Titan III B was launched on 15 April 1983, and operated for 128 days. It was followed by *Big Bird 18* on 20 June. The launch was the first to use the West Coast version of the Titan 34 D—an upgraded Titan III D with a longer first stage and solid-fuel, strap-on rockets, which increased payload from 24,255 pounds (11,000 kilograms) to 31,973 pounds (14,500 kilograms).[20] Between the Titan III B's launch

and its reentry on 21 August, the U.S. was operating all three types of reconnaissance satellites—the fourth and fifth KH-11s, the Titan III B, and *Big Bird 18*.

There was also a change in policy regarding the Goddard tracking data. Up to this point, daily reports were made about reconnaissance satellites' orbital behavior, which had allowed Kenden to make his analysis. Starting with 15 and 16 June 1983, however, no orbital elements for reconnaissance satellites were published.[21] Apparently, NORAD (North American Aerospace Defense Command) realized just how much could be learned from the data. Also, there was concern the photo targets could be determined. Although the Soviets have their own tracking network, it was decided not to give them the data "for free."

The satellites' photos again raised the issue of Soviet treaty compliance. In 1983, the Soviets began flight testing of two new solid-fuel ICBMs. The SS-X-24 carried up to ten MIRVed warheads and was designed for both silo and mobile deployment. The SS-X-25 was a single-warhead missile, similar to the Minuteman, carried on a large truck.[22] The SALT II treaty permitted only one new type of ICBM. There was also evidence the Soviets had deployed up to 200 SS-16s at Plesetsk, concealed under camouflage netting.

The Soviets were also pressing the limits of the ABM treaty. They were testing two new types of ABM missiles; testing the reloading of ABM launchers; producing ABM guidance and tracking radars; flight testing the SA-12 SAM against reentering warheads; and building two new phased-array, early-warning radars. The latter were an embarrassment. The radar at Pechora in northern Russia had not been detected for 18 months.[23] The radar at

Krasnoyarsk in Siberia had been under construction for a year before its detection in August 1983. It was apparently a defector who first warned the U.S. about the radars.[24] The Krasnoyarsk radar was a violation because it was located in the interior of Russia. Some believed its purpose to be defense of three SS-18 fields nearby.[25] Taken together, these moves seemed to indicate the Soviets were preparing to make an ABM breakout.

Dirty Pictures

The Reagan administration tried various means of countering the peace movement. One was to issue booklets containing previously classified data. Called *Soviet Military Power,* the first was published in September 1981, the second in March 1983, and others followed at yearly intervals. The booklets were criticized because they used sketches and paintings to depict Soviet weapons. Several European officials said they had little or even negative impact; the solution they advocated was the release of satellite photos. British Ambassador Sir Oliver Wright said, "Actual photos, if presented without much propaganda, would have impact."[26] There was historical support. After President Kennedy's speech, many in England had opposed the Cuban blockade; this turned around when the U-2 photos of the missile sites were released.[27] Satellite photos had been shown privately to allied officials. In 1982, the Japanese defense minister was shown a photo of the Russian aircraft carrier *Minsk* being repaired in a Japanese-built floating drydock. The drydock had been sold when the Soviets said it would be used for civil shipping. The photo was to underline U.S. concerns over Japanese exports being used by the Soviet military.[28]

There were security objections to public release. Technical details of the satellite's camera system could be determined from the photos. Also, CIA and DOD lawyers argued that release of even one satellite photo would cause all the others to be vulnerable to freedom-of-information requests.[29] These arguments prevailed and any plans to release photos were dropped. It was no surprise to observers. They had long decided that only another Cuban missile crisis would force the U.S. to release satellite photos. Short of this, all but the cleared few would be left forever curious about the photos. It was unimaginable that satellite photos would be released by accident. Yet, that is how it happened. CBS news reporter Bill Lynch was reading a volume of hearings on the DOD FY 1984 budget. In it was a chart dealing with two new Soviet fighters, the SU-27 Flanker and MiG-29 Fulcrum. Three lines had been deleted from the chart but the two photos remained; they had been taken from above—far, far above. CBS news had the photos in the book enlarged and showed them in a late February 1984 telecast.[30] The public had its first look at one of America's most carefully guarded secrets.

The second look followed 6 months later. The 11 August issue of *Jane's Defence Weekly* carried three photos of a Russian aircraft carrier under construction at the Nikolayev shipyard on the Black Sea.[31] The photos had been taken by a KH-11 satellite during 3 days in July 1984. On 1 October, Samuel Loring Morison, an analyst with the Naval Intelligence Support Center, was arrested. Morison admitted to FBI agents he had sent the photos to *Jane's.* The case took a year to reach trial. Beginning on 8 October 1985, Morison was charged under a 1917 espionage law that prohibited giving secret information to a person not authorized to receive it. Although leaks

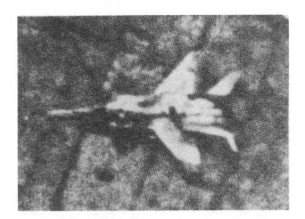
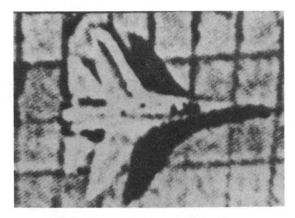

The first satellite photos seen by the public. The resolution is comparable to that of high-altitude aircraft photos (around 1 foot [.3 meters]). *Above left:* The Soviet MiG-29 (NATO code name Fulcrum). It was first spotted by satellite in 1977 and began entering service in early 1985. The MiG-29 is an all-weather interceptor with a look down–shoot down capability. The aircraft is broadly similar to the F/A-18 and is expected to eventually replace the MiG-21, -23, and -27. The black dot on the back of its fuselage is probably a ground crewman. Photo was taken from an altitude of 100–150 miles (161–241 kilometers). *Above right:* The Soviet Su-27 Flanker, an air superiority fighter equipped with long-range missiles and look down–shoot down radar. It is intended to defend against low flying bombers and cruise missiles. Deliveries began in early 1986 to air defense squadrons. *Below:* Chart showing the characteristics of the MiG-29 Fulcrum and the Su-27 Flanker. This chart was part of the congressional hearings on the Fiscal Year 1984 defense budget. It was subsequently published by accident.

One of the leaked KH-11 photos of the Soviet aircraft carrier *Leonid Brezhnev* at the Nikolayev shipyard, published in *Jane's Defence Weekly*. Visible are two large bridge cranes and several smaller ones. The resolution is probably several feet (sufficient to detect objects the size of trucks, but not people). Visible on the side of the ship's hull are scaffolds used by the shipyard workers.

The aircraft carrier is approximately 1,000 feet (300 meters) long and displaces 65,000 tons, somewhat less than a U.S. *Nimitz*-class carrier. It has an angled deck and what appears to be a ski jump bow (although it is not yet clear if it will have a full deck or antiaircraft and antishipping missiles on the forward deck). The two elevators and island superstructure are on the starboard side. The ship was launched in December 1985 and is expected to begin sea trials in 1989.

An oblique photo like this gives added information as it shows the side of objects rather than just a top view. The camera, however, must view the target through a greater thickness of the atmosphere. This is compensated for by large mirrors the Big Bird and KH-11 are believed to carry.

were common and investigations to find their sources equally so, Morison was only the second person charged with espionage for giving information to the press rather than a foreign power. (The other was Daniel Ellsberg for release of the Pentagon Papers in 1971. The case was dismissed because of prosecutorial misconduct.) Morison's American Civil Liberties Union defense team argued that his prosecution posed a grave threat to the press. Journalists would be criminals for receiving classified materials.[32] The argument was not accepted and, on 17 October, he was convicted on two counts of espionage and two of stealing government property.[33] Morison subsequently received a 2-year prison sentence. He was eligible for parole after 8 months.[34]

The lesson of the five ''dirty pictures''

is that the British ambassador overestimated their power, and the Cuban missile crisis was a misleading analogy. The photos do not seem to have caused any significant shift in opinion; minds were already made up. When two of the aircraft carrier photos were shown to a university class on U.S. foreign policy, they were regarded as an interesting novelty. The

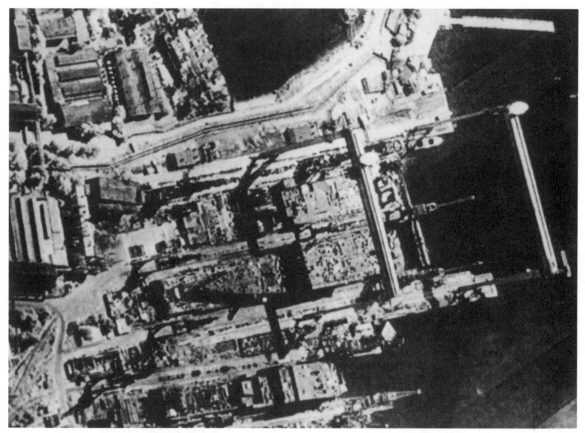

Another of the leaked KH-11 photos—this one looking directly down on the shipyard. This reduces the distance from the satellite to the target and the amount of atmosphere it must "look" through. Both improve resolution. This KH-11 photo has better resolution than many World War II high-altitude reconnaissance aircraft could achieve.

Photo interpreters can identify the function of each building in such a photo, measure floor space, and estimate the capabilities of the facility. The interpreters use stereo pairs of photos which overlap by 60 percent. When viewed through a stereoscope, the two photos merge into a single three-dimensional image. This resembles a model with a two- or three-fold vertical exaggeration. Details that otherwise would be hard to spot on a single print leap into prominence. Another technique is called "comparative covers." A series of photos, called "covers," are taken of a target over an interval of time. The photo interpreter can follow the construction step by step, which allows estimates of future forces to be made once the first signs of new construction are spotted.

Cuban missile photos were released at the height of the cold war, in a highly charged situation that tended to draw the West together. In the mid-1980s, that unity was gone.

The release of the photos in February and August came in the midst of the U.S. presidential campaign. The Democrats endorsed many of the peace movement's goals. Walter F. Mondale ran on a platform that included a nuclear freeze and cancellation of the B-1 bomber, MX ICBM, nerve gas production, and the Strategic Defense Initiative. Despite ''Another Freeze Voter'' bumper stickers, the issue was never in doubt. Mondale was seen as a symbol of worn-out liberalism, special interests, and the Carter years. Reagan seemed to reflect leadership, patriotism, and an optimistic future. On 6 November 1984, Reagan won forty-nine states.[35]

The Mondale disaster was a turning point for the peace movement. Its large demonstrations could not be turned into political success. The movement began to fade into the political background—present, but now one of the causes and movements on the American landscape no longer dominant. During the 1984 election, religious groups were more interested in banning abortion than the bomb.

Nineteen eighty-four was an eventful year for U.S. reconnaissance satellite operations. The first, a Titan III B, was launched on 17 April and continued to operate for 118 days. In June, the Air Force established a new naming system: military satellites would carry the designation ''USA.'' The first, USA-1, was a Navstar navigation satellite launched on 13 June. This was followed by *Big Bird 19* (USA-2) on 25 June. It deployed an electronic intelligence subsatellite designated USA-3. *Big Bird 19* (USA-2) operated for only 113 days—a disappointingly short lifetime for a Big Bird. USA-4, a Titan III B satellite, was launched

on 28 August and operated until the end of the year.[36,37] This was the fourth launched since the phaseout of the film-return satellites. Presumably, this used up the Titan III B satellites being held in reserve. The final reconnaissance satellite launch of 1984 was the sixth KH-11 (USA-6) on 4 December. The fifth KH-11 also operated through the year. The launch rate was twice that of previous years.[38]

Technically Fragile

Behind the high launch rate, there were problems. The KH-12 Ikon had run into technical difficulties. Originally scheduled for launch in 1984, by the start of the year this had slipped 2.5 years. It was planned for STS-62B, the second shuttle launch from Vandenberg, set for the fall of 1986. The satellite also had a $400 million cost overrun.[39,40] The first link in the chain was forged.

The first launch of 1985 was a surprise. USA-9 (8 February) was a Titan III B satellite—a species thought extinct.[41] Apparently, production had resumed or at least one or two ''extras'' had been built.

Winter would turn to spring and then to summer before the next reconnaissance satellite launch. Finally, on 28 August, a Titan 34 D stood ready. At its nose, reports would later say, was a KH-11. At about 2:20 P.M. PDT, the Titan 34 D's two solid-fuel rockets roared to life. Their thrust pushed the satellite into the California sky and ignited a brushfire (a standard occurrence with launches from the Space Launch Complex–4 pad). About 2 minutes into the flight, the two first-stage, liquid-fuel engines ignited. Several seconds later, the two solid-fuel, strap-on rockets were separated. The two first stage engines continued to burn, accelerating the Titan 34 D towards orbit. Then, 103.9 seconds into their burn, one of the engines shut down early. One possible reason was debris or fluids in the fuel lines. The failure occurred about 45 seconds before the normal shutdown time. The thrust from the remaining first-stage engine was not enough for the satellite to reach orbit. The booster impacted in the Pacific—a $150 million loss.[42]

The failure was unexpected. Of 133 Titan III launches, only 4 had failed. No Titan III D or Titan 34 D had ever failed.[43] The loss also disrupted the normal replacement sequence. The usual pattern was to have two KH-11 satellites operational simultaneously, supplemented by occasional high-resolution launches. Now only the sixth KH-11 was operating (USA-6, launched on 4 December 1984). The chain of events had grown longer.

The loss of the satellite took place against a confusing political environment. Defense spending was coming under new pressures. Congressmen were publicizing spare-parts overpricing—$7,000 coffee makers, $600 toilet seats, and an assortment of tools. It was a symbolic attack on Reagan administration defense policies—a potent mixture of serious questions, ridicule, and carnival sideshow. The constant battle over guns and butter also was changing. The pressure of a budget deficit around $200 billion had forced Congress into action. In January 1986, the Gramm-Rudman Act mandated automatic spending cuts to reach a balanced budget by 1991. The cuts were not evenly distributed. Social Security was untouchable.[44] Some believed Gramm-Rudman would accomplish what the peace movement could not—end the Reagan defense buildup.[45]

The foreign situation was equally confused. The issue of Soviet SALT compliance continued as before. On 22 October 1985,

Defense Secretary Caspar Weinberger announced that deployment of the SS-25 ICBM had begun. Two bases had been built; each had launcher garages with sliding roofs and several support buildings to house the trucks and other equipment.[46] A total of twenty-seven SS-25s were operational. A week later at a NATO meeting, Weinberger showed satellite photos of the new missile to counter Soviet claims that the SS-25 was not a violation; it was simply a modified SS-13. The photos showed the SS-25 was 10 percent longer, 11 percent greater in diameter, and had a 92 percent greater throw weight than the SS-13. This was more than the SALT II treaty allowed. Weinberger also had a model of the Krasnoyarsk radar and data on Soviet encryption of ICBM telemetry signals. The presentation had mixed results. Several of the NATO defense ministers publicly went on record as believing the Soviets were violating the SALT and ABM treaties. Yet at the same time, the Europeans' emphasis on arms control remained. Norwegian Defense Minister Anders Sjasstad said, "Because there are violations is no cause to stop negotiating." [47]

On 19 and 20 November, President Reagan met in Geneva with Soviet General Secretary Mikhail S. Gorbachev. Subsequently there seemed to be encouraging signals on Soviet compliance. Satellite photos showed SS-16 support equipment, such as warhead transporters, being loaded onto railroad cars at Plesetsk. The consensus was the equipment was being sent into storage, which lessened concern over the SS-16 issue. The Soviets were also dismantling old weapons. Fifty SS-11s were dismantled to make up for the forty-five SS-25s deployed by late November. The Soviets also retired more than thirty Mya-4 bombers to compensate for new cruise missile–equipped TU-95H bombers. During the summer, fifteen of the old Mya-4s had been lined up at an airfield and their tails cut off. The others were converted into tankers.[48]

The encouraging signals did not eliminate the problem. On 23 December, a report was issued on compliance. In addition to the old charges, it stated the Soviets had exceeded the 2,504 ceiling on launchers. It also said the Soviets were concealing the "association" of the SS-25 ICBM with its launcher to make it more difficult to determine which missile went with which launcher. Concealing was accomplished by draping material over the launcher and SS-25. The Soviets were also trying to hide the SS-24 and its launcher, but less information was available. The SS-25 was also being deployed at old SS-7 coffin sites. All were violations of the SALT II treaty. The report noted this "pattern of Soviet non-compliance" caused "grave concern regarding Soviet commitment to arms control." [49]

Other developments also attracted attention to the satellites. In early November 1985, the Soviets launched the aircraft carrier shown in the leaked satellite photos. Another was under construction in the Nikolayev shipyard.[50] In February 1986, a technology newsletter reported satellite photos had been taken of a prototype Soviet "stealth" fighter.[51]

It was cold and clear that Tuesday morning of 28 January 1986. Shuttle launches were considered routine. CBS Radio provided only 1 minute of launch coverage; they had already cut away when *Challenger* blew up. The explosion of *Challenger* had a devastating effect on both the U.S. civilian and military space program. The shuttle would be grounded for at least 2 years. In addition to the time lost, the schedule was disrupted and the number of launches that could be made per year was cut. This meant fierce competition for the available slots left over once military needs

had been met. It would take years to make up for the loss of *Challenger*.[52]

In the weeks and months following the tragedy, the ability of the U.S. space program to recover was in question. There was Gramm-Rudman's mandated spending cuts. Also, the political environment had changed from one of grief at the loss of the crew to that of back-biting, buck-passing, and scapegoat-hunting. There seemed to be no consensus about what to do. Some in Congress wanted to wait for a second-generation system—even though it would not be ready until the turn of the century. The White House was tied up for months by debate over how to pay for a replacement orbiter. Others in the military did not think the shuttle was worthwhile; Titan 34 Ds and the advanced Titan 34 D-7 would be sufficient. These expendable boosters, however, lacked the volume of the shuttle's payload bay and the flexibility of its human crew. A few, little more than vultures, wanted manned spaceflight to end. Given all this, the outcome would be problematical—whether or not the capabilities of the shuttle to expand man's horizons beyond the narrow limits of earth would be thrown away like the Apollo program.

The next Titan 34 D was ready for launch in April. At 10:45 A.M., 18 April 1986, it lifted off from Vandenberg. At 8.5 seconds into the flight, at an altitude of 700 feet (213 meters), a 12-foot (3.7-meter) in diameter fireball was seen erupting from one of the solid-fuel rocket motors. The rocket then exploded into a huge, mushroom-shaped cloud. Hundreds of glowing fragments rained down on the pad and surrounding area. The pad was damaged and several vehicles and two trailers were destroyed. The blast released a cloud of poisonous fuel; seventy-four people were treated, mostly for eye irritations. The cloud also caused the evacuation of tourists from the Channel Islands National Park and 120 oil workers from rigs in the Santa Barbara Channel.[53,54]

The initial suspicion centered on a casing defect or an accidental triggering of the destruct package. A leak in a joint (such as that which destroyed *Challenger*) was not believed to have been the cause of the Titan 34 D explosion.[55] The cause was subsequently determined to be a flaw in the insulation between the fuel and the casing.

Even as the cloud was dispersing, it was apparent the explosion had dealt a major blow to U.S. intelligence. Strategic photo reconnaissance was hanging on a single KH-11. It was then 16 months old and could be expected to operate into 1987. The question became whether or not it could continue to operate until a replacement could be launched. (The Titan investigation would require 6 to 12 months before another rocket could be launched. The pad would need about 5 months of repair.) If a prolonged gap should develop, either through a sudden failure or a long delay, the U.S. would be left without satellite coverage.[56] If this occurred, U.S. intelligence would no longer be fully confident of its count of Soviet weapons. Even the potential of such uncertainty would have a corrosive effect on U.S. security and arms monitoring and negotiations.

In the shorter term, one satellite meant limits on the amount of coverage. Obviously, two satellites could cover more targets than one. Still, high-priority targets can be photographed, as shown by use of the KH-11 for pre- and post-strike coverage of terrorist targets in Libya and photographs of the Chernobyl nuclear plant 24 hours after radiation was detected in Sweden.[57,58]

It was not certain what type of satellite was lost in the April Titan 34 D failure. Initial speculation centered on a KH-11. *Newsweek* claimed it was the last KH-11—an engineering test model refurbished for spaceflight.[59] Later, *Aviation Week & Space Technology* said it was actually the twentieth and last Big Bird.[60]

So the chain was now complete: Carter's decision to phase out film-return satellites and to build only four shuttles rather than the five planned; the lack of depth in the reconnaissance satellite and U.S. space programs that resulted; the delays in the KH-12 Ikon; and finally, the loss of both the Titan 34 D and *Challenger* that crippled the U.S. reconnaissance satellite effort in both the short and long term.

Throughout the spring and summer of 1986, the fortunes of the shuttle program continued to unravel. The Centaur G upper-stage program was cancelled for safety reasons; the hydrogen fuel was viewed as too dangerous to be placed in the shuttle cargo bay. This left the shuttle unable to place large payloads into geosynchronous orbit.

The *Challenger* disaster also reopened questions about the military use of the shuttle. Senator Jim Sasser (D-Tennessee), alleging safety problems, demanded the shuttle pad at Vandenberg be mothballed for 5 years; he claimed this would "save" $1.6 billion. He also asked the General Accounting Office to determine if a replacement shuttle was needed if Vandenberg operations were cancelled, and to make yet another study of manned military activities.[61]

On 31 July, Air Force Secretary Edward Aldridge made a wide-ranging statement on the future of military space activities. The shuttle pad at Vandenberg would be mothballed until 1992. He denied the reason was safety but, rather, logistics—with only three

The SLC-6 launch pad at Vandenberg AFB. Originally built in the late 1960s for the Manned Orbiting Laboratory program, it was mothballed in 1969 when the program was cancelled. In April 1971, NASA and the Air Force began looking at various shuttle launch sites. During 1974, three sites at Vandenberg were examined. Analysis indicated that using SLC-6 would save $100 million over a completely new shuttle pad. This was approved in 1975 and construction began in 1979. The shuttle is assembled on the pad in the center of the photo and the pad structures are rolled back. The first launch was scheduled for mid-1986. In the wake of the *Challenger* explosion, the SLC-6 facilities were mothballed again with the first launch put off until 1992. *U.S. Air Force Photo.*

shuttles it was better to keep them at the Cape than to move them back and forth. Although he supported Vandenberg shuttle missions, Aldridge stressed that the Air Force would not be limited to the shuttle. Expendable

boosters would be the workhorse of Air Force operations, not the shuttle. Thirteen additional Titan 34 D-7s (redesignated Titan 4s) would be built, bringing the total to twenty-three Titan 4s on order. Development of a medium-lift booster would begin and military satellites would be modified to fly on both the shuttle and expendable boosters. To reduce the backlog of military satellites, Aldridge said he had had discussions with the French about using the Ariane booster to launch nonsensitive payloads. (At the time, the Ariane, like the shuttle, Titan, and Delta, was grounded due to a launch failure.) The Aldridge statement was seen as yet another blow to an already beleaguered shuttle program.

The Road Back

Yet, even as the shuttle program was reeling, it was also pulling together. During the summer of 1986, the first KH-12 Ikon was switched to a Kennedy Space Center (KSC) launch, specifically, to the second mission after shuttle flights resumed. This would reestablish coverage, but with limitations. A shuttle launched from KSC cannot go into an orbit with an inclination greater than 57 degrees, which meant Russia north of Moscow could not be photographed. This included Leningrad, Plesetsk, and Murmansk.

On 15 August 1986, the real turning point came. The White House announced that President Reagan had decided to build a replacement shuttle for delivery in 1991. The new shuttle, it was understood, would be based primarily at Vandenberg for military launches. It would be lighter in weight and have better systems and higher performance than the previous shuttle, which would compensate for the payload penalties of a polar orbit. With this, the shuttle program could begin to travel the long, hard road back from the dark days of early 1986. Its route, only time will reveal; the destination—the stars.

In the area of future reconnaissance, several new programs are underway. One is a radar reconnaissance satellite. It will not use a camera but, rather, a radar unit to produce "pictures." Their resolution would be comparable to those of normal reconnaissance satellites. The primary advantage would be the ability to penetrate clouds and darkness. Some parts of Russia during the spring and summer have only 2 or 3 clear days per month. Once it took 4 months to photograph a target in Moscow.[62] With the radar satellite, the Soviets would no longer be able to hide their activities under clouds. The satellites' primary mission would be to watch for Warsaw Pact armor movements.[63]

Another project is the laser intelligence satellite (Lasint). It would be used specifically to find information on Soviet laser activities. Intelligence on this area is limited because laser light is hard to detect. Although details are lacking, certain speculations can be made. The satellite probably would be placed in geosynchronous orbit to scan the entire Soviet landmass. The satellite presumably would also have filters to determine the wavelength of the laser light, which would indicate the design of the laser. The Lasint satellite would cost $1 billion and be launched in the late 1980s.[64]

The Air Force is also studying the use of a network of small, relatively cheap reconnaissance satellites. The large number would make it more difficult for the Soviets to shoot them down.[65] This is a reversal of the trend of single, large, high-capability reconnaissance satellites.

Specific programs will change as emphasis changes and new needs arise. One thing is certain—in the future, as in the past, the watch will go on.

NOTES Chapter 8

1. "Space Reconnaissance Dwindles," *Aviation Week & Space Technology*, (6 October 1980): 18,19.
2. Deborah G. Meyer, "DoD Likely to Spend $250 Billion on C^{3}I Through 1990," *Armed Forces Journal*, (February 1985): 75.
3. "Space Reconnaissance Dwindles," *Aviation Week & Space Technology*, (6 October 1980): 18,19.
4. Craig Covault, "USAF, NASA Discuss Shuttle Use for Satellite Maintenance," *Aviation Week & Space Technology*, (17 December 1984): 14–16.
5. Adam B. Ulam, *Dangerous Relations*, (New York: Oxford University Press, 1983), 249,250,268,269.
6. Compton Yearbook, 1982, xxvii, 97–101,358.
7. Satellite Digest 147, *Spaceflight*, (August/September 1981): 227.
8. Satellite Digest 151, *Spaceflight*, (February 1982): 80.
9. Soviet Military Power 1985, (Department of Defense 1985), 28–30.
10. *Daily Californian*, 30 December 1981.
11. Anthony Kenden, "Was 'Columbia' Photographed by a KH-11?," *Journal of the British Interplanetary Society*, (February 1983): 73–77.
12. Compton Yearbook 1983, 84,125,290,291,391.
13. *San Diego Union*, 3 March 1982.
14. *San Diego Union*, 9 September 1982.
15. Satellite Digest 155, *Spaceflight*, (June 1982): 282.
16. "Unusual Military Satellite," *Spaceflight*, (November 1982): 410.
17. Anthony Kenden, "U.S. Military Satellites 1983," *Journal of the British Interplanetary Society*, (February 1985): 62–64.
18. "Washington Roundup," *Aviation Week & Space Technology*, (21 February 1983): 13.
19. Compton Yearbook 1984, xxvi,xxviii,84,153,154,158–160,290.
20. "New Titan Model," *Spaceflight*, (June 1985): 247.
21. Anthony Kenden, "U.S. Military Satellites 1983," *Journal of the British Interplanetary Society*, (February 1985): 62–64.
22. Soviet Military Power 1985, 29–31.
23. Clarence A. Robinson, Jr., "Soviets Accelerate Missile Defense Efforts," *Aviation Week & Space Technology*, (16 January 1984): 14–16.
24. *San Diego Union*, 23 October 1983, C-1.
25. Debrah G. Meyer, "Reaction to DoD's New Threat Analysis Ranges From Praise to Scorn," *Armed Forces Journal*, (May 1984): 8–13.
26. David Detzer, *The Brink*, (New York: Crowell, 1979), 203,204.
27. "Washington Roundup," *Aviation Week & Space Technology*, (12 April 1982): 15.
28. Benjamin F. Schemmer, "DoD Plans Soviet Military Power III in March," *Armed Forces Journal*, (February 1984): 34.
29. Deborah M. Kyle, "SACEUR General Rogers Urges U.S. to Release Threat Photos," *Armed Forces Journal*, (April 1984): 14.
30. Deborah M. Kyle, "Reaction to DoD's New Threat Analysis Ranges from Praise to Scorn," *Armed Forces Journal*, (May 1984): 8–13.
31. "Soviets Building Aircraft Carrier That Will Become Operational in 1990s," *Aviation Week & Space Technology*, (13 August 1984): 26,27.

32. *San Diego Union,* 6 October 1985, AA-8.
33. *San Diego Union,* 18 October 1985, A-27.
34. *San Diego Union,* 5 December 1985, A-20.
35. Compton Yearbook 1985, 125,290,291,374,375.
36. Craig Covault, "USAF, NASA Discuss Shuttle Use for Satellite Maintenance," *Aviation Week & Space Technology,* (17 December 1984): 14–16.
37. Satellite Digest 185, *Spaceflight,* (July/August 1985): 334.
38. Satellite Digest 182, *Spaceflight,* (April 1985): 184.
39. Clarence A. Robinson, Jr., "Soviets Accelerate Missile Defense Efforts," *Aviation Week & Space Technology,* (16 January 1984): 15.
40. *San Diego Union,* 26 February 1984, A-2.
41. Satellite Digest 184, *Spaceflight,* (June 1985): 284.
42. "Titan 34D Booster Failed Following Premature Shutdown of Aerojet Engine," *Aviation Week & Space Technology,* (18 November 1985): 26.
43. *San Diego Union,* 19 April 1986, A-1.
44. "Antinuclear Campaign Reawakens," *U.S. News & World Report,* (27 January 1986): 22.
45. "Budget Skirmishing Begins," *U.S. News & World Report,* (3 February 1986): 20,21.
46. *San Diego Union,* 22 October 1985, A-1.
47. *Los Angeles Times,* 30 October 1985.
48. *San Diego Union,* 24 November 1985, A-20.
49. *San Diego Union,* 22 December 1985, A-1.
50. *San Diego Union,* 16 January 1986, A-11.
51. *San Diego Union,* 8 February 1986, A-8.
52. "USAF May Increase Titan 34D-7 Production" and "NASA Asks Rockwell for Price of New Shuttles," *Aviation Week & Space Technology,* (17 February 1986): 26,27.
53. *San Diego Union,* 19 April 1986, A-1.
54. *San Diego Union,* 20 April 1986, A-1.
55. *San Diego Union,* 26 April 1986, A-3.
56. *Christian Science Monitor,* 21 April 1986, 3.
57. "Washington Roundup," *Aviation Week & Space Technology,* (21 April 1986): 17.
58. *San Diego Union,* 1 May 1986, A-13.
59. Walter Shapiro et al, "A Space Spy Gap," *Newsweek,* (28 April 1986): 45.
60. "Titan Explosion Cripples U.S. Launch Surveillance," *Aviation Week & Space Technology,* (28 April 1986): 16–19.
61. "KH-12 Reconnaissance Satellite Planned for Kennedy Launch," *Aviation Week & Space Technology,* (23 June 1986): 17.
62. Ralph Kenney Bennett, "U.S. Eyes Over Russia: How Much Can We See?," *Reader's Digest* (October 1985): 142–147.
63. "Space Reconnaissance Dwindles," *Aviation Week & Space Technology,* (6 October 1980), 18,19.
64. "Laser Intelligence Satellite," *Aviation Week & Space Technology,* (29 August 1983): 20.
65. "Washington Whispers," *U.S. News & World Report,* (17 February 1986): 17.

CHAPTER 9

Soviet Reconnaissance Satellites

In the late 1950s, the Soviet Union faced a completely different set of circumstances when it came to spying on the U.S. The reason was the openness of American society. The capabilities of U.S. weapons were freely discussed. To acquire similar information on Soviet weapons, the U.S. had to spend many millions of dollars per year.

To acquire classified information, the Soviets used two types of secret agents. The first, and potentially the most dangerous, were Americans recruited to pass information. The Soviets looked for individuals with a personal grievance or a need for money, who could be manipulated or had a personality defect (sex, alcohol, gambling, etc.). Simple greed was enough. The individual who spied out of ideological sympathy with Russia, common in the 1930s and 1940s, was by this time very rare. Spotting such an individual; developing a personal relationship with him; and finally, recruiting him was handled by either embassy personnel or by the other type of agent—the "illegal." An illegal is a Russian who lives in the U.S. under a false identity. The KGB expends a great deal of effort on their training and preparation. Along with the normal trade craft of espionage, these illegals are taught such seemingly trivial activities as how to act

at a party, check into a hotel, or apply for a job—how to be an American. He or she (married couples are sometimes used) is also given a complete life story or "legend," including not only an employment history but even the names, faces, and personalities of coworkers. This training may take several years and involve one or more foreign trips to practice posing as an American. Once in the U.S. (not by airdrop but by normal transportation), the illegal faces few of the operational difficulties of an agent in Russia. He can move about freely and meet anybody without raising suspicion.[1]

These agents could perform all of the functions for which the U.S. needed reconnaissance satellites. For this reason, some in the U.S. government doubted the Russians would build reconnaissance satellites.[2] Soviet actions during the first years after Sputnik seemed to support this. There were no military space missions or any overt signs that such a program was underway.[3]

Others felt the Soviets would use such satellites. Before Sputnik, a NATO group estimated that the earliest the Soviets could have a reconnaissance satellite would be 1962. In the wake of *Sputnik I* and *II*, this was moved up 2 years.[4]

A Vostok manned spacecraft on display. It is composed of two parts—the spherical reentry capsule and the service module which contains the batteries and retro-rocket. The Vostok was also used as the Soviets' first- and second-generation reconnaissance satellite. The porthole visible on the left edge of the capsule would make an ideal camera window. For the third-generation reconnaissance satellite, the service module was extensively modified to provide a longer 13-day lifetime versus the 8 days of earlier satellites. A maneuvering engine or supplementary payload was added to the top of the capsule. Over a quarter of a century after the first Vostok test flight, it continues to be used as the medium resolution satellite. *Soviet Photo via James E. Oberg.*

The Vostok Reconnaissance Satellite

The Soviets used a completely different approach in developing their reconnaissance satellite. Rather than building a completely new satellite, launch vehicle, and camera system, they used off-the-shelf hardware. The basic satellite was a modified version of the Vostok manned spacecraft. On 12 April 1961, Yuriy Gagarin, aboard *Vostok 1,* became the first human to venture into space. Four months later, in August, *Vostok 2* pilot Gherman Titov became the first to spend a full day in orbit.

The Vostok spacecraft had two parts. The first was the spherical reentry capsule, 7.5 feet (2.3 meters) in diameter. A sphere was chosen for several reasons. First, its aerodynamic properties were known over a wide speed range. Additionally, a sphere provides the maximum internal volume for a given outside surface area. Finally, by selecting the proper center of gravity, it could be self-orienting during reentry. There was no need for attitude-control rockets to hold its heat shield in the proper position.[5]

The sphere was attached to the service module shaped like two cones attached at the base. It contained batteries to supply electrical power; thrusters to orient the spacecraft; and the retro-rocket, which returns it to earth. Total spacecraft weight is about 10,360 pounds (4,700 kilograms). The launch vehicle was the A-1 booster. Use of a modified Vostok as a reconnaissance satellite had several advantages. Most important was reliability. The Vostok was designed to carry a man. Thus, it had a greater number of backup systems than an ordinary satellite would have. The Soviets were able to avoid the incredible string of failures the Air Force underwent with Discoverer. The Vostok was designed to function for up to 10 days in orbit—longer than Discoverer could. The reason for this design feature was, if the retro-rocket should fail, the on-board systems could keep the cosmonaut alive until the Vostok reentered naturally from atmospheric drag. Only a minimal amount of modification would probably be required. The Vostok was originally designed for fully automatic operation; the cosmonaut rode along as a passenger. Probably, the Vostok would need a more precise attitude-control system to avoid blurring the photos.

The biggest challenge the Soviets may have faced was not the satellite but the camera system. They had only very limited optical and camera manufacturing facilities in the late 1950s and early 1960s. Strategic reconnaissance did not play a very important part in Soviet doctrine up to this period. The U.S., in contrast, had begun developing reconnaissance capability in the 1930s as part of the buildup of strategic bomber forces. Ironically, the Vostok design got around this. The entire camera unit was fitted inside the reentry capsule, replacing the cosmonaut and his ejection seat. The complete package was returned to earth. After any damage from the touchdown was repaired, the camera could be reused. One of the windows in the sphere even faced the earth and was ideal as a camera view port. Both high-resolution and lower resolution area surveillance photos were returned to earth by capsule.[6]

The first attempt by the Soviets to launch a reconnaissance satellite came in the fourth quarter of 1961. The booster malfunctioned and the satellite did not reach orbit. It would be about 6 months before the next attempt.[7] *Cosmos 4,* the first successful Soviet reconnaissance satellite, was launched on 26 April 1962, and went into a 205- by 185-mile (330- by 298-kilometer) orbit, inclined 65 degrees. Three days later, the spacecraft fired its retro

and was recovered successfully.[8] The next Soviet reconnaissance satellite, *Cosmos 7,* went aloft 28 July 1962, and stayed in orbit 4 days.

Even though the program had barely started, Soviet reconnaissance satellites were about to face a critical milestone. During the summer of 1962, the Soviets stepped up weapons deliveries to Cuba to pave the way for the establishment of missile bases on the island. On 27 September, as the missiles were being delivered, *Cosmos 9* was launched. After its recovery 4 days later, examination of its photos showed no increases in U.S. military activity. At the same time, Soviet agents were being pressed by KGB headquarters for any signs of U.S. mobilization or preparations for an invasion of Cuba. The satellite's photos could have acted as confirmation of the agents' reports.[9] It was not until fall that the U.S. was able to sort through often-fanciful statements by Cuban refugees about missiles on the island.

On 14 October, a U-2 overflew the first medium-range missile base at San Cristobal, Cuba. Two days later, on the morning of the sixteenth, the photos were in the hands of President Kennedy. The next day, *Cosmos 10* was launched from Tyuratam. Like its predecessor, it orbited the earth for 4 days, returning on 21 October. If it had covered Florida, the photos would have shown the buildup of air and naval forces. One day after *Cosmos 10* returned, and even as its photos were being interpreted, President Kennedy went on television to announce a blockade of Cuba. It is impossible to say what part *Cosmos 10*'s photos played in Khrushchev's decision to back down and remove the missiles. Many factors—military, political, and diplomatic—probably went into his decision. It is worth noting, however, that preceding the crisis, Soviet reconnaissance satellites were launched at an accelerating pace. Three months separated the first and second, 2 months separated the second and third, and only 20 days separated the third and fourth. This could imply that Soviet reconnaissance satellites, like their human agents, were playing a supporting role in Khrushchev's Cuban adventure.

After the Cuban missile crisis, 2 months would pass before the fifth and final reconnaissance satellite of 1962. *Cosmos 12,* launched on 22 December, represented a major advance. Previous satellites had remained in orbit only 4 days. *Cosmos 12* stayed up 8 days— the duration planned for the operational satellite.[10]

Operations

After *Cosmos 12,* the Soviet effort went into hibernation. The reason was the difficulty of recovering capsules during the harsh Soviet winter. An error in retroburn timing could mean an overshoot of several hundred miles. Instead of parachuting onto the flat plains of Kazakhstan, the capsule would land in deep snow–covered forest. Flights resumed on 21 March 1963, with *Cosmos 13,* which stayed up 8 days. In April, the Soviets launched *Cosmos 15* and *16*—the first time two reconnaissance satellites had been orbited in a single month. *Cosmos 16* had the added distinction of remaining in orbit 10 days. Design, preparation, launch, and operation of reconnaissance satellites was handled by the GRU's Cosmic Intelligence Directorate, which controls research institutions, factories, a computer center, and the launch facilities. Its commander is a lieutenant general.[11] Although the Soviets made no public mention about their reconnaissance satellite effort, there were indications it was nearing operational status.

In July, on a picnic along the Dnieper River, Khrushchev and Belgian Foreign Minister Paul-Henri Spaak discussed a nuclear test ban treaty. Khrushchev, in a pleasant mood, said that on-site inspection was not needed to police an agreement. "Anyway," he continued, "that function can now be assumed by satellites. Maybe I'll let you see my photographs."[12] A more concrete indication came 2 months later. For the past 3 years, the Soviets had missed no opportunity to lambaste "American spy satellites." In the UN, they had tried to force an agreement banning reconnaissance from space. Because of its dependence on reconnaissance, the U.S. refused. On 9 September 1963, Dr. Nikolai T. Fedorenko, the Soviet delegate, omitted the obligatory attack in his speech to the UN Outer Space Committee. In this way, the Soviets signaled they were dropping their objections to reconnaissance satellites and were quietly acknowledging they were flying their own.[13] This was just as well, since on 16 November, the Soviets launched the first of a series of second-generation reconnaissance satellites. Unlike previous satellites, *Cosmos 22* used an A-2 booster, which had a more powerful second stage than the A-1 rocket. The satellite could carry a bigger film supply and larger high-resolution optics. It was recovered after 6 days. Before the winter halt, the Soviets had launched seven reconnaissance satellites, five of which stayed in orbit 8 days or longer. There were also two reconnaissance satellite launch failures in the last 3 months of 1963.

When flights resumed in 1964, the Soviet system was operational and set to begin an explosive growth. The first reconnaissance satellite of the new year was *Cosmos 28,* launched on 4 April 1964. On 10 June *Cosmos 32* went into a new orbital inclination of 51.3 degrees. Previously, 65 degrees was standard. This new inclination allowed coverage of targets in the northern hemisphere during the good lighting conditions of summer and fall. In all, twelve reconnaissance satellites were launched in 1964 (nine low-resolution satellites and three of the A-2–launched high-resolution satellites). The year's activity ended with *Cosmos 50* (28 October). A low-resolution satellite, it was the fourth placed into a 51–52 degree inclined orbit. After 8 days, preparation was made to return it to earth. When the attempt was made, the spacecraft suffered a retro-rocket failure, leaving it marooned in orbit. The Soviets had, however, allowed for the possibility this might happen. Western radars subsequently noted *Cosmos 50* had exploded into 98 fragments. To prevent the capsule and its enclosed camera and film from falling, literally, into the waiting arms of Western intelligence when its low orbit decayed, the Soviets had equipped their satellites with an explosive, self-destruct package.[14] This charge was so powerful, the vehicle was shattered. The debris subsequently reentered.

After a 2-month halt, possibly because of *Cosmos 50*'s failure, the Soviet reconnaissance effort resumed in January 1965. This was the first time winter launches had been made. Accordingly, there was an increase in the number of reconnaissance satellites launched in 1965—seventeen were orbited. Of these, ten were high-resolution satellites. In other words, the Soviets launched a reconnaissance satellite each twenty days in 1965 compared to one every thirty days the year before.

The previous 3 years of activity had revealed certain patterns in Soviet operations. The launches were made between 1 and 4 P.M. local time, which meant that objects in the photos would show shadows. The perigees, or low points, of both high- and low-resolution satellites were tightly clustered approximately 128 miles (205 kilometers) above the earth. This was a very narrow range. It was also a higher altitude than the perigees of U.S. reconnaissance satellites, probably to avoid atmospheric drag on the large Russian satellites.[15] In contrast to the narrow range of perigee values, the satellites' apogees (high points) could vary widely, from below 186 miles (300 kilometers) to over 217 miles (350 kilometers).[16] The satellites stayed aloft for 8 days, which allowed complete coverage of a target area on the earth's surface. The satellites' orbits and the earth's rotation caused the satellites' ground track to drift westward across a spot on the earth. The track after 8 days was the same as that after the first day. On the satellites' final orbit while northbound over Africa, they were oriented into the correct retrofire position. After the burn, the metal bands holding the capsule to the service module were released. Approximately 10 minutes after retrofire, the capsule reentered the atmosphere. Once the heating phase ended, a parachute was deployed.[17] Because of the size of the capsule, the Soviets did not use a midair recovery but, rather, a ground team equipped with special vehicles and helicopters, which were guided to the capsule by a radio beacon. Touchdown came in the recovery zone northeast of Tyuratam between 11 A.M. and 1 P.M. local time.

By the end of 1965, the U.S. had deemed Soviet reconnaissance satellites enough of a threat to take countermeasures. This was called SATRAN (Satellite Reconnaissance Advance Notice program). SATRAN was a joint effort of NORAD, the Defense Intelligence Agency, and the Air Force Systems Command. Presumably, its purpose involved calculating the orbit of a Soviet reconnaissance satellite, determining its ground track, then sending warnings to bases within camera range to conceal any sensitive equipment or activities during the satellite's pass. Although it would not stop the reconnaissance, SATRAN would lessen its effectiveness. SATRAN went into operation in early January 1966.[18]

Plesetsk

Following the introduction of a second-generation, high-resolution satellite in 1964 came the debut of a second-generation, low-resolution satellite in 1966. *Cosmos 120* and *124* were the first—both went into 51–52 degree inclination orbits. An A-2 booster was used. The orbital characteristics were identical with earlier satellites. Presumably, the added payload of the larger booster went into increased film capacity and batteries. Lifetime was still the standard 8 days.

The real development in 1966 had nothing to do with hardware; rather, it was how a group of 13- to 17-year-old English schoolboys, their teacher, and a war surplus radio discovered one of the Soviet Union's most closely guarded secrets. Beginning with *Cosmos 5* in 1962, Geoffery Perry, the senior physics master, and his students at the Kettering Grammar School had undertaken a systematic monitoring of the shortwave signals transmitted from Soviet satellites. On 17 March 1966 the Soviets launched *Cosmos 112*—a first-generation, low-resolution satellite. Its new inclination of 72.1 degrees attracted the Kettering group's attention. When they attempted to determine *Cosmos 112*'s ground track, they made an interesting discovery—*Cosmos 112* could not have been launched from Tyuratam. Instead, it appeared to have come from somewhere in northwest Russia. One early candidate was the southern tip of Novaya Zemlya—an island used in the Soviet 1961–1962 nuclear test series. The sec-

ond satellite from this new northern launch site was *Cosmos 114* in April 1966. Because of the very slight difference in inclination, the crossing of the two satellites' ground tracks was not precise enough to show the exact location of the launch site. This was finally pinned down with the launch of *Cosmos 129* (14 October). When the ground track of its 64.7 degree orbit was plotted, it intersected the other satellites' path near the small town of Plesetsk. The Kettering group had their launch site. The formal announcement came at the 3 November 1966 autumn meeting of the British Interplanetary Society.[19] The Soviets did not acknowledge the site's existence for satellite launchings. The site was used for military missions, which was something only "Yankee imperialists" did.

The Plesetsk launch site is the old SS-6 ICBM facility. The Soviets had several reasons for using it for satellite launchings. The expanding Soviet space program was putting a burden on the Tyuratam facility. It had only one A-booster pad operational until the late 1960s. Plesetsk had two pads. More important was Tyuratam's northerly location—62.9 degrees north latitude (equivalent to Anchorage, Alaska). A rocket will orbit its maximum payload if it is launched due east, because the earth's rotation provides an added boost. Tyuratam is at only 46 degrees north latitude; a reconnaissance satellite must be launched to the northeast if it is to cover western Europe and the northern U.S. This, however, entails a loss of payload. From Plesetsk, the rocket would be launched only a little north of east to reach the desired inclination.

During 1966, Plesetsk made six reconnaissance satellite launches out of twenty-one for the year. At this time, Plesetsk was a dual ICBM and space facility. In 1967, the four SS-6 ICBMs were retired from service and

Plesetsk undertook satellite launchings full time. These launches took place, like those at Tyuratam, between 1 and 4 P.M. local time. Because Plesetsk is farther west, its satellites make one less orbit before landing in the recovery zone. Of 1967's twenty-two reconnaissance satellite launches, fourteen came from the Plesetsk facility. Also in 1967—on 12 May—was the retirement of the first-generation, low-resolution satellites—*Cosmos 157*. Launches from the two sites could be readily separated by their orbital inclination—Tyuratam satellites going into 51–52 degree, 65 degree, or 69–71 degree orbits. The Plesetsk inclinations were 65–66 degrees or 72–73 degrees.

Third-Generation Reconnaissance Satellites

As Soviet reconnaissance satellites were becoming operational, development work began on a third-generation reconnaissance satellite. It would have the capability of remaining in orbit for 12 to 14 days. This doubling of the satellite's lifetime meant major modifications to the Vostok service module. The lower section of the new service module is a cone; attached to it is a short cylinder followed by a tapered segment and a final cylinder, to which the spherical reentry capsule is joined. The capsule is fitted to carry a supplementary payload on top. On low-resolution satellites, this is a large-diameter cylinder carrying piggyback scientific or military packages. Soviet reconnaissance satellites have long carried small scientific experiments. *Cosmos 4*, for example, carried a Geiger counter that detected four times as much radiation as *Sputnik V* did in August 1960. Several later reconnaissance satellites had weather-reporting equipment on board.[20] On the third-generation

satellites, this supplemental payload would separate before retrofire and would later decay.

On the third-generation, high-resolution satellites, this supplementary payload was a maneuvering engine, used to adjust the satellite's orbit to provide better coverage and resolution. Ideally, the spacecraft should pass directly over the target to be photographed. Because of differences between the satellite's actual orbit and the one planned, this is not always possible. The satellite must then view the ground target at an angle. This slanting path lowers resolution because the target is farther away and also because the camera is looking through a greater thickness of the earth's atmosphere. With the engine, the third-generation satellites can adjust their orbit and avoid the problem. The typical procedure was a two-part burn. First, the perigee was lowered from the initial 130 miles (210 kilometers) to an altitude as low as 109 miles (175 kilometers) to provide maximum resolution. The second-engine burn lowers the orbit's apogee, which causes the ground track to stabilize—repeating each day. The maneuvering engine can be used to finely adjust the orbit, causing the satellite's ground track to slowly drift in one direction. Another burn can send the ground track slowly drifting back the other way over the same limited area—a maneuver called the "Ali-shuffle."[21]

Because the stabilized ground track may be poorly placed for a landing in the recovery zone, the engine's final function is to once more adjust the orbit's ground track so it passes over the recovery site. Having thus completed its task, the engine is separated and the satellite makes a normal reentry.

The first third-generation satellite was *Cosmos 208*—a low-resolution model launched on 21 March 1968. After 8 days of

Launch of a Soviet A-2 booster from Tyuratam. The A-booster is based on the SS-6 Sapwood ICBM first flown in 1957. The A-1 is the SS-6 with a small upper stage; the A-2 uses a larger upper stage. The A-2 has launched the Soviets' second-, third-, fourth- and fifth-generation photo reconnaissance satellites. The first stage consists of four strap-ons attached to a center core. These carry a total of 20 main engine nozzles. Steering is accomplished by 12 small rocket engines which pivot (two on each strap-on and four on the center core). All 32 are burning at lift-off. The second stage on the A-2 has four nozzles. *NASA Photo.*

flight, it separated the supplementary payload. Four days later (12 days after launch), *Cosmos 208* deorbited. The supplementary payload decayed 4 days after the main satellite's recovery. The second third-generation satellite, *Cosmos 228,* followed 3 months later, in June. Still another was orbited in September (*Cosmos 243*). The first high-resolution type, *Cosmos 251,* went into orbit on Halloween and

operated for 12 days. The remainder of the twenty-nine reconnaissance satellites orbited in 1968 were the older, second-generation satellites, two of which—*Cosmos 210* and *214*—went into a new inclination of 81 degrees from Plesetsk. In subsequent years, two or three such satellites would be orbited—usually in the spring. It is believed they are used to photograph the breakup of winter ice on the northern sea route across the top of Russia. If true, this was the first use by the Soviets of reconnaissance satellites for earth resources missions.

War with China

As the new, third-generation satellites were being phased into operation, a crisis developed that was potentially as dangerous to the Soviet Union as had been the Cuban missile crisis 6 years before. Soviet-Chinese relations had been deteriorating throughout the 1960s. In the spring of 1966, the Soviets moved military units into Mongolia and strengthened their forces in other locations along the border. The conflict remained limited to verbal abuse. In 1969, however, it turned violent. The events began on 25 and 26 February, when *Cosmos 266* and *267* were launched. On 2 March 1969, a week after they went up, a Soviet patrol was ambushed on Damanski Island on the Ussuri River in the Soviet Eastern Maritime Province—one of the areas taken from China in the nineteenth century that were in dispute. Thirty-one Soviet border guards and an unknown number of Chinese were killed.[22] The first incident was followed on 14–16 March by a larger clash.

These events sent Moscow into a panic. Despite the years of worsening relations, the attack came as a complete surprise. The Soviets now faced the nightmare of millions of Chinese invading the disputed area. In Politburo discussions, the possibility of a nuclear attack on China was seriously raised. Defense Minister Marshall Andrei Grechko called for wide-scale, unrestricted use of multimegaton weapons to remove any threat from China—permanently. Millions of Chinese would be killed and fallout would spread to the Soviet Far East and countries bordering China. The

plan met opposition within the military; they argued it would lead to world war. A limited strike on Chinese nuclear facilities was also too dangerous. The Soviets could find themselves bogged down in an endless guerrilla war. The Politburo found itself deadlocked, unable to decide what action to take.[23]

These events caused an explosive upsurge in reconnaissance satellite activity. Between 25 February and 23 April, the Soviets launched ten satellites, twice the normal two-per-month rate. These flights were probably to see if the Damanski Island battles were a prelude to a full-scale Sino-Soviet war. They would also provide an updated target list should full-scale hostilities break out. Their photos were apparently reassuring. As conditions settled down, the launch rate of reconnaissance satellites also returned to normal during May.

On 10 June 1969, there was a minor border incident in western Mongolia at the Dzungarian Gates. Five days later, *Cosmos 286* went up. It was followed by four more satellites in the next month and a half. The second increase may well have put a strain on the supply of reconnaissance satellites and boosters. After the launch on 22 July of *Cosmos 290* (a second-generation, low-resolution satellite) and its recovery 8 days later, there was a halt in flights. This could not have happened at a worse time. On 13 August, there was a larger battle on the Mongolian border. On 16 August, *Cosmos 293* (a third-generation, low-resolution satellite) was orbited. Three days later, *Cosmos 294* (a second-generation, high-resolution satellite) followed it. Another reconnaissance satellite went up on 29 August. This once-a-week launch schedule continued through September.

The debate over a nuclear strike on China continued. Grechko argued the U.S. would stand aside. To test this, the KGB began circulating reports in mid-August that the Soviets were considering a preemptive nuclear attack on China and asked the U.S. government's response. It soon became clear the U.S. would not stand aside, and any Soviet attack would lead to a crisis. This forced the abandonment of any plans for an attack. Instead, Brezhnev ordered a buildup of Soviet forces (with nuclear weapons) on the Chinese border along with diplomatic efforts to settle the dispute.[24] Between the threat of nuclear attack and offers of talks, the crisis eased. Low-level formal discussions began on 20 October 1969.

The Soviet launch rate had, in the meantime, returned to normal.[25,26] Soviet reconnaissance satellite activities during the China crisis highlight a maturity in the program. There is a certain irony in the Soviet use of satellites to cover China. The reason the U.S. turned to reconnaissance satellites was the difficulty of gathering information inside Russia—problems KGB agents did not face in the U.S. With the emergence of China as a military threat, the Russians came up against a police state that matched their own. The solution was once again found in space.

In all, thirty-two reconnaissance satellites were launched in 1969—three more than the previous year. Seven were third-generation satellites. Although there had been an increase in launches, the rapid growth of the 1960s was tapering off. Subsequent years would show minor variations.

After the crisis year of 1969, the early years of the new decade were rather anticlimactic. Only twenty-nine reconnaissance satellites were launched in 1970—the first time the yearly total had gone down. This may have been because satellites intended to be launched in 1970 had been used during the crisis. Eighteen of the satellites were third-generation models. The second-generation

satellites were being phased out; by 1971, they had completely disappeared. The third generation was undergoing improvements. In 1969, *Cosmos 264* (launched 23 January) was the first to stay aloft for 13 days. *Cosmos 399*, launched 3 March 1971, had a lifetime of 14 days.

Since the discovery of Plesetsk, the Kettering group had followed its activities. They were able to show that there were actually five versions of the satellite—three low-resolution and two high-resolution types. The difference is the telemetry the satellites transmit—pulse duration modulation, a Morse code, or a two-tone signal. Another variation between the satellites has to do with their recovery beacons, used to guide the ground team. The Kettering group noted that 7 minutes after reentry, at the point the parachute would open, the capsules began to transmit a two-letter Morse code signal. The first noted was TK. After another 7 or 8 minutes, the capsule touched down and the beacon's strength diminished, although it continued to transmit until the recovery crew arrived and shut off the beacon. Over the years, other letter combinations have been noted—TG, TF, and TL.[27] What exactly these differences mean in terms of equipment or other hardware is not known. On occasion, these beacon signals have ceased abruptly in less than 7 minutes. A possible reason is a parachute failure—ending with a hard landing.

The October War

On 6 October 1973, at 2:05 P.M., the Egyptian and Syrian armies, backed by tanks and artillery, attacked Israel. The Israelis, celebrating Yom Kippur, were caught by surprise. Against this violent background, the Soviets began their most concentrated reconnaissance effort. When the war began, *Cosmos 596* was in orbit, having been launched 3 days before from Plesetsk. The low-resolution satellite was in position to photograph the battle area both before fighting began and for the first 2 days of battle. On 8 October, after only 6 days in space, *Cosmos 596* deorbited and was recovered. Clearly, the Soviets were eager to see its photos. They were understood to have revealed the heavy losses on both sides. On 10 October, the Soviets began to airlift in supplies. As *Cosmos 596*'s film was being analyzed, the watch on the battlefield was taken over by *Cosmos 597*. This satellite, a high-resolution, maneuverable type, was launched from Plesetsk less than an hour after the fighting began. *Cosmos 597* made its first pass over the area on 8 October. It fired its engine to stabilize its ground track.[28]

On the ground, the Israelis had suffered staggering losses—forty aircraft and fourteen fortified positions along the Suez Canal in the first few hours. The Israelis' first priority was the Syrians. Their 900 tanks were pressing to within 12 miles (19 kilometers) of Israel itself. Despite their losses, the Israelis were able to knock out the computer that controlled Syrian SAM missile defenses. This opened a wedge in the coverage into which the Israelis could send their Phantom and Skyhawk bombers to destroy the Syrian tanks and SAMs.[29]

All the while, *Cosmos 597* kept watch. As its ground track drifted slowly westward, it passed over the Golan Heights, the Sinai Peninsula, and Israel. Finally, on 12 October, *Cosmos 597* had moved beyond the area and was recovered. Flight time was again only 6 days. *Cosmos 598,* its replacement, had been launched 10 October. With an inclination of 72.9 degrees, it made a pass over the battle

area the day after launch. Thus, coverage was continuous. On 12 October, shortly after *Cosmos 597* had landed, *Cosmos 598*'s maneuvering engine adjusted its orbit. As with previous high-resolution satellites, *Cosmos 598*'s passes were tightly clustered over the battle area. The satellite was brought down on 16 October after 6 days. When the Soviet photo interpreters examined the film, they found the situation had changed. In the Golan Heights, the Israelis had wrested control of the air from Syrian SAMs and had begun to bomb the Damascus airport. As *Cosmos 598* was making its final pass over the Middle East on 15 October, the Syrian armored thrust had been turned back and the Israelis had begun a drive into Syrian territory. With Israel's first goal achieved (stopping the Syrians), attention could be turned to Egypt. On 14 October, an attempt by the Egyptian armor to break out of their beachhead and drive eastward was blunted.

Cosmos 598 was recovered on 16 October. That same day, two different groups arrived in Egypt: the first was Soviet Premier Alexei Kosygin for 3 days of secret talks; the second group had come in the early morning hours. It was a unit of Israeli paratroopers crossing the Suez Canal on rubber rafts. They were soon followed by tanks and armored personnel carriers. This as yet small force immediately fanned out across the desert and attacked Egyptian SAM sites, which were only lightly defended against ground attack.[30] Photos of the crossing were taken by *Cosmos 598,* probably during its last two passes on 14 and 15 October. The photos would have shown the bridging equipment being moved up and the armor being assembled for the crossing. In any event, by 20 October at the latest, the Russians knew the Egyptians were in trouble.[31]

Also on the eventful day of 16 October, *Cosmos 600* was launched. It made its first pass over the area the next day, when it performed the Ali-shuffle maneuver. On the second day, the orbit's apogee was lowered, which caused the ground track to drift eastward. On the flight's fourth day, the process was reversed: the apogee was raised and the satellite's ground track was moved westward. Finally, on 23 October, 7 days after launch, the capsule was recovered.

The watch was continued by not one, but two satellites. *Cosmos 599* was a low-resolution satellite. Unlike the others, it was not specifically intended to photograph the battle. Although the satellite was launched on 15 October, its orbital path was not correctly placed over the area until the last 3 days of its 13-day mission. Apparently, the Middle East coverage was an add-on to a general survey mission. About 70 minutes after *Cosmos 599* passed overhead, another satellite appeared. *Cosmos 602* was a maneuverable type, launched on 20 October, with its orbit stabilized with no lateral drift at all. *Cosmos 602* passed directly over the south end of the Suez Canal. It was recovered on 29 October after a 9-day flight.

The flights of these three satellites coincided with the closing stages of the war. The Israelis continued to pour across the Suez Canal. On 22 October, a UN cease-fire resolution was accepted by Israel and Egypt. The truce broke down almost immediately in the Sinai. On 23 and 24 October, the Israelis were able to cut off Suez City and the 20,000 men of the Egyptian Third Corps. Alarmed, the Egyptians asked for U.S. and Soviet troops to enforce the cease-fire. The U.S. refused but the Soviets began moves that indicated they were about to send in seven parachute divisions, two mechanized divisions, and 6,000 naval

infantry. To forestall this, on 25 October, U.S. forces went on an increased alert level. A few hours later, the UN Security Council passed another truce resolution, which included authorization to send a UN emergency force, to be followed by a larger group that would monitor the truce. This time the cease-fire held. By the time *Cosmos 602* was recovered, the war was virtually over. The Third Corps was receiving relief supplies and there were only sporadic violations.[32,33]

The final reconnaissance satellite of the war, *Cosmos 603,* was launched on 27 October. Although it made a pass over the target area the next day, its orbit was not stabilized until 1 November. As with *Cosmos 602,* the southern canal area was the target for its cameras. It stayed aloft for the normal 13 days, being recovered on 9 November. Two days later, Israel and Egypt signed the six-point peace agreement.[34]

The 1973 Middle East war demonstrated both the high priority the Soviets gave to space reconnaissance in times of crisis and the ability to cope with fast-breaking situations. It seems likely that the Soviets keep a reserve of satellites and boosters on hand for just such a situation. It is also worth noting that, despite the step-up, the Soviets were still able to cover their routine targets (*Cosmos 599*). Reflecting the increased launches, the 1973 total was thirty-five. The war's effects continued into the next year. Only twenty-eight satellites would be launched in 1974. The situation parallels that in 1969–1970 after the Sino-Soviet crisis.

Reconnaissance in the 1970s

After the disruption of the 1973 war, the Soviet reconnaissance satellite program returned to normal. In the last half of the 1970s, the number of launches per year would stabilize at thirty-two to thirty-five. The Soviet effort has been called "clocklike" and with good reason. Launches occur on Wednesdays, Thursdays, or Fridays; recoveries are made on Tuesdays or Wednesdays.

Despite such regularity, the program was not static. One major change came in 1974. Up to this point, both the high- and low-resolution satellites were placed into a typical orbit of 205 by 130 miles (330 by 210 kilometers). The two types were distinguished by launch vehicle or whether or not they maneuvered. Starting in 1974, the high-resolution satellites were placed into an initial orbit with a low point of about 114 miles (185 kilometers), which would improve the resolution (resolution improves as the distance between camera and target is reduced). The low-resolution satellites remained in the higher orbit, which would allow them to survey a wider area.

Two years later, the replacement for the low-resolution satellites appeared. The so-called "medium-resolution" satellite combines several features of both types. Like the low-resolution satellites, they go into a higher altitude orbit. The medium-resolution satellites also carry a maneuvering engine to adjust their orbits. The first medium-resolution satellite was *Cosmos 867,* launched on 23 November 1976, from Plesetsk. It went into a 250- by 155-mile (402- by 250-kilometer) initial orbit, inclined 62.8 degrees. After 3 days, the maneuvering engine fired and *Cosmos 867* went into a 244- by 219-mile (401- by 352-kilometer) orbit.[35] After going into this high, final orbit, the maneuvering engine made one or two "tweak" burns, which corrected the orbital decay caused by atmospheric drag. Orbital lifetime for the medium-resolution satellites is the standard 13 days.

Some of these satellites carry earth resources packages, which could be similar to the MKF-6 camera carried on the *Soyuz 22* mission. This package had six cameras to simultaneously photograph the earth in different wavelengths, four of which operate in visible light and two in infrared. On the *Soyuz 22* flight, each cassette carried enough film to photograph six million square miles (16 million square kilometers). The MKF-6 camera is equipped with a motion compensator and weighs 450 pounds (205 kilograms). The photos can be used by specialists at the Priroda Earth Resources Center to determine the condition of crops (diseased crops will be a different color in the photos than healthy ones), soil humidity, the amount of snow cover, and the amount of water available.[36] A modified MKF-6 was carried aboard the *Salyut 6* space station.[37] It should be stressed such earth resources are only a supplement to the medium-resolution satellites' primary function of military reconnaissance. Subsequent launches shifted to an 82.3 degree inclined orbit.

Introduction of the medium-resolution satellite was slow. Only one or two were launched per year during the late 1970s. In 1979, the rate underwent an explosive increase. Medium-resolution satellites would provide about a quarter of the year's total. With this, the phaseout of the older, low-resolution satellites began in earnest. By the early 1980s, only one or two low-resolution satellites would be orbited per year. They used an inclination of 82.3 degrees. Previously, the complete range of inclinations was used. It is believed their mission is military mapping.[38] Presumably, they carry equipment similar to that in U.S. Titan III B reconnaissance satellites during the late 1960s. Their photos would be used to precisely locate military targets in the U.S., western Europe, and

China. This updating was necessary because of the Soviets' development of super-accurate ICBMs.

The Soyuz Reconnaissance Satellite

Looking over the history of the Soviet effort in space reconnaissance, it can be seen that one disadvantage is the satellites' short lifetime. The Soviets must launch ten times as many reconnaissance satellites as the U.S. to achieve the same time in orbit. It is not unusual for the Soviets to have four reconnaissance satellites in orbit simultaneously. This level of activity is very expensive. Clearly, a reconnaissance satellite with a lifetime longer than 13 or 14 days would be desirable. To develop this long-duration reconnaissance satellite, the Soviets turned to the Soyuz manned spacecraft. Work had begun on it in the early 1960s. Unmanned orbital tests were made in 1966 and 1967. The first manned flight, *Soyuz 1,* was launched on 23 April, 1967. Due to hardware failures, it was brought down early. After reentry, the parachute tangled and its pilot, Vladimir Komarov, was killed. After redesign work, manned flights resumed 18 months later.

In the 1970s, the Soyuz was used to transport cosmonauts to Salyut space stations. The improved version, the *Soyuz-T,* has been extensively updated with new, more sophisticated systems. The Soyuz is also the basis for the unmanned Progress resupply spacecraft. The Soyuz is a three-part spacecraft. The cylindrical service module contains the prime and backup rocket engines, fuel tanks, electronics, and the two solar panels. Attached to it is the bell-shaped command module, which houses the cosmonauts during launch and reentry. At the front is the spherical orbital module, the working and living area for the crew.[39]

There is no precise information on the modifications made to the reconnaissance Soyuz. According to one account, the orbital module is replaced by two, small reentry capsules or "buckets," which are deorbited periodically. At the end of the mission, the descent module, with its camera and film, is recovered.[40] An alternative modification is that the reconnaissance Soyuz is similar to the Progress spacecraft, in which the descent module is replaced by a cylindrical unit that contains fuel tanks. Once it has docked to a Salyut station, the fuel is pumped into the Salyut's tanks. Additional supplies of food, water, and mail are contained in the orbital module.

In this design, the cameras would be in the cylindrical module. The multiple reentry capsules would be fitted to the nose—again replacing the orbital module. The service module would remain the same. The camera would not be recovered but, rather, allowed to burn up during reentry. One bit of supporting evidence for this is the geometry of the recovery orbits of the Vostok and Soyuz reconnaissance satellites. The geometry and timing of the Vostok satellites have a very narrow range, which is necessary if the capsules are to land within the limited confines of the recovery zone. With the reconnaissance Soyuz, this narrow set of values does not exist, which implies that when all the film has been transferred to the capsules, the spacecraft is commanded to make a destructive reentry. Any debris would sink safely beneath the northern Pacific. Since there was no need for a precise impact point and since a day or night landing made no difference, there would be no need for the limitations seen in the earlier satellites.[41] Unfortunately, there are no open

An orbital view of the Russian *Soyuz 19* spacecraft. The Soyuz is composed of three parts. At the forward end is the spherical orbital module where the crew works during the flight. In the middle is the bell-shaped descent module, which can seat up to three cosmonauts in cramped conditions. The crew is launched and lands in this section. The final part is the service module. This contains electronics, the fuel tanks, and the two rockets used to maneuver and deorbit the spacecraft. Power for the spacecraft is provided by the two solar panels. Like the Vostok, the Soyuz was also adapted for use as the fourth- and possibly fifth-generation Soviet reconnaissance satellites. The specific modifications are not known but may be extensive. It is believed by some that two reentry capsules replace the orbital module and that the descent module is replaced by the camera unit. *NASA Photo.*

sources to explain which modifications were actually made to the reconnaissance Soyuz.

The reconnaissance Soyuz program got off to a shaky start. The first satellite, *Cosmos 758,* was launched from Plesetsk on 5 September 1975, into a 218- by 112-mile (351- by 181-kilometer) orbit. Its inclination of 67.2 degrees was a tip-off that this was a new type of reconnaissance satellite. Its low orbit indicated it had high-resolution optics. For the first few days, *Cosmos 758* appeared to function well. It made several maneuvers. On the ninth day of the mission, it went into a 359- by 121-mile (578- by 194-kilometer) orbit.

Seventy-six objects were tracked with it. Whether this was an intentional self-destruction of a malfunctioning satellite or the result of a rocket system problem is unknown. Clearly, *Cosmos 758* was a failure. The largest piece of debris decayed and reentered after 20 days.

The second satellite, *Cosmos 805,* launched on 20 February 1976, was more successful; it was recovered after 19.6 days. Any optimism the Soviets may have felt was dashed by the next flight. *Cosmos 844* was launched on 22 July and blew up 3 days later. The wreckage reentered 39 days after launch. The roller coaster–like history of the program continued with *Cosmos 905* (26 April 1977), which was successfully recovered after 29.5 days. This was the first reconnaissance Soyuz satellite to reach the operational lifetime.[42] It was the only reconnaissance Soyuz launched in 1977. Only one Soyuz was launched the following year—*Cosmos 1028* on 5 August 1978. It stayed up for 29.5 days.

There was a step-up in activity—four launches—in 1979. The reconnaissance Soyuz became operational in 1980. This pattern of several years of sporadic tests, followed by a sudden step-up, is the same as that of the medium-resolution satellite. One significant launch was *Cosmos 1177* (29 April 1980), which stayed in orbit 44 days. Beginning in 1981, the reconnaissance Soyuz satellites would make about one-fourth to one-third of each year's flights. The expansion in activity was not limited to just numbers. Previously, only Plesetsk had been used for reconnaissance Soyuz launches. With the beginning of Soyuz operations, launches were also made from Tyuratam. These used inclinations of 65 degrees and 70 degrees. There is, as yet, no sign the Soviets intend to phase out the older Vostok high-resolution satellites.

An improved Soyuz reconnaissance satellite made its debut in 1982. Rather than staying aloft for 29–31 days, these new satellites remained in orbit 40 or more days. This increased lifetime would require additional fuel and film. It has been suggested that the satellites use Soyuz-T systems rather than the mid-1960s technology of the original Soyuz. By 1983–1984, this improved Soyuz would be standardized at a 44-day lifetime.

The introduction of the reconnaissance Soyuz does seem to indicate a change in procedure. The Vostok high-resolution satellites appear to be used to photograph only a single target area. The ground track is stabilized to pass over a given area until recovery. The October 1973 war is an example, but there have been enough others, over the years, to indicate a general trend. The reconnaissance Soyuz is maneuvered as needed to cover multiple targets,[43,44] which can be seen in the behavior of satellites during 1983:

- *Cosmos 1454*
Lebanon war	27–29 April
	9–11 and 19–22 May

- *Cosmos 1457*
Lebanon war	23–26 May

- *Cosmos 1471*
Global surveillance	28 June–3 July
El Salvador	4–7 July
Global surveillance	7–28 July

- *Cosmos 1489*
Faya-Largeau (Chad)	10–13 August
Lebanon war	3–4 September

- *Cosmos 1504*
Grenada invasion	28 October–1 November
Lebanon war	1–3 December

Identification of specific targets is possible because the satellites' ground track is stabilized over the area.[45]

The launches of the reconnaissance Soyuz satellites appear to be seasonal: the year's first launch comes in the spring, the bulk of the flights are made in the summer months, and only two launches are made in the fall. The reason may be the need for very clear weather. The photos from the satellites' high-resolution optics would be adversely affected by the shadows cast by clouds, even if the target area itself were clear. The satellites are placed into a 217- by 106-mile (350- by 170-kilometer) orbit. Because of this low orbit, atmospheric drag is a problem. Accordingly, orbital maneuvers are made at weekly intervals, which returns the satellite to an orbit close to the initial one. It may be at this point that the reentry capsules are separated.[46]

A fifth-generation reconnaissance satellite also may have made its debut at this time. On 28 December 1982, *Cosmos 1426* was launched from Tyuratam and went into a 218- by 127-mile (351- by 205-kilometer) orbit inclined 64.9 degrees. Unlike previous satellites, it signals could not be picked up by Western amateur listeners. *Cosmos 1426* operated for a record-breaking 67 days, until 5 March 1983.[47] A second example of this new type was *Cosmos 1552,* launched on 14 May 1984, also from Tyuratam; its orbit was 200 by 113 miles (322 by 182 kilometers). The orbits were marginally higher than the standard Soyuz reconnaissance satellites.[48] What marked this satellite as unusual was its lifetime—173 days. There are three other satellites—*Cosmos 1576, 1585,* and *1599*—which also showed unusually long lifetimes and may be members of this new generation. *Cosmos 1643* set a record for the fifth generation by remaining aloft for 207 days; this is equivalent

to the best performance by the Big Bird but considerably less than the lifetime of the KH-11.

It has been suggested the satellites use electronic digital imagery like the KH-11, making them the first Soviet radio transmission satellites. It is also possible that once the photos have been transmitted, the tapes are returned to earth via capsule.[49] Radioing the photos to earth would have the advantage of getting them into the hands of the interpreters faster; there would be no need to wait for the next capsule recovery. If this is true, the Soviets have a system comparable to the Big Bird or KH-11. Since the spacecraft is launched by the same A-2 booster as earlier satellites, the weight of this new type would also be approximately 13,230 pounds (6,000 kilograms).

The fourth-generation Soyuz reconnaissance satellites, like the Titan III Bs, exhibited a continuing growth. *Cosmos 1504* remained in orbit a record 53 days, which became the new standard lifetime. In 1984, *Cosmos 1616* orbited for 54 days; the next year, *Cosmos 1647* remained aloft 53 days.

The number of launches remained stable. In 1983, there were twenty-seven reconnaissance satellites and ten medium-resolution, earth resources missions. There was one less earth resources flight in 1984. The earth resources satellites have proven extremely valuable to the Soviets, resulting in the saving of millions of rubles per year. Their cost was ten to fifteen times less than aerial photos. There was also intelligence value to the photos: crop assessments of other countries could be made, and knowledge of world harvests and reserves would aid the Soviets in negotiating grain purchases.[50]

Within the static numbers, there are continuing new developments. The most important in 1984 was the demonstration of in-orbit storage. After *Cosmos 1587* and *1613* were launched, they were shut down and allowed to drift in orbit. After 10 days, ground controllers turned them back on; they then flew a normal, 14-day mission and reentered. This was seen as enabling the Soviets to improve survivability under wartime conditions.[51]

The Future

One can make certain projections about the future of the Soviet reconnaissance satellite effort. First, its high priority will continue. The Soviets would not launch more than thirty satellites per year unless they felt the effort was worthwhile. Second, as more and more of the reconnaissance Soyuz satellites are flown, the total number of launches will start to drop. This would mirror the trend of the U.S. program. Looking over the past 20 years, one surprising aspect is how long it took the Soviets to develop a radio transmission satellite. The Soviets have built such systems for their lunar and planetary probes; yet it appears they are only now flying this type of reconnaissance system. Clearly, this would have been an advantage in the October 1973 war.

The Soviets had to wait until the satellites were recovered before they could examine the film, which meant much of the information would be at least several days old. With a radio transmission–type satellite, they could have photos on a daily basis.

In comparing the U.S. and Soviet reconnaissance programs, one must remember the different design philosophies. However, it seems clear that on a technological level, the U.S. is superior. The Soviets are only now flying a spacecraft similar to the Big Bird. The resolution of Soviet reconnaissance photos is also believed to be inferior to those of the U.S. This applies to both the optical systems and the film. On the other hand, under combat conditions where satellites are being attacked, the higher Soviet launch rate is an advantage; it would be easy for them to replace destroyed satellites.

Whatever form the future takes, the orbital watch goes on. The satellites will make their endless rounds through the cold, silent vacuum, their mechanical eyes following activities on the green fields of earth below—that beautiful, blue sphere where Allen Drury once observed, ". . . there was, as always, little peace and less goodwill, toward men who had conquered almost everything but man."

NOTES Chapter 9

1. John Barron, *KGB*, (New York: Bantam, 1974), 262–269,362–378.
2. Carrollton Press DoD 39A 1981.
3. Carrollton Press PSAC 181B 1981 and NASC 72C 1981.
4. Carrollton Press DoD 58B 1980.
5. Nicholas L. Johnson, *Handbook of Soviet Manned Space Flight*, (San Diego: American Astronautical Society, 1980), 10.
6. Philip J. Klass, "Soviets Trying Mid-Air Satellite Recovery," *Aviation Week & Space Technology*, (4 October 1971): 18.
7. Memo: Bromley Smith to Bill Moyers, Forecast of Soviet Space Spectaculars in the Balance of 1964, Mandatory Review Case #NLJ 84–225, Document #2, Lyndon Baines Johnson Library, Austin, Texas.
8. *Soviet Space Programs 1971–1975*, (Library of Congress 1976).
9. Barron, *KGB*, 414,415.
10. Philip J. Klass, *Secret Sentries in Space*, (New York: Random House, 1971), 125.
11. Viktor Suvorov, *Inside Soviet Military Intelligence*, (New York: Macmillan, 1984), 55,60.
12. *New York Times*, 15 July 1963.
13. Klass, *Secret Sentries in Space*, 127.
14. William J. Normye, "Cosmos 57 Believed Destroyed by Soviets," *Aviation Week & Space Technology*, (12 April 1965): 34.
15. Philip J. Klass, "Soviets Hike Reconnaissance Satellite Pace", *Aviation Week & Space Technology*, (4 July 1966): 16–18.
16. *Soviet Space Programs 1971–1975*, 447,448.
17. Kenneth Gatland, *Manned Spacecraft in Color Second Revision*, (New York: Macmillan, 1976).
18. DoD Weekly Report 25 January 1966, Lyndon Baines Johnson Library, Austin, Texas, Mandatory Review Case #NLJ 84–221 Document #1.
19. G. E. Perry, "The Soviet Northern Cosmodrome", *Spaceflight*, (August 1967): 274–276.
20. "Cosmos Spacecraft: First Pictures," *Flight International*, (4 June 1964): 946.
21. *Soviet Space Programs 1971–1975*, 444,464,468, 470.
22. Compton Yearbook 1970, 181,182,192,193,480,481.
23. Arkady N. Shevchenko, *Breaking With Moscow*, (New York: Knopf, 1985), 164,165.
24. Klass, *Secret Sentries in Space*, 160,161.
25. Shevchenko, *Breaking With Moscow*, 165,166.
26. Philip J. Klass, "USSR Accelerates Recon Satellite Pace," *Aviation Week & Space Technology*, (6 April 1970): 72–74,79.
27. *Soviet Space Programs 1971–1975*, 441–445, 459–462.
28. *Soviet Space Programs 1971–1975*, 465,466.
29. Bryce Walker, *Fighting Jets*, (New York: Time Life Books, 1983), 147–150.
30. Peter Allen, *The Yom Kippur War*, (New York: Scribners, 1982), 236–247.
31. *Soviet Space Programs 1971–1975*, 473.
32. Allen, *The Yom Kippur War*, 288–296.
33. *Israel & The Arabs: The October 1973 War*, Lester A. Sobel, Editor, (New York: Facts on File 1974).

34. *Soviet Space Programs 1971–1975,* 471–473.
35. Satellite Digest 104, *Spaceflight,* (May 1977): 200.
36. Gordon R. Hooper, "Soyuz 22—Mission Report," *Spaceflight,* (February 1977): 61–63.
37. James E. Oberg, *Red Star in Orbit,* (New York: Random House, 1981), 194.
38. Phillip S. Clark, "Aspects of the Soviet Photoreconnaissance Satellite Programme," *Journal of the British Interplanetary Society,* (April 1983): 171.
39. Johnson, *Handbook of Soviet Manned Space Flight,* Chapter 7.
40. "Soviet Union Emphasizes Military Spaceflight," *Flight International,* (2 January 1982): 27.
41. Phillip S. Clark, "Soviet Spacecraft Recoveries," *Journal of the British Interplanetary Society,* (April 1983): 186–191.
42. Satellite Digest 108, *Spaceflight,* (October 1977): 362.
43. *Soviet Space Programs 1971–1975,* 463–465, 473–475.
44. Nicholas L. Johnson, "Soviet Satellite Reconnaissance Activities and Trends," *Air Force Magazine,* (March 1981): 90–94.
45. Nicholas L. Johnson, "The Soviet Year in Space: 1983," *Space World,* (October 1984): 19–21.
46. Phillip S. Clark, "Aspects of the Soviet Photoreconnaissance Satellite Programme," *Journal of the British Interplanetary Society,* (April 1983): 177.
47. Satellite Digest 163, *Spaceflight,* (June 1983): 284.
48. Satellite Digest 177, *Spaceflight,* (November 1984): 428.
49. Letters to the Editor, *Aviation Week & Space Technology,* (8 October 1984): 104.
50. Nicholas L. Johnson, "The Soviet Year in Space: 1983," *Space World,* (October 1984): 19–21.
51. "Soviets Develop Heavy Boosters Amid Massive Military Space Buildup," *Aviation Week & Space Technology,* (18 March 1985): 120.

CHAPTER 10

Independent Reconnaissance Satellites

After the road to space had been opened by Russia and the U.S., other nations were quick to follow. The area of reconnaissance satellites, however, remained a U.S.-Soviet monopoly. Although Western Europe and Japan have long had the technical capability to build reconnaissance satellites, they had no need. The U.S. shares the intelligence the satellites gather.

With Russia, the situation is the reverse. The Soviets share nothing with the other Warsaw Pact members. Indeed, the intelligence services of the various Eastern European countries are little more than appendages of the KGB. Information is so tightly controlled by the Soviets that if an Eastern European country wants to know the Soviet order of battle, they consult Western data.[1]

For a country to develop an independent reconnaissance satellite, it must meet two basic requirements. First, it must have the capability, which includes not only the necessary technology but also the industrial base to build them and an economy large enough to pay for the effort. This eliminates most of the world's countries. Second, the country must face a military threat that cannot be assessed by conventional espionage (spies, reconnaissance aircraft, or cooperating with other nations). So far, only one country has met these requirements—the People's Republic of China (PRC).

In many ways, the intelligence situation facing China resembled that of the U.S. in the early 1950s. The Chinese would need to know the location of Soviet forces and any changes that might signal an attack.

The Chinese Space Program

China's space program, like most of their military technology, goes back to aid supplied by the Soviets in the 1950s. Before breaking off assistance in 1959 and 1960, the Soviets supplied copies of the SS-2 Sibling, a large battlefield rocket based on V-2 technology and similar to the U.S. Redstone. It was not much; but from this small seed would grow the world's third military power.

There is also an American dimension to the Chinese effort. One of the important people in their rocketry program is Dr. Chien Hsueh-Sen, director of the Dynamic Research Institute. Born in Shanghai, he obtained an American scholarship to Caltech in 1935. He became associated with Dr. Theodore Von Karman, who established the Jet Propulsion Laboratory at Caltech. Dr. Hsueh-Sen worked on the WAC-Corporal rocket and studied the

V-2's performance. In 1949, he made a study of a winged rocket. During the early 1950s, he fell victim to Senator Joe McCarthy's witch hunts and left the country. He first went to Canada to teach at Toronto University; finally, in 1955, he returned to China.[2]

Four years later, work began on an improved version of the Soviet SS-2. This is the CSS-1, a 600-nautical-mile (1,111-kilometer) range liquid-fuel missile.

At the same time, the CIA was supplying U-2s to the Nationalist Chinese Air Force for overflights of the mainland. Particular attention was paid to the PRC's nuclear and rocket programs. On one of the early overflights, a U-2 obtained photos of the Chinese rocket test center at Shuang Chen-Tzu in northwestern China. When the photos were analyzed, it was apparent that the Soviet-Chinese technical cooperation had been on a high level. However, with the Soviet pullout, the work had come to a virtual halt.[3]

Reflecting these difficulties, the CSS-1 was not ready for flight tests until 1966. One of these launches, however, involved a live nuclear warhead. On 27 October, a CSS-1 was launched towards the Lop Nor nuclear test site. Upon reaching the target, the warhead exploded with a yield of 20 to 30 kilotons.[4]

Three years later, the second Chinese rocket was ready for testing. The CSS-2 was a single-stage IRBM with a 3-megaton warhead and a 1,500-nautical-mile (2,778-kilometer) range. The next year, a two-stage version was being flight-tested—the CSS-3 limited-range ICBM. From silos in China, it could strike targets in European Russia, possibly including Moscow. The CSS-3 also meant the road to space was now open for the Chinese. A small, third stage was added to make the Long March-1 satellite launcher (U.S. designation CSL-1).[5]

The first Chinese satellite launch attempt may have been made on 1 November 1969. A poster reportedly appeared on Peking's Chang An Street that said "warmly acclaim launching of China's first man made satellite."[6] It is understood that on the same day, radio Peking said that an important announcement would be made that afternoon. Yet, nothing was forthcoming.[7] Either way, it would be another year before success was theirs.

On 24 April 1970, *China 1* went into a 1,481- by 273-mile (2,384- by 439-kilometer) orbit. The satellite was a multisided sphere 3.28 feet (1 meter) across. Attached to it were four 9.8-foot (3-meter) antennas.[8] With a total weight of 380 pounds (173 kilograms), this was the largest first satellite launched by any nation. *China 1* had twice the weight of *Sputnik I* and about thirteen times the weight of the U.S. *Explorer I*. *China 1* is also remembered for broadcasting the song "Tungfang-hung" ("The East Is Red").

The second Chinese launch came on 3 March 1971. *China 2,* which also used a Long March-1 booster, was slightly heavier than *China 1;* it weighed 487 pounds (221 kilograms) and contained scientific equipment.

At roughly the same time, the Chinese reconnaissance satellite program was underway. The booster was the CSS-4 ICBM, a large, two-stage vehicle with a 4-megaton warhead and a 6,500-nautical-mile (12,038-kilometer) range.[9] When used as a satellite launcher, it is referred to by the Chinese as the FB-1. Final development work was concluded around 1973, but the program was halted for a year. This was a time of great internal turmoil. The Chinese satellite program was, according to some reports, going through its own difficulties; three straight FB-1 launches failed.[10] Finally, on 26 July 1975, *China 3* went into a 288- by 116-mile (464- by 186-kilometer) orbit. The satellite weighed approximately 7,718 pounds (3,500 kilograms) and was understood to carry a television camera for weather and reconnaissance observations. Two months after launching, it was apparently deorbited into the Pacific. In an article related to the launch, the Chinese referred to their satellite program as for "preparedness against war." It appears that *China 3* was the prototype for a radio-transmission satellite.

There was no question about the mission of the next satellite. *China 4* was the first Chinese recoverable satellite. On 26 November 1975, it went into a 300- by 108-mile (483- by 173-kilometer) orbit. This low orbit would be excellent for high-resolution photos. The satellite also used a new inclination—63 degrees versus the 69 degrees used for the previous Chinese satellites. After 6 days, a reentry capsule was separated from the main satellite, deorbited by an on-board retro-rocket, and recovered by the Chinese. The new inclination, along with a pass over the desert areas in northern China, made the recovery easier. China thus became only the third country to both launch and recover satellites. Details about the satellite are lacking. The only description is a "sphere cylinder." It is known the satellite has two parts. After the capsule is recovered, a module remains in orbit until atmospheric drag brings it down. This might be some type of service module carrying the camera, electrical power supplies (solar cells or batteries), attitude-control rockets, and the guidance system. The other part is the reentry capsule and its retro-rocket,

which makes up the greater part of the spacecraft's weight of 7,718 pounds (3,500 kilograms).[11]

The Chinese have published photos of the capsule after reentry. It is bell-shaped, about 5 by 6 feet (1.5 by 1.8 meters), and resembles a slightly smaller version of the Soyuz command module.

The photos are used for both reconnaissance and earth resources. The spacecraft has a multispectral capacity. Photos shown to Western engineers can resolve small farm fields.[12]

The final flight in 1975, *China 5,* was the same type of satellite as *China 3*—a suspected radio-transmission type. It went into orbit on 16 December.

The next satellite, *China 6* (30 August 1976), was a small Long March-1–launched satellite, like *China 1* and *2*. It is believed to have carried a small television camera for weather photos rather than for reconnaissance.

The second Chinese recoverable satellite, *China 7,* was launched on 7 December 1976 into a 298- by 107-mile (479- by 172-kilometer) orbit. The inclination was only 59 degrees—less than its predecessor, *China 4.* Two days later, the capsule was separated and recovered; a module remained in orbit as before. The short operating lifetime meant only a few specific targets could be covered. *China 8,* the next recoverable satellite, went into orbit on 26 January 1978. The 5,292-pound (2,400-kilogram) capsule was recovered 5 days later; the module decayed after 12 days. With this, the Chinese space program went into a hiatus. According to a Russian Tass dispatch, a Chinese launch attempt failed in late July 1979.[13] This rocket was carrying a 3.28-foot (1-meter) in diameter sphere powered by three solar panels. The satellite was carrying a scientific and engineering payload.[14] It would be another 2 years before a Chinese satellite entered space. *China 9, 10,* and *11* (19 September 1981) were actually three scientific satellites launched on a single booster. One was a prism-shaped satellite 3.29 by 3.9 feet (1 by 1.2 meters), the second was cone-shaped, and the third was a balloon attached by a wire to a metal ball. They carried experiments on magnetic fields, infrared and ultraviolet radiation, charged particles, X rays, and atmospheric drag.[15]

Another year would pass before the next Chinese spacecraft was orbited. *China 12,* the first reconnaissance satellite in more than 4 years, was launched on 9 September 1982; its reentry capsule was recovered 4 days later.[16]

The long gap between these missions could have a number of explanations—to allow time for analysis of the photos, to correct technical difficulties (although the "public" part of the flights, launch, and recovery seemed to have gone well), or budget constraints. Continuing the one-launch-per-year rate, *China 13* went up on 19 August 1983; capsule recovery came 5 days later. These two satellites used 63 degree inclinations.[17]

The next two Chinese satellites, *China 14* and *15* (29 January and 8 April 1984), were development flights for geosynchronous communications satellites. Reconnaissance satellites resumed with *China 16,* launched on 12 September 1984. It orbited for the standard 5 days before the capsule was returned to earth. Inclination was increased to 67.9 degrees. All the Chinese recoverable satellites had low points around 108 miles (174 kilometers).[18] *China 17* followed in the fall of 1985.

With only one mission per year, the amount of photo intelligence that could be gathered was limited. The satellites could update China's target list but could not indicate day-to-day changes.

These achievements are all the more remarkable in light of the PRC's difficult economic condition. The military and social demands it faces are great and the available resources are severely limited. For this reason, the Chinese have decided to concentrate on military and applications missions, to avoid disbursing the available money and personnel over too many projects. In terms of technology, the Chinese seem fairly solid. They are very proud that their space program's electronics and digital computers are domestically designed and built. It is not an "imported" space program. The Chinese also have several research institutes to develop the needed technology.[19]

As for the future, the Chinese have begun to fly the Long March-3, an FB-1 equipped with a liquid hydrogen–fueled third stage. Although intended for geosynchronous orbit satellites, the Long March-3 would also allow a greater payload into low earth orbit. This could include larger military satellites as well as manned spacecraft and deep space probes, two areas that figure in China's long-term planning. It is doubtful the Chinese effort will achieve the level of even the small-scale European program for a long time to come, but the Chinese will continue to leave their mark in space.

France

Although the People's Republic of China is the only nation, as yet, to fly an independent reconnaissance satellite, France is actively developing a similar system. France has long pursued an independent military policy. In the 1960s, it pulled out of NATO and has maintained an independent nuclear deterrent of submarines, land-based missiles, and aircraft. In space affairs, France was a driving force behind the development of the Ariane booster. Given this background, the decision to have an independent intelligence capacity is to be expected. Ironically, the French reconnaissance satellite is to be developed from a civilian satellite—the SPOT earth resources satellite (Satellite Probatoire d'Observation de la Terre). The SPOT is box-shaped, 10 by 7 feet (3 by 2 meters), with a solar array spanning 51 feet (15.6 meters); it weighs 3,859 pounds (1,750 kilograms). It carries two identical high-resolution imaging systems, which observe the ground in either visible or infrared light. The imaging systems have two operating modes—panchromatic (across the complete spectrum) or multispectral (at several selected narrow wavelengths). The resolution in panchromatic is 33 feet (10 meters) and in multispectral, 66 feet (20 meters).[20]

The military version of SPOT is the Satellite Militaire de Reconnaissance Optique (SAMRO), which will share structure, thermal control, solar array, and attitude-control systems with SPOT. The major change will be a new, high-resolution optical system; details are secret, but it will have to be at least two to ten times better than SPOT. The SAMRO would also be placed into a sun-synchronous orbit by an Ariane booster. Several satellites would be required for complete coverage and to handle any failures.[21] The work was temporarily halted in 1982 because of military spending constraints. Active work resumed in 1986; the program was renamed Helios and a first launch is envisioned in 1992.

The satellite would have greater capabilities than the original SAMRO.

PEACE SAT

An alternative to an independent national reconnaissance satellite program is one under the control of an international body. The satellites would be used to enforce the peace on a worldwide basis; they would keep watch on the various trouble spots. The purpose would be to give warning of any buildup or military moves that could signal an attack. With this foreknowledge, diplomatic pressure could be brought on the would-be aggressor. To continue the aggression would mean defying both international opinion and the now-alerted intended victim.

The idea goes back before the first reconnaissance satellite went into orbit. The originator is Howard Kurtz, an Air Force officer turned management consultant. Since the early 1950s, Kurtz has argued that conventional arms control efforts are flawed. U.S. and Russia's mutual distrust would generate pressure to continue weapons development, treaties notwithstanding. Kurtz admits he is too cynical about human nature to believe disarmament has much chance. Unilateral disarmament would be foolhardy. A one-world government, in Kurtz's view, is an illusion because of the vast differences between peoples and regions. The solution, he believes, is security by satellite. This would be accomplished through a "global information cooperative"—a set of multiple agencies that would undertake not only military reconnaissance but also weather, mapping, search and rescue, and communications. The global information cooperative would receive continuous updates from its network of satellites. This data would give the military status of every nation in the world. It would be the basic guarantor of the peace. The global information cooperative satellites would do for the entire world what U.S. and Soviet satellites already do on a more limited basis.

After a quarter century of publicizing his ideas, Kurtz, in the late 1970s, was able to gain the interest of various congressional and administration figures. Olin E. Teague (D-Texas), chairman of the House Science Committee; Senator Adlai E. Stevenson (D-Illinois), chairman of the subcommittees on space science and, more importantly, intelligence; Senator Edward M. Kennedy (D-Massachusetts); President Carter's science advisor, Dr. Frank Press; and NASA administrator Dr. Robert A. Frosh all voiced interest.[22,23]

It was the French, however, who took the lead in setting up such an agency. At a 25 January 1978 press conference, French President Valery Giscard d'Estaing proposed the establishment of an International Satellite Monitoring Agency (ISMA). He subsequently noted, "The two biggest powers are the only ones that have this equipment right now. . . . Other countries such as France will acquire it in the next five years." This International Satellite Monitoring Agency would be used to enforce arms treaties between the superpowers as well as regional agreements. It could also be used to follow arms shipments. It would be paid for by a tax on what the French president called "overarmed" countries.[24]

The UN took up the French proposal and in February 1981 issued a report. It found a UN International Satellite Monitoring Agency was feasible, although the working details of access to data and reports; the confidentiality of the material during transmission, process-

ing, interpretation, and storage; and how action would be initiated were open questions. The study proposed a three-phase program to establish the ISMA. Phase I would be the relatively small step of establishing a photo interpretation section to provide training and experience for the personnel. The photos, to be supplied by member countries, could include weather satellite photos, earth resources data, and archive material such as photos from the Skylab and Salyut missions. At this point, the ISMA would have a very limited function. Its ability to do any verification or crisis monitoring would depend on the U.S. and/or Soviets providing reconnaissance data. In Phase II, the ISMA would no longer be dependent upon donated photos. By building a worldwide network of tracking stations, ISMA could get its photos directly by tapping into U.S. and Soviet satellites. Building and operating the stations would increase costs. Although the ISMA would be a full-fledged organization at this point, it still could not undertake significant reconnaissance activities. To become a true "monitoring" organization, the ISMA must make the enormous leap to Phase III and its own independent reconnaissance satellite network. The study presumed the UN would have to build its own satellites. This means big money—a minimum of several hundred million dollars per year to operate the satellites, to provide launch replacements, and to pay for the ground-based segments (tracking, interpretation, and storage). To provide complete area coverage, multiple satellites would be required. The UN study spoke of a three-satellite network carrying optical, infrared, and radar systems. There would also be a need for a high-resolution satellite with a maneuvering capability to cover specific trouble spots. An alternative possibility would be to maintain the high-resolution satellite on alert, ready for launch in a crisis, which would save about $120 million per year in operating costs. The specific details as to sensors and operations had to be studied further.

Problems and Prospects

The UN proposal was all very nice, but the world is filled with wonderful peace programs, from superpower proposals to slogans on protest signs. What about the nasty little details of the real world that so often ruin such high hopes? Although most members of the UN were favorable about ISMA, both the U.S. and Russia were opposed and did not participate in the study. The other delegates believed the U.S. was worried about disclosure of reconnaissance technology and its capabilities. Since U.S. security depends upon satellite monitoring of Russia, this is not a trivial concern. The study group felt, however, that there were factors that might cause the U.S. to moderate its opposition. First, the U.S. may come to feel that information about world trouble spots would best be made available through an independent source; an example given was the Sino-Soviet border. Second, if an ISMA was established, U.S. participation would prevent domination by French and Western European technology. Finally, an ISMA may be seen as preferable to multiple independent reconnaissance programs. The U.S. position was felt to be one of wait and see rather than flat opposition. Not so with the Soviet Union. Its opposition was much stronger and there does not seem to be much chance of it lessening.[25]

It must also be asked how well the UN could perform in this role. Many critics argue that the UN has become dominated by Third World countries, which, in the view of the critics, have shown a continuous anti-Western bias. What guarantees would there be that the ISMA would not be similarly used to condemn the West while excusing more favored nations? And how supportive would various nations be once the idea is no longer theoretical. Most of the world's countries are authoritarian to varying degrees. One of the basic principles of such rule is control of information. How willing would they really be to have a network of orbital ''snoops'' prying into their activities? Although they may want information on their enemies and rivals, having those other countries receive the same data is another matter. To add insult to injury, they would also have to pay part of the costs of supplying it.

Finally, an ISMA could cause a space versus welfare controversy similar to that which has crippled the U.S. civilian space program. The ISMA would probably be the largest item in the UN budget. Most of the other UN agencies provide welfare-type services (famine relief, refugee aid, etc.). The large sums given to the highly visible ISMA could generate jealousy. The welfare agencies might argue that the funding ''would do more good'' if it were funneled to them.

Given these difficulties, the best solution may have been suggested by Arthur C. Clarke, science fiction writer, past president of the British Interplanetary Society, and originator of the term ''PEACE SAT.'' He suggested that an ISMA not be formally established but gradually and silently evolved. Already, earth resources satellites are nearing resolution of military interest. With future advances in sensor technology, data processing, and image enhancement, they could cross over the line. Additionally, the roles these satellites would play (resource and pollution monitoring, disaster warning, search and rescue) are not simply national but global in scope. Some type of international coordinating agency would be an advantage. This agency would have its own interpreters, processing equipment, and archives. With no fuss or bother, the ISMA would, in effect, have been set up. The only thing that remained would be to add the military intelligence experts and analysts.[26]

Indeed, this last step may not be needed; once the photos reach sufficient resolution, intelligence personnel would start checking them for anything of interest. Once the benefits are realized, all countries who feel the need would be making such independent photo interpretation. This is not a projection of the future but something that has already happened. In the summer of 1978, the People's Republic of China ordered $105,000 worth of Landsat earth resources photos. The 2,800 photos, which covered much of Russia, have a resolution of 264 feet (80 meters). The smallest object that can be seen is the size of a football field; this is sufficient, however, to show the location of roads, airfields, bases, and cities—information that the Chinese would be interested in. Landsat photos of an airfield, for instance, would show not only the runways but also taxiways, aircraft revetments, and antiaircraft defenses.

Similarly, television networks have used Landsat photos to illustrate developments in world trouble spots. Thus, NBC, CBS, ABC, CNN, and major magazines have their own independent reconnaissance capability, although one with an extremely low resolution. This capability was used extensively during the Chernobyl nuclear accident in April 1986. The earth resources photos showed the plant area, the "hot spots" created by the burning graphite core, and the heated water in the reservoir.

The most ironic user of this capability is the U.S. Department of Defense. In the 1987 edition of the booklet "Soviet Military Power" were SPOT photos of the Pechora phase array radar, an airbase on Etorofu Island, and the Chernobyl reactor site. [27] Needless to say, much higher resolution photos have been taken of these targets, but security rules out their public release. Thus, the SPOT photos become a ready substitute.

NOTES Chapter 10

1. Col. William V. Kennedy, *Intelligence Warfare,* (New York: Crescent, 1983), 49,50.
2. Theo Pirard, "Chinese 'Secrets' Orbiting the Earth," *Spaceflight,* (October 1977): 355–361.
3. Dr. Ray S. Cline, *The CIA Under Reagan, Bush & Casey,* (Washington, D.C.: Acropolis Books, 1981), 201.
4. *The Chinese War Machine,* Dr. James E. Dornan, Jr., Editor, (New York: Crescent, 1979), 174.
5. Ken Gatland, *The Illustrated Encyclopedia of Space Technology,* (London: Salamander Books, 1981), 44,45.
6. Ken Gatland, *Missiles and Rockets,* (New York: Macmillan, 1975), 217,220.
7. Theo Pirard, "Chinese 'Secrets' Orbiting the Earth," *Spaceflight,* (October 1977): 355–361.
8. Correspondence, *Spaceflight,* (June 1979): 283,284.
9. Robert Berman & Bill Gunston, *Rockets & Missiles of World War III,* (New York: Exeter, 1983), 64,65.
10. Gatland, *The Illustrated Encyclopedia of Space Technology,* 44.
11. Theo Pirard, "Chinese 'Secrets' Orbiting the Earth," *Spaceflight,* (October 1977): 355–361.
12. Craig Covault, "Austere Chinese Space Program Keyed Toward Future Buildup," *Aviation Week & Space Technology,* (8 July 1985): 17,19.
13. Gatland, *The Illustrated Encyclopedia of Space Technology,* 44.
14. Craig Covault, "U.S. Team Tours China Space Facilities," *Aviation Week & Space Technology,* (25 June 1979): 77–79, 81,82.
15. Satellite Digest 152, *Spaceflight,* (March 1982): 136.
16. Satellite Digest 161, *Spaceflight,* (February 1983): 84.
17. Phillip S. Clark, "The Chinese Programme," *Journal of the British Interplanetary Society,* (May 1984): 201–203.
18. "Chinese Satellite," *Aviation Week & Space Technology,* (24 September 1984): 16.
19. Craig Covault, "U.S. Team Tours China Space Facilities," *Aviation Week & Space Technology,* (25 June 1979): 77,79,81,82.
20. "Earth Observation," *Flight International,* (14 May 1983): 1330.
21. "France Approves Military Reconnaissance Satellite," *Flight International,* (20 March 1982): 666.
22. *Los Angeles Times,* 22 May 1978, Section II, 7.
23. Arthur C. Clarke, "War and Peace in the Space Age," *Spaceflight,* (February 1983): 50–53.
24. *New York Times,* 26 January 1978, 9.
25. *Outer Space: A New Dimension of the Arms Race,* Bhupendra Jasani, Editor, (London: Taylor & Francis Ltd, 1982), 280–295.
26. Arthur C. Clarke, "War and Peace in the Space Age," *Spaceflight,* (February 1983): 50–53.
27. *Soviet Military Power 1987* (Department of Defense 1987), 49, 68, 115.

CHAPTER 11

U.S. ELINT

The least known category of intelligence gathering is the monitoring of electromagnetic signals such as radio transmission, radar, and telemetry signals. Electronic intelligence, or ELINT, sometimes called Ferret activities, is unfamiliar for a number of reasons. Human intelligence gathering is familiar from stories of World War II and spy movies. Photo intelligence is known from aerial photos of World War II, Korea, Vietnam, and various other wars. In contrast, comparatively little has been written about ELINT. Another reason is the nature of ELINT—even an untrained individual can recognize an airplane in an aerial photo. The analysis of ELINT data, however, is extremely complex and technical, involving the measurements of such unfamiliar quantities as peak power and beam width.

Beginnings of Airborne ELINT

The roots of ELINT satellites, like those of photo reconnaissance satellites, extend back to the years just after World War II. In those early years of the cold war, it was found that monitoring Soviet radar signals and military communications was an excellent source of intelligence. The signals did not stop at the border but could be picked up from Western territory. In addition to ground stations, both the U.S. and Britain began to outfit aircraft to monitor Soviet signals. The first U.S. aircraft modified to perform ELINT flights was an RB-29. It carried five receivers and crew positions fitted into the tail section behind the bomb bay. The aircraft was delivered in December 1948; by July 1949 it was flying missions along the Soviet coast from Wrangel Island down to the southern tip of the Kamchatka Peninsula. These early flights were pathfinders—no rules or procedures existed to guide them. There were also many technical problems. The early wire recorders were unreliable. The wires broke easily, usually after catching an important signal. Once broken, the recording was useless. Even if the wire survived, the problems were not over. There were no specialized analysis teams, so after flying a 12- to 18-hour mission, the "crows," as the ELINT crewmen were known, would have to spend another 12 hours analyzing what they had caught.

Slowly, often painfully, the problems were worked out. In the early and mid-1950s, the first RB-29 was followed by specialized ELINT aircraft such as the RB-50, RB-47, C-130, and Navy P-2V and P-4M. The necessary aircrews, crows, ground personnel, and analysts were trained. They faced a violent existence; for all intents and purposes, the crews were flying combat missions. The So-

Navy PB4Y-2 Privateer patrol bomber, a modified version of the Army Air Force B-24. In 1950, a similar PB4Y-2, stripped of its guns, was the first ELINT aircraft to be shot down. It was lost over the Baltic near the Russian coast. *U.S. Navy Photo.*

viet response to the ELINT flights was to push their airspace outward—at first from 3 miles (5 kilometers) to 12 miles (19 kilometers); then to 50 miles (80 kilometers); and finally, to as much as they could get away with.[1]

The first "incident" was on 8 April 1950, when the Soviets shot down a U.S. Navy PB4Y-2. All ten crewmen were lost. The second casualty was also a Navy plane—a P-2V shot down on 6 November 1951, off Vladivostok; again, all ten aboard were killed. The next victim was a Swedish Air Force C-47. On 13 June 1952, it, along with the eight-

man crew, "disappeared" over the Baltic. Several days later, a Swedish Catalina amphibian, searching for survivors and wreckage, was shot down by Soviet MiGs. Four months later, on 7 October, a U.S. Air Force RB-29 was destroyed by Soviet MiGs while flying over Japanese territorial waters. All eight crewmen were killed.

Nineteen fifty-three was an especially violent year. It began with the 18 January shooting down of a P-2V by Chinese shore batteries. Two of the thirteen crewmen were killed when the aircraft crashed into the sea; four

A P2V-5 Nepture taking off from Atsugi, Japan, in September 1952. Between 1951 and 1955, three P2Vs were shot down while flying patrol missions off the Russian and Chinese coasts. Another was damaged but made a successful forced landing. *U.S. Navy Photo.*

more died when their rescue plane crashed after picking them up.[2] Then, on 12 March, a Royal Air Force Lincoln bomber was shot down over East Germany. Six of the seven crewmen were killed in the crash. That same day, another Lincoln was subjected to mock attacks by several MiG 15s. On 15 March, the tables were turned—an RB-50 off the Kamchatka Peninsula was jumped by two MiGs. The RB-50 shot back, sending them

running. This was followed on 29 July 1953 by the loss of an Air Force RB-50. Only the copilot survived.

The incidents continued. On 4 September 1954, another Navy P-2V was shot down off the Siberian coast by two MiGs. Nine of the ten aboard survived. Two months later, on 7 November, an RB-29 was attacked. Despite damage, it was able to crash-land in Japan. One crewman lost his life. Still another Navy

P-2V was damaged on 22 June 1955, when flying above the Bering Strait between Alaska and Russia.[3] The two MiGs, which attacked without warning, set the right engine on fire. The P-2V was able to make an emergency landing on St. Lawrence Island. Investigation indicated that although it had drifted west of its planned flight patch, the aircraft was well outside Soviet airspace at all times.[4] The sixteen-man crew of a Navy P-4M was less fortu- nate. They were shot down by Chinese aircraft over the Formosa Strait on 22 August 1956. All were lost.

The next loss was especially bitter because it was a deliberate ambush. On 2 September 1958, a C-130 was lured across the Soviet border by use of false radio navigation beacons. These transmitted on frequencies that mimicked beacons inside Turkey. Once inside Soviet airspace, the slow-flying transport was

An RB-29 taxies out for a predawn takeoff. In 1948, an RB-29 became the first Air Force aircraft converted for Cold War ELINT missions. In 1952, an RB-29 was shot down by Soviet MiGs; in 1954, another was damaged and made a crash landing in Japan. *U.S. Air Force Photo.*

On September 2, 1958, a C-130A, similar to this one but fitted with ELINT equipment, was lured across the Soviet border and shot down. The Soviets denied any involvement with the crash, even after a tape of the Soviet pilots' conversations was made public. Only six bodies were returned out of the 17-man crew, although an eyewitness report that reached the CIA indicated all on board were killed in the crash. *U.S. Air Force Photo.*

jumped by five MiGs. They methodically shot it down, continuing to fire even after the C-130 was burning.[5] Witnesses inside Russia saw the aircraft's right wing blown off. The C-130 crashed in Soviet Armenia, killing all seventeen crewmen.[6]

The incidents continued with dreary regularity. In the 10 years between 1959 and 1969, three more aircraft were shot down and two damaged, with the loss of thirty-seven crewmen. Some areas, such as the Caspian Sea, were considered too dangerous, and operations had to be temporarily suspended.[7,8,9] However, the casualties in the crows' 20-year war were not in vain. One early achievement was determining that Soviet aircraft had been fitted with radar to perform interceptions in darkness or bad weather. As late as 1950, the question had been argued endlessly between the Navy, Air Force, and CIA. There were only vague rumors and little ELINT data. The Navy tried to find out by sending aircraft over the Black Sea. It was a British aircraft that came through with a 20-second recording of a ''Scan Odd'' fighter radar. End of argument.

Another early operation, which stretched over a year, was determining the status of a new Soviet air defense radar called ''Token.'' This was the first Soviet-developed radar and was comparable to U.S. systems then in use. The prototype Token radar was spotted in the fall of 1951 at a test area near Moscow. Once the specifications were determined by a ground listening station, the ELINT aircraft set to work. Six months later, the Navy had found six to eight Tokens in the area of the

Black Sea. Numerous others were located in Eastern Europe. By June 1952, Tokens had been spotted in the Soviet Far East. After a year, it had been confirmed that the Soviets had more Tokens than the U.S. had equivalent radar. The determination was based almost entirely on ELINT data.[10]

ELINT Missions

The hunt for Tokens is a typical example of a primary mission of ELINT aircraft—locating the position of enemy radars, finding their characteristics, and determining their area of coverage. One difficulty in describing the characteristics of radar is their invisible transmissions; all one sees is the rotating antenna. If one could see a radar beam, it would look like a searchlight sweeping across the sky. The radar signals are detected by an array of antennas. The RB-47 and the aircraft that replaced it, the RC-135, seemed to be covered

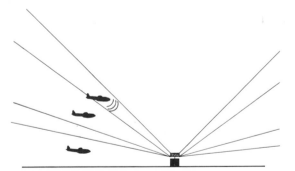 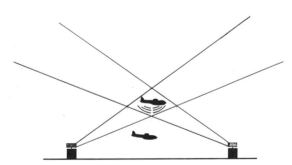

Two drawings to illustrate the unusual nature of radar. *Left:* Radar signals are transmitted in beams and only aircraft in the beams can be tracked. The first aircraft is below the beam and is not detected. The middle plane is between the radar's two beams and also is not spotted. Only the third aircraft can be tracked. *Right:* The disadvantages of high altitude vs. low altitude flight. The aircraft at high altitude is being tracked by two radar sites, while the plane at low altitude has not been spotted.

ELINT aircraft and spacecraft must determine the frequency of the radar signal, the number of beams, their coverage, blind spots, and detection range. In this way aircraft can avoid them, or if necessary, use countermeasures to render them ineffective.

with lumps and bulges, which are the antenna housing. The detected signals are displayed on an oscilloscope. One important feature to be measured is the transmission frequency. A particular type of radar will transmit in a certain narrow range of frequencies. Each individual radar is set at its own specific frequency (if they were the same, they would jam each other). Determining the frequency can be a complex question, since a modern radar may have multiple beams operating at different frequencies. Moreover, some radars can also change their frequencies. The radars transmit a series of short pulses. In the earlier analogy, the searchlight would seem to flicker rapidly. The number of pulses can be counted to determine the rate.

Another instrument can display a tracing of a single pulse, which can tell the ELINT operator much. The pulse of an early warning radar is comparatively long. That of a radar used to aim antiaircraft guns is very brief. (Early warning radar needs strong echoes at long range. Gun-aiming radars need precise positional data, thus the difference in pulse characteristics.) The polarization of the beam can also be determined. Small irregularities in the shape of the pulse can even be used to identify individual radar sets. These are caused by peculiarities in the electronics of each radar. The location of a radar site is determined by taking several bearings at different points on the aircraft's course.[11]

Finally, the coverage of the radar can be determined. A radar sends out one or more vertical fan-shaped beams that sweep out a circle as the antenna rotates. If the plane is below the beam, skimming the ground, it will not be detected, because the beams travel in straight lines rather than curving along the earth. Additionally, it is impossible to watch every inch of sky. Some areas would not be covered simply because radars were too far away. A final factor affecting coverage is terrain; a mountain can block out one area.

The tape recordings from an ELINT mission are a collection of beeps, chirps, whistles, screeches, buzzes, and tones.[12] This seemingly meaningless collection of noise, however, once processed can give a picture of Soviet air defenses. This, in turn, is used to plan the attack routes for U.S. bombers in the event of war. The Strategic Air Command uses a three-part philosophy concerning enemy radar. The preferred option is to avoid them by flying through gaps and under the radar. The air defenses would have no indication of U.S. aircraft in the area. If this is not possible, for example during the run into a heavily defended target, the bomber would jam the radar. Bombers carry very complex arrays of electronic countermeasure equipment (ECM), which is constantly updated to match changes in Soviet radar. An example of the level of sophistication, even as early as the 1960s, can be found in the ECM of the B-58. It automatically cycled through the frequencies used by Soviet radars (much like a police scanner). When the equipment spotted a signal, it determined if it represented a threat. It looked at such things as signal strength or the number of times the aircraft was scanned. If it was deemed a threat, the ECM equipment would put out an extremely strong transmission at the same frequency. This would overload the offending radar and cause its protective relays to open, knocking it off the air for half an hour or more. Operators of U.S. weather radars near training areas had to be careful or they would be zapped by a passing B-58. The ECM now carried by B-52s and B-1s are much more complex. Finally, if the radar cannot be jammed, it would be destroyed. The bombers would need location data to program their Short-Range Attack Missiles and cruise missiles.[13,14]

A typical ELINT mission involves flying near the Soviet border while remaining over international waters or friendly territory. Close approaches to the border are not always necessary, since the aircraft receivers can detect a radar signal at up to twice the distance the radar can pick up the aircraft. The flight may be a general survey or be targeted against a specific installation. Many flights may be needed to find the information. A single negative flight proves nothing. The crows must assemble their understanding of Soviet radars over a period of time. In the back of the aircraft, the crows listen to their assigned frequency. As each new signal is detected, a crow logs the instrument readings of the radar's characteristics and takes bearings on its position. A mission lasts many long hours as the aircraft flies endless circles in the sky.

In the early 1960s, RB-47s based at Thule, Greenland, had to fly across the North Pole to reach their patrol area. Only then could the crew begin their mission (they didn't mind—anything to get away from Thule). The "Common Cause" patrols by RB-47s off Cuba, made just after the missile crisis, lasted 12 hours. RC-135 missions are even worse—they take off from their home base, fly a mission lasting for 24 hours or more, then land at an overseas base. The crew fly several "local" missions lasting 12 to 20 hours, then make another 24-hour return flight home.

As the long hours drag by with little activity, crows get bored. They have been known to record kazoos, buzzers, duck calls, and even a live cricket as "intercepts." The analysts, however, are never fooled. ELINT missions can best be described as involving hours of numbing boredom interrupted with moments of either fear or the excitement of discovery.[15]

The Soviets, of course, do everything they can to deny the aircraft any useful data. They can jam reception or transmit false signals to mislead analysts. The simplest procedure, however, is not to transmit anything. A silent radar cannot be picked up and so goes undetected. It becomes a battle of wits between the crows and the Soviet radar operators. To draw out the Soviet radars, the aircraft may make a feint—flying towards the border as if it was going to cross, then veering away at the last moment.[16]

One RB-47 crew found that by dropping bundles of chaff, they could generate a lot of activity. Unfortunately, headquarters objected. The crew soon found a ready substitute—after takeoff, several cases of empty beer cans were thrown into the forward wheel well. At the proper location, the doors were opened and the radars "went crazy."

A more subtle method was to prevent the Russians from realizing there were ELINT aircraft flying off the coast or based at a particular location. One approach was to use standard security measures. The ELINT crews would avoid saying or doing anything that could identify them or their unit. Markings were also removed from their aircraft. A crow, based in Turkey, recalled how a Russian propaganda station greeted their arrival by name and crew number. Amid the threats, "Moscow Molly" warned them about loose steps and floor tiles in various buildings and a broken window crank on a crew vehicle. Another, hopefully more successful, procedure is radio silence. Rather than have the tower radio takeoff clearances, green lights were used to signal the aircraft to start engines, taxi, and take off.[17]

The silent takeoff highlights another aspect of ELINT missions. The Soviets monitor the tower frequencies and know the call signs used by ELINT aircraft. If they hear the clearance, they would have several hours warning of the aircraft's approach—time enough to shut things down. To monitor such radio communications, some ELINT aircraft carry crewmen trained in the target country's language. Using the on-board equipment, they can determine the frequencies used, record the transmission, and take bearings on the station's location. The real value of such intercepts is in their content. Soviet interceptors are closely directed from the ground. The pilots are told where to fly, when to climb or descend, and when to open fire. By listening to training exercises, the tactics and procedures used and the capabilities of the aircraft, weapons, and ground controllers can be determined as well as the meaning of code phrases found. The same applies to any military activity that relies on radio traffic. The analysis is conducted by the National Security Agency, which has responsibility for both protecting U.S. communications and intercepting those of other countries. From their analysis, capabilities and weaknesses in Soviet forces can be found. One obvious weakness is the radio link between ground controllers and the aircraft. With these links disrupted, the fighters would have difficulty finding the target aircraft. To do this, of course, the frequencies would have to be known. In the early 1960s, British bombers carried a device called "Green Palm"—a VHF jammer tuned to the four frequencies then used by Soviet ground controllers. The MiG pilot would get an earful of what sounded like a cross between a police siren and bagpipes.[18]

New weapons and procedures require extensive testing. Such testing requires radio links between the participants. By listening in, the analyst can figure out what is going on; thus, early warning of new dangers is gained. During World War II, British ELINT teams monitoring brief, cryptic messages from German interceptors were able to reconstruct new procedures being tested for use by German night fighters. British bomber crews were thus warned before the tactic could be made operational. Such analyses rely on various things—the discovery of new call signs, patterns, and frequencies as well as the inevitable slips made by the test team; sometimes they need something repeated or clarified. All the while, the enemy is listening.[19]

The most important function of ELINT or any other type of intelligence is that of early warning—providing information that indicates a Soviet attack is imminent. This can take many forms. A sudden step-up in radio traffic between the high command and field units, indicating they are being placed on a war footing, is one. Another is the disruption of routine activities or the sudden movement of equipment—for example, radars "disappearing" from central Russia and reappearing in East Germany and Poland.

On a less cosmic level, such information can indicate upcoming Soviet actions that require a U.S. response. Starting in late 1958, the U.S. and Russia refrained from any nuclear testing. By 1961, however, the pressure on both governments to resume testing had grown. President Kennedy refused, still hopeful the stalemated Geneva talks could produce a comprehensive agreement banning all tests. Khrushchev had other things in mind. At about 1:15 P.M. eastern daylight time (EDT) on 30 August 1961, a technician at a ground ELINT station on Cyprus was wading through a mass of intercepted teletype copy. It was not military signals but the text for newspapers in the Caucasus. In this unlikely source, he found the text of a Tass announcement for release at 7 P.M. EDT that the Soviets were resuming nuclear testing. About 2 hours later, the State Department had been notified and soon after, the president was told. Thanks to nearly 4 hours of early warning, the White House was able to prepare a harsh statement condemning the Soviet action and shifting the blame onto their shoulders.[20] Although the early warning could not stop the tests, the U.S. was able to avoid the political consequences of being caught unprepared.

Higher and Farther

Airborne ELINT is tremendously valuable. It does, however, have one fundamental limitation—radar signals and some radio transmissions travel in straight lines. They cannot be picked up from beyond the horizon. For an aircraft flying at 35,000 feet (10,668 meters), the horizon is 228 miles (367 kilometers) away. ELINT can pick up any signal within this range but not beyond, which meant that most of the vast landmass of Russia was out of reach. One way around this is to fly higher. U-2s were equipped with ELINT receivers. Flying at 70,000 feet (21,336 meters), they could detect a radar 323 miles (520 kilometers) away. The U-2s in Turkey would fly along the Soviet border as far east as Pakistan before returning. Other missions were made over the Black Sea and as far west as Albania. At all times, the U-2s would stay outside Soviet airspace. These flights were far more numerous than the overflights. The U-2s, because of their higher altitude capability, added one-third to the area that could be covered by a border mission. Despite this, large areas remained out of reach, particularly in western Russia, since it was shielded by the Eastern European buffer states. This lack of information was dangerous because the radars inside Russia might use different frequencies from those the ECM were designed for. Additionally, the bombers would have to fly

blind—not knowing where the dangerous areas were. The only way to determine the degree of radar coverage of the interior of Russia was to go in. The Moby Dick reconnaissance balloons were originally planned to carry both photo and ELINT packages. The ELINT receivers were not, however, ready in time. U-2s subsequently carried ELINT packages on overflights. By examining the strength of the recorded signals, it could be determined if the Soviets had tracked the aircraft.

The amount of overflight ELINT data was limited by the same factors mentioned earlier—the risks of the overflights and Eisenhower's reluctance to run them, and the range limitations of the U-2 and the small payload it could carry. The solution, as with photo reconnaissance, was to go into space.

The First ELINT Satellites

Ironically enough, the first concept for space ELINT did not use an artificial satellite but the earth's natural one. The concept behind Project Moonbounce was that Soviet radar and radio signals traveling out into space would be reflected from the moon back towards the earth. They would then be picked up by a 600-foot (183-meter) dish antenna to be built at Sugar Grove, West Virginia. Studies made in the mid-1950s involved both theoretical calculations and the construction of a small test telescope. Funding was approved in 1959 but the project immediately ran into technical problems. The telescope was the largest movable structure ever built, weighing 30,000 tons. Engineering calculations required the use of the most advanced computer then available. Costs increased from $79 million to $135 million to $200 million, and the completion date was pushed back from

1962 to 1964. With the entire project looking increasingly questionable, it was cancelled in July 1962.[21]

Despite Project Moonbounce, more conventional ELINT satellites were also examined. It was not until the launch of Sputnik that action could be taken. On 10 November 1957, the Army made a proposal for a wide-ranging military space effort. Part of this was a 500-pound (227-kilogram) ELINT satellite, which was to be launched by a modified Jupiter rocket. The satellite would record radar transmissions along with the time the signals were intercepted. The recordings would later be played back to a U.S. ground station. The satellite could locate a radar with an accuracy of 25 miles (40 kilometers). Once the general area was determined, reconnaissance photos could be used to pin down the radar's location exactly. The Army estimated the first ELINT test package would be ready in November 1958. The 500-pound (227-kilogram) operational satellite would be launched in June 1959. Later missions would be equipped to record both Soviet radar and military communications.[22]

Although the Army launched the first U.S. satellite, *Explorer I,* it was soon edged out of significant space activities by the Air Force and NASA. When President Eisenhower approved the WS-117L program, it included not only photo reconnaissance but also an ELINT development effort called ''Pioneer Ferret.'' A 30 June 1958 schedule envisioned launches beginning in August 1960, with subsequent launches to be made at 2-month intervals. The launch vehicle was to be an Atlas. The various changes in the WS-117L program also affected the Pioneer Ferret. Instead of a separate satellite, the ELINT package would be carried aboard the first three SAMOS satellites. The first combined photo and ELINT payload was

undergoing ground tests by February 1960.[23] There was no mention of any ELINT package aboard SAMOS. The only comment on the payload was that it contained ''photographic and related equipment.'' The first SAMOS launch attempt failed on 11 October 1960, preventing any ELINT data from being obtained.

The next attempt, *SAMOS 2,* was successful, going into polar orbit on 31 January 1961. *SAMOS 2* operated for a month. If the ELINT package operated successfully, this was the first such data from deep inside Russia since the previous April. Analysts apparently had difficulty interpreting the early satellite tapes. According to one account, the tapes were a confusing jumble of multiple signals, which is understandable when one considers the situation in which the tapes were made. A receiver in a 300-mile (483-kilometer) orbit such as *SAMOS 2* used could pick up a radar as far away as 1,500 miles (2,414 kilometers). A satellite over central Russia could simultaneously detect radar signals on both the northern and southern borders. Given the number of radars and SAM sites in such a large area, the difficulty in sorting them out is not surprising.

The operating altitude of 300 miles (483 kilometers) is a compromise—the higher the receiver, the larger the area that can be covered. This, however, means the radar signals will also be weaker because of the greater distance they must travel (if the distance is doubled, the signal will be one-fourth as strong).

The third ELINT package aboard *SAMOS 3* was lost on 9 September 1961, when the Atlas Agena blew up on the pad. Available documents refer only to these three ELINT packages. However, it would seem likely that additional ELINT packages were flown on first-generation reconnaissance satellites. In any event, the Air Force and Defense Department must have been encouraged, because work was begun on an operational system. As with photo reconnaissance, activities were divided between two types of ELINT satellites.

The initial survey was performed by a small satellite carried piggyback aboard a photo reconnaissance satellite. After reaching orbit, the ELINT subsatellite would separate and its on-board engine would fire, which would place the satellite into a 300-mile- (483-kilometer-) high circular orbit. From radar measurements, the subsatellites are estimated to be about 3 feet (0.9 meter) in diameter and 1 foot (0.3 meter) high, with a weight of approximately 125 pounds (57 kilograms).[24] Being this small, the satellite could carry only an extremely limited amount of ELINT equipment. It could be as simple as an antenna, a receiver, a tape recorder, and a clock. The rough position of a Soviet radar would be determined from the exact time a signal appeared and then disappeared. This would be combined with the satellite's orbital ground track. This approach would eliminate the need for sophisticated correction-finding equipment.

Once a new radar was located, its exact characteristics would be determined by a heavy ELINT satellite. A TAT Agena booster would put it into a 300-mile- (483-kilometer-) high orbit. The Agena would carry a 2,000-pound (907-kilogram) ELINT payload. With this large package, the satellite could perform a detailed survey, including the radar frequencies, pulse shape and rate, the number of beams, and the transmission power—in short, the same characteristics the aircraft would measure. The satellite would also carry a precise direction-finding system; one possibility

would be a dish antenna. In theory, a radar's location could be determined from orbit to within a few hundred yards.

The contract to build the ELINT receivers was awarded to Airborne Instruments Laboratory, a company with many years' experience with ELINT and ECM equipment.[25]

It is not clear when the first of these operational satellites went into orbit. For a long time, it was believed the first heavy ELINT satellite was launched on 15 May 1962, by a Thor Agena B. It was subsequently determined that this was actually a film-return reconnaissance satellite.[26]

Accordingly, the launch of the first heavy ELINT now appears to have been on 18 June 1962. A Thor Agena B propelled the satellite into a 255- by 230-mile (411- by 370-kilometer) orbit inclined 82.14 degrees.[27] The orbit was higher than typical for photo reconnaissance satellites and would be better suited for ELINT. Presumably, the flight would have the dual purpose of gathering intelligence and testing the equipment. This was the only suspected heavy ELINT launch in 1962 (a suspected launch on 1 September 1962 was also later shown to be a film-return spacecraft).[28]

The heavy ELINT program got off to a quick start in the new year. On 16 January 1963, a Thor Agena D–launched satellite went into a 331- by 285-mile (533- by 459-kilometer) orbit—higher than the first to give it better coverage. It was followed on 29 June by another heavy ELINT, which used a new launch vehicle—the TAT Agena B. This implied that a larger, more complex payload was carried.

In the meantime, the development of the ELINT subsatellites had been underway. During the summer, the first was prepared for launch; it was orbited on 29 August 1963, by a Thor Agena D. The area surveillance satellite went into a 201- by 181-mile (324-

by 292-kilometer) orbit. The subsatellite later separated and placed itself into a higher, 268- by 193-mile (431- by 310-kilometer) orbit, inclined 81.9 degrees. It was followed by two more on 29 October and 21 December. Both rode piggyback on TAT Agena D boosters.

In 1964, operations were stepped up. Heavy ELINT satellites were launched in February, July, and November. The increased number, and the use of a standardized booster (TAT Agena D), implied a move to operational status. The smaller subsatellites also became operational. Although two were launched (July and October 1964), they were switched to Atlas Agena D boosters carrying high-resolution reconnaissance satellites.[29]

These events took place against a background of steadily increasing Soviet air defense threat. The U-2 notwithstanding, SAM coverage was spotty in the early 1960s. The bombers could still get through to their targets by flying at high altitude. By 1964, however, the Soviet SAMs were too numerous for a high-altitude attack to be successful. Bomber tactics would have to change. The solution was to send the bombers in at low altitude—only 1,000 feet (305 meters) or lower. Although the SAM coverage from separate sites overlapped at high altitude, there were gaps at low levels. The bombers could slip through, masked by the terrain. Even if the bombers were detected, the Guideline SAM couldn't hit them. The missile's minimum effective altitude was 4,921 feet (1,500 meters). Below this, the guidance radar could not direct it to the target. The data from the ELINT satellites would be important in pinning down the location of SAM sites and the gaps between them—doubly important, since low-altitude flight used more fuel, which meant less range for evasive flight paths.[30]

Apparently, the 1964 flights mapped out the basic pattern of Soviet and Chinese air defenses. Subsequent flights would look for changes and the development of new types of radar. The Soviets would be expected to close the low-altitude gap in their defenses—a supposition reflected in the launch activities in 1965. Only one of the heavy ELINT satellites went into orbit (17 July), while three of the smaller subsatellites were orbited (28 April, 25 June, and 3 August). The single heavy ELINT satellite used a new inclination, 70.2 degrees versus the 82 degrees previously seen. This pattern was to continue through the 1960s and into the early years of the 1970s.

Two heavy ELINT satellites were launched in 1966 (9 February and 29 December) along with three of the smaller subsatellites (14 May, 16 August, and 16 September). The first of the heavy ELINT satellites used an 82 degree inclination; the other used a 75 degree orbit. All of the subsequent flights would use this lower inclination.

The first changes in the ELINT satellite program came in 1967. The LTTAT Agena D had been introduced to carry a third generation of area surveillance satellites. Accordingly, the ELINT subsatellites also made the changeover. The first was the 9 May 1967 launch; it was followed by two more in June and November. Only one heavy ELINT satellite was launched in 1967; this came in July. In 1968, there was a considerable step-up in the launch rate: five ELINT subsatellites were orbited. One of these, the 12 December 1968 launch, went into a new orbit. Previous satellites had used a roughly circular 300-mile- (483-kilometer-) high orbit. This launch, however, went into a 912- by 864-mile (1,468- by 1,391-kilometer) orbit—three times higher than the standard. The first subsatellite launch of 1969, made on 5 February, also went into a similar 900-mile (1,448-ki-lometer) circular orbit. These two satellites were intended to monitor the Soviet ABM radars around Moscow that were then going into operation. Because the radar beams extended several thousand miles into space, the higher altitude was better suited to monitoring their characteristics and coverage. The data would be used to assess the system's effectiveness and to design U.S. decoys and warheads.

A New Heavy ELINT Satellite

Nineteen sixty-eight was also a pivotal year for the heavy satellite program; two were launched that year. The first, on 17 January, used a TAT Agena D; the second, launched on 5 October, used an LTTAT Agena D. This was the debut of a more advanced heavy ELINT satellite. The larger booster meant a 20 percent increase in the payload. In the early 1960s, before the secrecy curtain was lowered, Air Force spokesmen had indicated that future ELINT satellites would be able to analyze the signals on board, comparing them with previously recorded signals and automatically noting new or unusual radars. The satellite would also intercept military communications and analyze them on the basis of modulation, linguistics, and semantic patterns. The satellites would also be able to receive transmissions over a wider range of frequencies.[31]

One of these new satellites was launched

in each of the next 3 years. The last launching, on 16 July 1971, brought the heavy ELINT satellite program to a close. No subsequent launches showed their orbital characteristics. At the time, it was believed that a new heavy ELINT satellite was being developed by Hughes Aircraft. Called Program 711,[*] it was to be placed in a high elliptical orbit by a Titan III. It was expected the Program 711 satellite would handle both the survey and detailed analyses. Its computer would carry a complete listing of all Soviet radar sites and their individual characteristics. The satellite would compare the signals it received to the catalog. If a new signal or location was detected, it would then automatically make a detailed survey. The first launch was expected in late 1970 or early 1971, but the Program 711 satellite never appeared.[32,33] Either the reports were incorrect or the satellite was cancelled, or possibly it was a misinterpretation of the Hughes Aircraft Satellite Data System, a highly classified communications satellite that actually uses an elliptical orbit.[**]

The ELINT subsatellites continued their clocklike operation:

1969 Three radar   One ABM subsatellite
1970 Three radar
1971 One radar

After 1971, the LTTAT Agena D booster was retired and the ELINT subsatellites were transferred to the Big Bird. Their launch record follows:

---

[*] Apparently, a reflection of Howard Hughes's extensive holdings in Las Vegas.
[**] It has been claimed that some SDS launches were of an ELINT satellite called Jumpseat. The first was launched on 10 March 1975 and a second on 6 August 1976. This has not been confirmed.

1972	Two radar	One ABM
1973	One radar	One ABM (Both launched on same Big Bird)
1974	Two radar	
1976	One radar	
1978	One radar	
1979	One radar	
1980		One ABM

The 1976, 1978, and 1979 subsatellites were apparently a new generation. Rather than the traditional 300-mile (483-kilometer) orbit, they went into higher, 390-mile (632-kilometer) circular orbits.[34]

The pattern appears to be a more complex one than that used in the 1960s. Rather than several launches per year, there is a flurry of activity followed by several years without launches, followed in turn by another increase. One possible explanation is that the U.S. establishes a network of such satellites that operates for several years. Then, as the satellites begin to fail, replacements are launched, reestablishing the network for several more years of listening. In any event, no subsatellites were launched for 2 years. Then a single subsatellite was launched in each of the years 1982, 1983, and 1984 (the latter was designated *USA-3*). The 1983 launch was an ABM ELINT. None were launched in 1985.

A Soviet Lie

The Soviets call the Big Bird ELINT sub-satellite a "Ferret-D."[35] The term would first come to public attention as part of the Russian campaign to justify their destruction of Korean Airlines' flight 007 on 31 August 1983. Two weeks after the airliner was shot down, Soviet Marshal of Aviation Pyotr S. Kirsanov claimed that the flight of the 747 was timed to coincide with the passage of a Ferret-D ELINT subsatellite. The specific satellite was later identified as 1982-41C, the subsatellite launched aboard *Big Bird 17* on 11 May 1982. It went into a 439- by 436-mile (707- by 701-kilometer) orbit inclined 96 degrees with a period of 98.87 minutes.[36] Marshal Kirsanov claimed that Ferret-D first passed over Chukotka at 6:45 P.M. Moscow time and was east of Kamchatka and the Kuril Islands for about 12 minutes. This was before the plane entered Soviet airspace. The Marshal said, "On that revolution the satellite had the opportunity . . . to monitor the Soviet radioelectronic means on Chukotka and Kamchatka, working in the normal regime of combat duty to determine their exact location and the level of activity. . . ." He continued that the satellite's next pass took place at 8:24 P.M. Moscow time over Kamchatka as Soviet air defense stepped up its activities to track the airliner. The third pass of Ferret-D occurred as the airliner was over Sakhalin Island; radars on Sakhalin, the Kuril Islands, and the Primorski Territory were switched on in response. Marshal Kirsanov concluded by saying, "It is doubtless that the moment of the penetration by the intruder plane . . . had been carefully planned in advance so as to assure the gathering of the maximum information by the [Ferret D]."[37]

Others were not so sure. James Oberg, an engineer with the space shuttle program and an expert on the Soviet space program, spent several months examining the shoot-down, including the marshal's claims. He found that Kirsanov's claims had several holes. First, the Soviet map showed the satellite as being in an orbit with an inclination of less than 90 degrees. Yet the only U.S. ELINT satellites to use such an inclination were the heavy satellites of the 1960s. Their use had been discontinued after the 1971 flight. Moreover, all of these satellites had reentered years before the incident.[38] ELINT subsatellites launched aboard Big Birds use an inclination of 96 degrees. When 1982-41C's ground track is compared to the Ferret-D's ground track, the two differ by as much as 500 miles (800 kilometers). The timing of the pass is also changed. When the Soviets claimed Ferret-D was due east off the Kamchatka Peninsula, 1982-41C was over the east coast of Siberia. Satellite 1982-41C was, in reality, poorly placed for "gathering the maximum information," to quote the marshal. It was far to the west. The Soviet radars were pointed east towards the airliner and away from the satellite, which could pick up only transmitter "leakage" or 360 degree scans of search radar. Any highly directional radar signals would not have been detected. Furthermore, the presence of a satellite at the time of the incident is not proof that the airliner was on a spy mission. With a dozen satellites in orbit, the odds of one being in the vicinity of a given area at a random time were ten to one.[39]

Moreover, the Ferret-D ground track shown in the Soviet drawing is a physical impossibility. First, Ferret-D went east of south; then due south; then finally, east of south. A real satellite's ground track shows no such erratic changes of course. There are also internal inconsistencies in the satellite's timing. The Soviets claim that Ferret-D had an orbital period of 96 minutes (1982-41C's period was 98.8 minutes). Yet, the times given for specific points on the orbit give a period of 98 to 100 minutes (the Soviets claim

an accuracy to the minute). Further, the separations between the three passes contradict all the timing. If these were correct, the satellite period was only 80 minutes, which would require an orbit so low that the satellite would reenter and burn up after only a few hours. Lastly, the range of 932 miles (1,500 kilometers) the Soviets gave is incorrect for the given orbital period. It should be twice as great.[40]

These inconsistencies are symptomatic of the Soviets' handling of the entire affair. First, the Soviets claimed the only other aircraft involved was an RC-135. Several days later, the Ferret-D, a second RC-135, two P-3s, an E-3A AWACS aircraft, and a Navy frigate "appeared." In later versions, the E-3A "vanished," the RC-135 was moved to a new location, and a third P-3 "appeared." When this is combined with the Soviet lies about weather conditions and attempts to contact the airliner, a pattern becomes apparent.

In a courtroom, whether a legal one or a court of public opinion, each side presents the evidence that supports its position. In this case, the Ferret-D is part of the Soviet alibi to justify and explain their actions, but it is an alibi filled with contradictions and outright physical impossibilities. Ironically, this may not matter, since the Russians have already caused a segment of the world's population to wonder if there might be some validity to their claim.[41–43]

Rhyolite

The term ELINT covers other things besides radar and radio communications. It also includes the interception of such things as telemetry signals from long-range missiles, ground guidance signals that direct their flight, and the various transmissions related to the countdown. The intelligence value is obvious. The telemetry package aboard a rocket is designed to record how its many thousands of systems functioned. The data is used to find the reasons for any failures and to suggest the needed changes. In the hands of an intelligence analyst, the same data can give an understanding of the missile's design; such things as the general level of technology, accuracy, number of warheads, throw weight, and other capabilities can be determined. Watching the test flight's progress can show if the development effort is going well or is running into trouble, which, in turn, can indicate whether the missile will be abandoned or the U.S. must face a new threat. It is almost as if U.S. personnel were in the blockhouse, looking over the shoulders of Soviet engineers.

ELINT activities against the Soviet missile program began early. The Kapustin Yar test site for short- and intermediate-range missiles was identified in the spring of 1947. After the launch of the first captured V-2 by the Russians in the fall, western activities were stepped up. In 1948, a British team posing as archeologists monitored the V-2 tests from Iran. The CIA and British SIS (Secret Intelligence Service) worked closely on monitoring Soviet missile activities, including both sharing information and writing joint reports. In September 1952, a joint U.S.-British team also examined the reports of former German missile scientists who had worked in the Soviet program. The Germans gave details of the V-2 test flights, the modifications to the missiles, and design studies they had made. The German engineers also warned that the Soviets were working on a 240,000-pound- (108,864-kilogram-) thrust rocket engine, which could be used to power an IRBM able to hit western Europe from Soviet territory.[44]

Concern about the pace of the Soviet effort increased. By 1954, the intelligence community began to look for innovative ways to gain more information. A long-range radar emerged as the way to gain this added information. Previously, radar had been used only for air defense rather than for intelligence gathering. Perhaps because of the novelty, there was opposition, primarily in the Air Force. Supporting the radar was Trevor Gardner, the Air Force assistant secretary for research and development, who played a major part in accelerating the U.S. ICBM program. He believed that any feasible means had to be used; his support overcame the objections. The plans were approved by President Eisenhower and work soon began. The radar was designed by General Electric and built near the Turkish resort of Samsum on the Black Sea. About 700 miles (1,126 kilometers) to the northeast was the Kapustin Yar test site. The radar began operation in the summer of 1955. Its echoes provided such data as height, direction, speed, and size of the Soviet IRBMs.[45]

During the early months of the radar's operation, only sporadic IRBM test launches were tracked. There was no pattern, which indicated the rocket was still in the research and development phase. In 1956, the pattern changed from the irregular tests to a regular, five-per-month series. These were training firings for the operational crews. Soon the IRBMs would be deployed and the Soviets would have the ability to destroy U.S. bases in western Europe.[46]

The following year, as the Soviets began ICBM test flights from Tyuratam, the Turkish radar was upgraded. Previously, it had a range of only 1,150 miles (1,852 kilometers). This was sufficient to cover Kapustin Yar, but Tyuratam was about 1,400 miles (2,253 kilometers) away, so coverage was, at best, marginal. Advanced antenna design extended the radar's range to just over 4,000 miles (6,436 kilometers), which would cover the first part of a Tyuratam-launched missile's trajectory.

U-2s were also used. Some of the border flights carried equipment to monitor the Soviet tests. Gary Powers would later write about these flights: "One unit came on automatically the moment the launch frequency was used and collected all the data sent out to control the rocket." By the late 1950s, through such monitoring, the U.S. knew several days in advance when a Soviet launch was scheduled.[47] Such early warning allowed ground stations, the Turkish radar, and the U-2s to be geared up to monitor the tests.

As the Soviet ICBM program progressed and the missile gap controversy became more heated, the U.S. stepped up its intelligence-gathering effort. In 1958, plans were approved to build another intelligence radar.[48] It was located on the desolate shores of Shemya in the Aleutian Islands. Cold and overcast the year around, Shemya can experience high winds and fog simultaneously. The hostile environment puts great strain on both the equipment and the people who operate it.[49] What it lacked in climate, Shemya more than made up for in intelligence potential. The radar beam could look out across the Kamchatka Peninsula and track the Soviet missiles in the final part of their flight before impact. The added coverage would provide information on the accuracy of the missile, which is defined as the area around the target into which half the warheads fall. Plotting the radar track of the missile could indicate this. Knowing the accuracy of a missile is important, since it allows its effectiveness to be assessed, i.e., whether the missile could destroy a hard target such as a missile silo or if it would be restricted to "soft" targets such as cities. The first would require a very accurate missile, while the second would need only limited accuracy.

In 1959, the U.S. and Pakistani governments reached agreement on establishing an ELINT ground station at Peshawar near the Khyber Pass, only 150 miles (241 kilometers) from the Soviet border. This position was ideally suited to monitoring signals from Tyuratam, the ABM test site at Sary Shagan, the nuclear weapons test site at Semipalatinsk, as well as Chinese missile and nuclear tests in Sinkiang Province. The facility was ready in November 1960 and operated until 1969, when the 10-year lease ran out and Pakistan refused to renew it.[50]

By this time, however, replacements had been built. In 1963–1964, a new radar had been built at Diyarbakir in eastern Turkey. Its position was closer to Tyuratam than the earlier site.[51] Also in the 1960s, two stations were built in Iran. The most valuable of these was the Tacksman 2 facility at the mountain village of Kabkam. It was less than 700 miles (1,126 kilometers) from Tyuratam and was equipped with several large dish antennas. Tacksman 2, its capabilities, and ultimate fate had a major bearing on the directions taken by the U.S. ELINT program in the late 1970s and early 1980s.

The foregoing stations are all oriented toward monitoring Kapustin Yar and Tyuratam. This is not to say that Plesetsk was ignored. U.S. stations were built at Bodo and elsewhere in Norway.[52] Plesetsk is a particularly important target, since it is the world's busiest spaceport, a major ICBM test center, and is near an operational ICBM field.

The most sensitive of the U.S. monitoring activities was the use of nuclear submarines inside Soviet waters. This was known as Holystone—a Navy ELINT project that operated from the early 1960s until the mid-1970s. One part of its activity was monitoring telemetry from Soviet submarine–launched missiles. Such launches are often made from the White Sea in northwestern Russia.[53]

Ground stations, however, have a limitation. A radar station can track a rocket only when it is above the horizon. Similarly, microwave telemetry signals are also strictly line of sight. Since the stations are several hundred miles away from the launch pad, the initial part of the missile's flight cannot be picked up. Some intelligence experts have been quoted as saying that these first moments are critical in determining the details of the missile and its throw weight.[54]

A low-orbit satellite can pick up the signals but, again, only when the satellite has a direct line of sight with the missile. Since a satellite's pass over the launch site lasts only 10 or 15 minutes, it is a relatively easy matter for the Soviets to schedule their test launches around them.

There is, however, a way to pick up the signals. If the ELINT satellite is placed in a 22,300-mile- (35,881-kilometer-) high orbit, it will take 24 hours to complete one revolution. Thus, it appears to stay fixed at one point and can keep a constant watch on one hemisphere of the earth. This ability contains a problem: the signals received by a geosynchronous-orbit satellite are more than 5,500 times as weak as those picked up by a low-orbit ELINT satellite. This technical challenge was undertaken by the Rhyolite program. TRW was selected as the prime contractor, with work apparently beginning in the mid-1960s. The heart of the Rhyolite satellite is a 70-foot- (21-meter-) wide dish antenna, which picks up the weak telemetry signal of the ICBMs from the background hash of other

VHF transmissions. The difficulty of the task is reflected by its code name, "Rhyolite," a volcanic rock made up of light gray ash. Embedded in the bland mass are small, colorful crystals of quartz and feldspar. The Rhyolite satellite's job was to find these small crystals of knowledge.[55] The dish was kept rigid by a framework grid that resembled the supports for stadium bleachers. The Rhyolite also had several additional smaller antennas, which picked up microwave, radio, radar, and telephone signals. The data would then be relayed to a ground station at Alice Springs, Australia, and from there to the various analysts. The communications tapes, for instance, would be processed by trained linguists, who would listen for certain key words or phrases. With only a year of language school, the linguists could not understand the complete conversation word for word. Once the important conversations were spotted, they could then be translated in full. The electrical power to run the various receivers was provided by several large panels of blue solar cells.[56] Despite its large size, the entire package folds up into a 19.7-foot- (6-meter-) long by 4.7-foot- (1.4-meter-) diameter Atlas Agena payload shroud and weighs only 606 pounds (275 kilograms).[57,58] Use of a geosynchronous orbit provides a major advantage. The low-orbit ELINT satellites were over the Soviet Union for only a relatively brief period each day. Also, once the Soviets knew the satellites' orbits, the times and locations of the passes could be predicted. When the satellites were overhead, radars and radios could be shut down and false signals transmitted. Rhyolite, from its high vantage point, constantly monitored Soviet electronic signals day in and day out.

Yet, this advantage also pointed out the fragile nature of ELINT intelligence. Rhyolite would be fully effective only as long as the Russians were unaware of the satellite's function. If they knew Rhyolite was an ELINT satellite, precautions could be taken to deny it the signals.

The first Rhyolite was launched on 6 March 1973. The Atlas Agena placed it into a 22,280- by 22,171-mile (35,855- by 35,679-kilometer) orbit. It was stationed above the Horn of Africa. From this position, it could monitor ICBM tests from Tyuratam. The launch coincided with the beginning of flight tests of a new generation of Soviet ICBMs—the SS-16, 17, 18, and 19. The satellite had a life expectancy of 2 years. As it was being tested and returned its first ELINT data, other Rhyolite satellites were being built. The success of the first, still operating after nearly 4 years, meant the replacement launches were put off. The completed Rhyolite satellites were stored. Once or twice a year, one of these satellites would be unfurled so its systems could be examined. At one of these checkouts, in TRW's high bay area, a spectator was Christopher John Boyce—Special Projects employee; holder of a Top Secret clearance, a Strategic Intelligence–Byeman clearance, and a Crypto clearance; and KGB spy.[59]

Boyce was born in 1953 and grew up on the Palos Verdes peninsula, California, south of Los Angeles. It was not an easy time or place to grow up. One reason was the drug problem. At Palos Verdes High School, drugs were as much a part of school life as classes and football. During the late 1960s and early 1970s, Boyce used marijuana and cocaine regularly. Another factor was the general self-destructive pessimism of American society in those years. The films of Vietnam battles on the evening news and his reading of history books from the school library had an effect

on Boyce. He came to reject the affluent life around him, patriotism, nationalism, and his country. By the time he turned 21, some have argued, he was already so cynical and rebellious that the leap to becoming a traitor was small.[60,61]

After graduating from high school in 1971, Boyce drifted, dropping out of three colleges and seven jobs over the next 3 years. In July 1974, he was hired by TRW as a general clerk. In November, he was transferred to Special Projects, the section responsible for the various intelligence satellites TRW built. Here Boyce first learned about reconnaissance satellites. His job was to relay coded messages sent to TRW from CIA headquarters and other facilities.[62] He saw his work, TRW, and the people around him as representing all he hated about America. In January 1975, at a drug party at his friend Andrew Daulton Lee's house, Boyce talked Lee into taking secret documents to the Russian embassy in Mexico City. Boyce said they would be worth thousands of dollars. With this inducement, Lee decided to go into dealing secrets as well as drugs. Lee was soon making regular trips to Mexico. One delivery in July was a stack of twenty or so sheets of paper; on one, Lee noted the word "Rhyolite." With its delivery, the Soviets learned that their rocket telemetry was being monitored; Rhyolite was "blown." Between then and December of 1976, Boyce photographed thousands of documents, many detailing the capabilities of U.S. reconnaissance systems.

His theft was aided by appallingly lax security at TRW. Personnel in the "black vault" where Boyce worked often drank on the job and even had a potted marijuana plant.[63] Lee was caught trying to make one last delivery on 6 January 1977. He was seen throwing a note through the fence of the Soviet embassy.

Mexican police, thinking he was a terrorist, arrested him. After several days of brutal interrogation, he implicated Boyce. Ten days after Lee's arrest, Boyce was picked up. Both were subsequently convicted. Lee was given a life sentence; Boyce received 40 years.

As their cases were going through the court system, Rhyolite launches resumed. The second went up on 23 May 1977. It was placed in geosynchronous orbit above Borneo, allowing coverage of Plesetsk launches. Two more Rhyolite launches followed on 11 December 1977 and 7 April 1978. They were positioned near the first two and were to be on-orbit spares. Subsequently, the Atlas Agena pads at Cape Kennedy were torn down. The complete set of four satellites was now in orbit, but their usefulness was compromised by the leak at TRW.[64]

On 29 July 1978, the Soviets launched a single-warhead SS-18 ICBM. When analysts examined its telemetry, they found it was encrypted to an extent not seen before; in particular, information related to the critical matter of the missile's throw weight was coded.[65] The Russians, indeed, knew about Rhyolite.

Argus and the Iranian Ground Stations

As Andrew Daulton Lee was making his deliveries to the Soviet embassy, events were underway that further damaged the U.S. ability to monitor Soviet missile tests. The early successes of Rhyolite suggested the possibility of a more advanced satellite. With a dish antenna nearly twice the diameter of Rhyolite's and with more advanced systems, it could rival the capabilities of the Tacksman 2 ground station in Iran.[66] Studies of this new satellite were underway at TRW by the early 1970s. The new satellite was code named "Argus,"

or AR for advanced Rhyolite. Argus was, in Greek mythology, a giant with 100 eyes chosen to be an ever-watchful guardian of Io.[67] The new Argus would be a vigilant observer of Soviet missile tests. The price for Argus, however, was very high, which was a problem. At the same time, the KH-11 photo reconnaissance satellite was running into technical problems. According to some reports, a fight developed over whether to build Argus or use the funding to keep the KH-11 on schedule. The Defense Intelligence Agency (DIA) wanted Argus, but Defense Secretary James Schlesinger rejected the request. CIA director William E. Colby, protesting this decision, wrote a letter to President Ford requesting the National Security Council to review the whole question. Ford agreed. Argus's problem was not simply the amount of money but also what it would buy. Argus was a fallback in the event the ground ELINT stations in Turkey, Iran, or elsewhere should be lost. Some argued that the cost was too high to simply duplicate something the U.S. already had. After examining both sides, the NSC endorsed Argus. Congress, however, would have to appropriate the money. The House Appropriations Committee noted the disagreement over whether or not Argus was needed and decided not to fund the project. By the end of 1975, Argus was dead. Had Argus gone ahead, it would have been ready for launch in about 4 years.[68]

At the same time, a study called Project 20,030 was underway; it was an estimate of what would be feasible in the 1980s in the way of satellite ELINT. Both the large amount of preliminary work done on Argus before Congress shut it down, and the Project 20,030 study, were done by Special Projects at TRW—the same section for which Christopher Boyce worked. Among the documents photographed by Boyce were ones relating to Argus and Project 20,030. Through these, the Soviets learned both the current state of U.S. ELINT technology and what would be possible in the future. In a way, the code name Argus was quite fitting. In Greek mythology, Argus was killed by Hermes, the god of commerce, cunning, and theft—the god of thieves . . . and spies.[69]

Even as Argus was being both cancelled and betrayed, the possibility of losing the ground stations became a reality. The U.S. had cut off arms shipments to Turkey after it had invaded Cyprus. In retaliation, Turkey, in July 1975, shut down the listening operations. Despite this, there was no move to develop orbital replacements other than launching the three Rhyolites in 1977 and 1978. In both years, attempts were made to revive Argus, but these were stopped by CIA director Stansfield Turner. Finally, in 1978, the arms embargo was lifted and the ELINT stations in Turkey were reopened.[70] The problem continued to be the seeming duplication of the ground stations in Iran. The CIA estimate was that the Shah of Iran was secure; and so it seemed, with his large, well-equipped armed forces and secret police. What nobody had counted on was the power of one hate-filled old man and the conflict he unleashed. With the return of the Ayatollah Ruhollah Khomeini to Iran in January 1979, the two ground stations' fates were sealed. The station at Behshahr on the coast of the Caspian Sea was closed in December 1978. Since the Shah's government was disintegrating, Tacksman 2 secretly remained open. The U.S. hoped some arrangements might yet be made with Iran's new rulers.

In late February, the new Iranian chief of staff, Maj. Gen. Mohammed Wali Qaraneh, believing Tacksman 2 had closed, announced that U.S. ELINT stations would not be tolerated in Iran. The Iranian Air Force personnel who worked at Tacksman 2, on hearing the broadcast, mutinied and took the U.S. personnel hostage. U.S. diplomats at the embassy in Tehran ransomed them for $200,000 in severance and back pay.[71]

Ironically, if Argus had gone ahead, it would have been ready to take over from the now-lost ground stations. At best, Rhyolite was a partial substitute. What everyone forgot in the fight over Argus was that the most valuable thing a government spends is time.

The loss of the Iranian ground stations coincided with the final stages of the SALT II negotiations. The SALT II treaty was quite different from its predecessor. SALT I had simply set limits on the number of ICBMs and submarines each side could have. SALT II, however, was more specific, dealing with such things as the throw weight of a particular type of missile, the maximum internal volume of a silo, or the number of warheads a particular type of missile could carry. It also limited each side to one new type of ICBM. This represented an obvious danger: what if the Soviets developed two new ICBMs, claiming that one was just a modified version of an old missile? To forestall this, the agree statements and common understandings to the treaty (which were more than twice the length of the treaty itself) stated that if the missile had a different number of stages or used a different type of fuel, or if the largest diameter, the launch weight, or throw weight were changed by more than 5 percent, it would be considered a new type.[72]

Clearly, monitoring of telemetry would be critical in policing the 5 percent rule. If the U.S. lacked the means of picking up the signals, the agreement would be worthless. The loss of the Iranian ground stations brought this long-smoldering issue of verification into the open. A related issue was the coding of telemetry. The data the U.S. could pick up would be useless if it could not be translated. The SS-18 test on 29 July 1978 and a second one on 21 December had been extensively encrypted. The issue of encryptment was one of the first raised by the Carter administration and one of the last settled. Some in the administration, particularly CIA director Turner, wanted a complete ban. Others felt this was not possible. It was not until 7 April 1979 that the question was finally settled. The SALT II treaty banned encryptment "whenever such denial impedes verification of compliance." [73]

A related question was how long it would take to replace the capability lost in Iran. Some estimates were that it would take 3 or 4 years to fully recover. Since the SALT II treaty expired on 31 December 1985, it would be only the last year of the treaty that would be fully verified. [74] Others stated that it would take only a year to improve the existing system enough to verify the agreement. This was the narrow area of verification rather than the total spectrum of intelligence-gathering activity the two stations in Iran had undertaken.

The Carter administration initially proposed to replace the Iranian stations with U-2 aircraft. While flying over the Black Sea, they would carry a 1,000-foot (305-meter) wire antenna that would be unreeled from the aircraft. From high altitude, the antenna could pick up the missile's telemetry. [75] This would be supplemented by an upgrading of the ground stations in Turkey. The plan met with immediate criticism. The Turkish stations' effectiveness was limited by the Caucasus Mountains, and the U-2s could not stay on station constantly or carry the huge loads of equipment necessary to duplicate the Iranian

stations. [76] Whether or not they could was rendered academic when the Turkish government said that they might need Soviet permission before allowing the flights to go ahead.

It was at this time that the full dimension of what KGB spies Boyce and Lee had done began to be made public. At their trials, Rhyolite, Argus, and the rest of the satellites were not mentioned. It was not until 8 December 1978, that the first oblique references were made. William Clements, the deputy secretary of defense during part of the time Boyce was selling documents to the Russians, said, "A major satellite intelligence system, developed and deployed at a cost of billions of dollars over the past decade, without Soviet knowledge, has been compromised. . . ." [77]

Several months later, a brief *New York Times* article first mentioned Rhyolite and Argus. This was soon followed by a more detailed article in *Aviation Week & Space Technology* magazine, which described Rhyolite, the saga of Argus, and the importance of the Iranian ground stations.

One month later, on 18 June 1979, Jimmy Carter and Leonid Brezhnev signed the SALT II treaty in Vienna. The disclosures about Rhyolite made the issue of verification a central one in the Senate debate. Despite an intensive SALT selling campaign, even by early December, the votes to ratify were still unsure—sixty-seven were needed. The administration felt it had the support of forty-seven senators; twenty-one had announced they were opposing it. The remaining thirty-nine were undecided. The opposition seemed more confident. Very soon, however, such arithmetic would not matter.

In April 1978, a Marxist coup had overthrown the government of Afghanistan. Soon, Soviet personnel were advising the government and army. This Soviet involvement, however, sparked rebellion by Muslim tribes- men. They proved to be a difficult foe and the Afghan government seemed threatened. Moscow was becoming impatient with the Af- ghans' failure to put down the revolt. In Sep- tember 1979, the Soviets replaced President Nur Mohammad Taraki with Hafizullah Amin; but they were quickly disenchanted with Amin and came to the conclusion more drastic action was needed. On the night of 27 December 1979, Soviet troops invaded Afghanistan. As they did, a special KGB unit killed Amin. As a replacement, the Soviets installed Babrak Karmal, the leader of a rival Marxist faction then living in exile and a person Moscow felt would be more obliging than Amin had been. The invasion represented the first time Soviet troops had moved outside the boundaries es- tablished at the end of World War II. President Carter felt betrayed and responded angrily. The U.S. ambassador to Moscow was re- called; sales of grain and technology were cut back; and in January 1980, Carter withdrew the SALT II treaty from Senate consid- eration.[78,79] Although Afghanistan had tempo- rarily brought an end to SALT II, it was the question of verification that had really killed it. Boyce and Lee's role in this was summed up in a "60 Minutes" broadcast in November 1982. Senator Daniel Patrick Moy- nihan said, "With respect to the satellite sys- tems that were compromised, they made them, temporarily at least, useless to us. Be- cause the Soviets could block them. And the fear that that would happen, had happened, permeated the Senate and, as much as any one thing, was responsible for the failure of the SALT treaty. And if you think, as I do, that the breakdown of our arms negotiations with the Soviets is an ominous event, then nothing quite so awful has happened to our country as the escapade of these two young men."

A CIA official put it more simply: "What he did was a national calamity."[80]

Aquacade

In the wake of the failure of SALT II and the invasion of Afghanistan, there re- mained the task of replacing the lost stations in Iran. The interim solution was one that would have been unimaginable a decade be- fore. The ground station would be built in the People's Republic of China. The U.S. first proposed the idea in 1978 before formal diplo- matic relations with China were established. The Chinese were reluctant, apparently fear- ing the consequences of cooperating too closely with the U.S. Soon after the closing of Tacksman 2, the U.S. renewed its request. This time the Chinese were interested but with provisions. The station was to be manned by Chinese technicians and built and operated in secret. Final agreement was reached in 1980. Two sites were surveyed, but only one was built; it is located in the mountains of the Xinjiang Uighur Autonomous Region in western China. The equipment is supplied by the U.S. and operated by the Chinese. U.S. advisers visit periodically. The intelligence is shared between the U.S. and China.

With the start of operations in 1980, the Chinese station became the U.S.'s most sensi- tive and important facility. It was ideally lo- cated to cover both Tyuratam and the Sary Shagan ABM test center. It could monitor ICBMs from the early part of launch to the separation of the warheads. The facility also goes to show that a common enemy makes for unusual allies.[81]

Once the warheads separate from the final stage, they pass out of range of the Chinese facility. U.S. attempts to upgrade its ability to cover this part of the missile's trajectory predated the loss of the Iranian station. The first of the new radars—the Cobra Dane phased array radar on Shemya Island—went into operation in early 1977. Housed in a building that looks like an Egyptian temple, the radar can track the warheads down to an altitude of 400,000 feet (121,920 meters).[82] For coverage at lower altitudes, nearly to impact, a shipboard system called Cobra Judy is used. The phased array radar is mounted in a turret on the deck of the USNS Observation Island; it went into operation in the early 1980s.[83] The third radar is an airborne system. The Cobra Ball radar is carried aboard a modified RC-135, which flies off the Soviet coast when Soviet missile tests are expected. Like Cobra Judy, it has the advantage of mobility and was also operational in the early 1980s.

As valuable as the Chinese ELINT station is, the bitter experience of Iran taught the U.S. the vulnerability to political upheaval of such facilities on foreign soil. As the personnel were being evacuated from Iran, work began on a new satellite system called Aquacade, which began as an updated version of Argus. Some of the modifications were to take advantage of the added capabilities of the space shuttle. One such advantage was that Aquacade could be larger. The shuttle's payload bay is 60 feet (18 meters) long and 15 feet (5 meters) wide—vastly larger than a conventional rocket shroud. Additionally, the shuttle's crew could check out the satellite's systems before it was released into space. All this, plus the improvements in technology in the decade since Rhyolite was designed, meant Aquacade would be a much more versatile system. It would, however, take several years to develop.

As work began on Aquacade in 1979, it seemed that use of the shuttle was more of a problem than an advantage. The shuttle was to be the world's most advanced flying machine. Yet, it was to be developed on a restricted budget. When problems appeared, as they must in any complex project, NASA either shifted money from one budget to cover them or put off testing and later purchases. By 1978, the limited funding had caught up with NASA; schedules began to slip and costs increased. Questions began to be asked about whether the shuttle was worth it.[84] The uncertainties alarmed Congress and in early 1979, after Iran, the Senate Select Committee on Intelligence directed Aquacade to be redesigned for launch on a Titan 34 D/IUS. Although this could limit its capability, the committee felt that redesign was less important than having Aquacade delayed by the continuing shuttle problems.[85]

At the same time as this snub, the shuttle program received a major boost. NASA needed $220 million in added funds or it would have to put off building additional shuttles to cover increased development costs. When NASA administrator Dr. Robert Frosch met with Carter administration officials to discuss the problems, he expected a bad time. He was lectured all right, but much to his surprise it was on the need to get the program back on schedule and how NASA did not understand the importance of the shuttle. The reason for this turnaround from the disinterest of previous years was the Carter administration's realization that the shuttle was critical to the next generation of reconnaissance satellites. If it were further delayed, the satellites would also be delayed, which would adversely affect U.S. intelligence and monitoring of arms control agreements. Moreover, the Air Force needed the complete fleet of shuttles to assure a reliable launch service.[86] The military need for the shuttle reversed the program's technical decline and may well have prevented an eventual cancellation.

On the night of 21 January 1980, a bit of old business returned to public attention: Christopher Boyce escaped from prison. After their conviction, Boyce and Lee had been sent to the federal prison at Lompoc, California, near Vandenberg Air Force Base. After several unsuccessful escape attempts, Boyce finally made his break, cutting through the wire and vanishing into the night. For the next 19 months, U.S. marshals pursued him, going on wild-goose chases as far away as Central America and South Africa. Boyce was actually in the northwestern United States, supporting himself by robbing banks. He had also learned to fly and thought about escaping to the Soviet Union. Finally, on 21 August 1981, Boyce was arrested at a drive-in restaurant. He was later sentenced to 25 years for sixteen bank robberies and an additional 3 years for escaping. To prevent any further escapes and to protect him from attacks by other inmates, he was sent to the federal prison at Marion, Illinois, where he is in solitary confinement. If he serves his full term, Boyce will be released in the year 2047, when he will be 94 years old.[87]

First Launch

This time was also an eventful one for both the shuttle and Aquacade. In April 1981, the space shuttle made its first flight. Aquacade was also progressing; development was

probably completed and construction of the first satellite underway. The crew for the first launch was selected in October 1982. The members were Capt. Thomas K. Mattingly (USN), Lt. Col. Loren J. Shriver (USAF), Lt. Col. James Buchli (USMC), Maj. Ellison Onizuka (USAF), and Maj. Gary Payton (USAF).[88]

The need for Aquacade was increasing when, in the early 1980s, it became apparent the Soviets were violating a number of SALT II provisions. When the Soviets signed the SALT II treaty, they were in the process of developing not one new ICBM but three. The first was the SS-X-24, a medium-sized, solid-fuel ICBM with better accuracy than the older SS-18s and 19s, to be deployed in both silos and on a mobile launcher. The Soviets notified the U.S. that this was the new missile the treaty permitted. After doing so, they began flight tests of the SS-X-25, a solid-fuel missile about the same size as the Minuteman. It had a single warhead and was carried in a massive, wheeled transporter. Both missiles were test flown from Plesetsk. Then, in 1984, U.S. intelligence detected a third new ICBM. Its solid-fuel rocket motors were being tested at Pavlograd in the Ukraine. Also under test was its container/ejector launch system. Test silos were also spotted at Plesetsk. The third new ICBM represented a violation, not only because of its existence but also because its launch weight and throw weight exceeded the 5 percent limit, being a 15 to 20 percent increase over the SS-19.[89]

The Soviets were also improving their older missiles. In late 1983 and early 1984, tests were made of an ICBM warhead meant to carry biological warfare materials. It was observed to tumble, spraying its contents out at high altitude.[90] It was subsequently claimed

The crew of mission 51C, the first military shuttle mission. *Front row, left to right:* Loren J. Shriver (pilot) and Thomas K. Mattingly II (commander). *Back row:* Gary E. Payton (payload specialist), James F. Buchli and Ellison L. Onizuka (mission specialists). Mattingly was command module pilot on *Apollo 16* and commander of STS-4. Payton was the first manned spaceflight engineer to fly, one of a group of 27 air force and navy engineers selected to fly on military shuttle missions. Onizuka was killed in the *Challenger* explosion a year later. *NASA Photo.*

that the warhead would be carried by the SS-11 Mod 4 Sego, which is described as carrying three to six reentry vehicles rather than the one or three carried by the earlier versions.[91] It is probable that the warheads would carry biological agents, such as Anthrax spores, rather than chemical weapons like nerve gas.

Although .75 milligram of Sarin nerve gas can kill an adult, it would take 250 tons to kill everybody in a city the size of Paris.[92] Anthrax spores, on the other hand, could start epidemics that could spread outside the target area, so the amount of material carried by each rocket could be relatively small.

Clearly, U.S. intelligence needed all the information it could muster. The Soviets knew this too and made a concerted effort to deny the U.S. the necessary data. In addition to continued encrypting of telemetry, they began, in January 1983, to jam the Cobra Dane, Cobra Judy, and Cobra Ball radars. They also jammed reception from the Rhyolite satellites. In the spring of 1984, they further refined the process by switching to low-power transmitters aboard the rockets. Their telemetry signals can be picked up only by aircraft within the test area.[93]

A final problem was one of age. By early 1984, the first Rhyolite satellite was 11 years old, while the other three were between 6 and 7 years old and sure to be degraded to some extent.

Yet despite the need, the first Aquacade launch was delayed repeatedly from mid-1983 until late 1984. The reason was apparently not problems with the satellite but in the two-stage Inertial Upper Stage (IUS), used to boost the satellite into geosynchronous orbit after it has been released from the shuttle. The problem appeared during the STS-6 mission in April 1983. The IUS carried the first NASA Tracking and Data Relay Satellite (TDRS). The first-stage firing went well, but 83 seconds into the planned 107-second burn of the second stage, the IUS went out of control. The TDRS was left in a useless orbit. It was salvaged only by repeated firing of small, on-board engines, which nudged it into geosynchronous orbit. To prevent a repetition, all IUS launches, both NASA and Air Force, were stopped until the reason was found and a solution developed. This took 20 months and four ground test firings. The problem was found to be in a seal on the engine nozzle, which allowed it to pivot for steering control. The seal had overheated and collapsed, throwing the stage out of control. Insulation was added and other changes were made, and the IUS was cleared for flight in late 1984.[94] The Aquacade launch was then scheduled for 8 December 1984, aboard space shuttle *Challenger*. The mission was designated 51-C. During preparations for launch, an inspection found 4,202 of *Challenger*'s heat-shield tiles might have been loosened by repeated re-entries and use of a waterproofing agent. The problem could be corrected but would take several months. Rather than wait, NASA switched shuttles, using *Discovery* instead and rescheduling the launch for 23 January 1985.[95]

DOD versus the Media

In 1972 when the shuttle was approved, it was stated that it would carry both civilian and military satellites. Military launches would account for about 30 percent of all shuttle flights. This was never hidden, but as the 51-C launch approached, the news it was to be military seemed to have taken some in the media by surprise. Additionally, some details of the payload were beginning to leak out to television networks and wire services. Defense Secretary Caspar Weinberger was soon on the telephone asking them to ''kill'' the story. Although management agreed, the decision left the involved reporters increasingly hostile towards the Air Force, which showed at the T minus 30-day press conference. Because of the launch delay, the press conference took place on 17 December 1984—during the traditionally slow news period around Christmas. The Air Force's chief of public affairs, Brig. Gen. Richard Able, announced that the restrictions normally applied to Air Force launches were being eased. For unmanned launches of military satellites, past practice was that no launch date or time was made public beforehand, to make it more difficult to track the satellite. The 51-C, he said, would be launched between 1:15 P.M. and 4:15 P.M. EST. He continued, "Our intention is to make the maximum information available to you consistent with national security. We are working to deny our adversaries any information which might reveal the identity of the mission or payload." Since the media wanted such information, reports of the press conference centered on what would not be made public—no interviews with the crew, no press kit, no release of communications to or from the shuttle. General Able further antagonized the press when, in answer to a question about press speculation, he said that it might be investigated to find the source.[96]

The reaction to the press conference was uniformly hostile, one commentator saying that he thought the shuttle was carrying a load of chocolate chip cookies but then, he continued, he didn't think much of the military. Given the venomous atmosphere and the number of people who knew about the payload, it was only a matter of time before the story would be published. The secrecy lasted 2 days. On 19 December the *Washington Post* printed a front-page story saying that the shuttle cargo was an ELINT satellite. Secretary Weinberger reacted angrily, calling it ". . . the height of journalistic irresponsibility."[97] Many in the media supported the *Post*. The editor of the Traverse City, Michigan, *Record-Eagle* went so far as to criticize the news agencies that held back as "frightening examples of how close the government sometimes comes to producing news for its own pur-

poses. Thank God this unnecessary exercise in censorship [failed]."[98]

Over the next days and weeks, details about Aquacade leaked out. It had two large dish antennas: one picked up telemetry, radio, radar, and phone calls—the entire radio spectrum coming from Russia, Asia, and Africa; the other antenna relayed the data to the Alice Springs ground station. The artist's concept showed the two antennas as being equal in size. When the antennas and solar panels were unfurled, the satellite was 100 feet (30 meters) wide. According to one report, Aquacade was so large, only the shuttle could carry it. If correct, this would imply that Congress's order to fit Aquacade on a Titan 34 D could not be met. The satellite could also be maneuvered in orbit to reposition itself. Aquacade weighed about 5,000 pounds (2,268 kilometers)—roughly eight times that of Rhyolite. It cost $300 million.[99,100]

Despite the controversy, preparations went smoothly, the only delay being a 1-day slip caused by freezing temperatures that went as low as 19 degrees. NASA officials were concerned that ice could build up on the external tank; at launch, it might shake loose and damage the heat shield tiles.

At 2:50 P.M. EST, 24 January 1985, the shuttle's engines ignited and it roared into the Florida sky. Sixteen hours after launch, at about 6 A.M. EST the next morning, the satellite and two attached IUS stages were released from the shuttle. The shuttle then maneuvered away. Fifty-five minutes later, as Aquacade crossed the equator, the first stage fired; it burned for 146 seconds. Then the second stage and attached Aquacade separated from the first stage. When the burn was complete, the velocity was low by 50–55 feet per second (15.2–16.8 meters per second). To make this up,

Discovery clears the launch tower, 24 January 1985. This was the first military shuttle mission. The orbiter carried an Aquacade ELINT satellite and IUS upper stage to boost it into geosynchronous orbit. Reflecting its military role, launch and landing times and mission duration was not publicized beforehand. Inflight conversations and photos were not released. The news media was predictably upset.

At left is part of the structure CBS claims was used to hide the shuttle; in fact, this is its standard launch position. The 51C shuttle mission brought to public attention the bad relations between press and military. One reporter, in discussing their adversary relationship, would say while it was the duty of the government to keep secrets, it was the role of the media to make them public. *NASA Photo*.

Discovery lands on the Kennedy Space Center runway, bringing the 51C mission to a close. The January 27, 1985, landing was a day earlier than planned due to a forecast of bad weather for the next day. *NASA Photo.*

small liquid-fuel rockets on the second stage fired briefly, insuring that Aquacade would reach the exact point in geosynchronous orbit at the right time. The long climb to geosynchronous orbit took 5 hours, during which time the second stage of the IUS and attached satellite were turned so ground stations could receive telemetry. Other maneuvers kept the temperature equal.

Reaching an altitude of 22,300 miles (35,881 kilometers), the IUS second stage oriented itself, then fired for 103 seconds, circularizing the orbit. The solar panels extended and Aquacade separated from the IUS. Under the new system, Aquacade was designated after launch as USA-8.[101,102] The shuttle continued to orbit. The mission had apparently been planned to last 4 days, but a bad weather forecast caused the mission to be called down a day early. Touchdown on the Kennedy Space Center runway was at 4:23 P.M. EST, 27 January 1985.[103]

Aftermath

The saga of *Aquacade 1* is much more than the story of one satellite. It highlights the mutual contempt between the U.S. military and the news media, dating from Vietnam, the military being seen by the press as liars, fools, and criminals and the press being viewed by the military as biased to the point of disloyalty—the enemy behind the lines. Aquacade raises questions about the role of the press in a technological society. Is the media to be the final arbitrator of what will remain secret? As *Aviation Week & Space Technology* magazine noted, the press could find itself in the hypocritical position of condemning the Air Force for buying $7,000 coffee makers, yet by their own actions, lessening the effectiveness of a satellite costing hundreds of millions of dollars.[104]

Despite all the controversy, when the launch came the networks seemed indifferent. The countdown was made public at T minus 9 minutes.[105] Yet, there were only 7 or 8 minutes of live television coverage. Network news cut away even before *Discovery* reached orbit. The landing time was announced 16 hours beforehand; yet, even with this notice, not one network was willing to interrupt its regular programming to cover it. Worse, the secrecy was exploited by the media strictly for sensationalism. CBS said it would ''try'' to cover the launch; then, in their coverage of the launch, they said that photographers were kept 3 miles away from the pad for security reasons and that the launch tower was used to hide the vehicle.[106] In fact, nobody is ever allowed closer than 3 miles for safety reasons in the event of an explosion. As for the other charge, the press site was built during Apollo. When the pad tower was rebuilt for the shuttle, part of the structure blocked the view of the shuttle as seen from the press site—for *all* launches.

The controversy over Aquacade was repeated two years later. In January 1987, the British Broadcasting Corporation was preparing a film on government secrecy and discovered the existence of an independent British ELINT satellite program code-named Zircon. The satellite was described as having a 100 foot (30.5 meter) dish antenna for picking up Soviet radio and radar signals. Work had begun in 1983. The satellite was being built by British Aerospace and GEC Avionics with technology transfer from TRW. It was based on the technology developed for the Rhyolite, Argus, and Aquacade satellites. The Zircon would be launched from the shuttle and placed in a geosynchronous orbit at 53 degrees east longitude. This would allow it to view central Russia.[107] The data would be relayed to the Government Communications Headquarters at Cheltenham, England. Zircon would give the British an independent ELINT satellite capability (although the NSA provides satellite data to the British, some is withheld). This capability had great political importance to British relations with the U.S.

Knowledge of Zircon was closely guarded. Parliament was not told of its existence, and the 500 million–pound budget was hidden. The cover story was that the satellite was a third Skynet 4 communication satellite added to the two previously announced. (Later increased to five Skynets.)

When the British government became aware of the leak, it reacted differently than the U.S. did with Aquacade. The documentary was banned, police Special Branch officers raided BBC offices, and members of Parliament were prohibited from privately viewing the film.[108,109]

Assuming the Zircon story is accurate,

particularly the Skynet 4 cover story, the launch schedule isn't clear. Before the *Challenger* explosion, *Skynet 4A* was set for launch aboard the shuttle in July 1986, *Skynet 4B* scheduled for a January 1987 shuttle launch, and *Skynet 4C* was to be launched by the French Ariane booster in 1988.[110,111] After *Challenger, Skynet 4B* was rescheduled for a late 1987 launch on an Ariane booster. *Skynet 4C* was to be launched on a 1989 Ariane flight. *Skynet 4A* remained on the shuttle and was rescheduled for a 1990 launch. A fourth Skynet 4 was set for a 1991 shuttle mission.[112,113] A fifth Skynet 4 is also planned. Security and payload considerations would seem to rule out an Ariane launch; therefore, the Zircon must be one of the three shuttle launches.

The Aquacade and Zircon controversies raise issues about the role of secrecy in a free society—specifically, whether classifying everything, in fact, means that nothing is secret and whether this breeds disregard for the very secrecy that classifying is meant to enforce. It is an old question. In 1957, the Air Force refused to say if the first Atlas ICBM had been launched—even as photos of its brief flight were appearing in magazines and newsreels. After Sputnik, a more open attitude was established, one that preserved the real secrets without taking things to absurd limits. Unfortunately, the technological precision with which the satellites were created cannot be applied to the issue of secrecy versus openness.

NOTES Chapter 11

1. Lt. Col. Bruce M. Bailey, USAF, (Retired), *We See All-A Pictorial History of the 55th Strategic Reconnaissance Wing,* (Privately Printed, 1982).
2. "Soviets May Have Lured RB-47 to Bolster Confidence in Defenses," *Aviation Week & Space Technology,* (18 July 1960): 30.
3. Dick van der Aart, *Aerial Espionage,* (St. John's Hill: Airlife Publishing Ltd., 1985), 13,57,58, 110,111.
4. Research Publications, Department of State 1641, 1984.
5. John M. Carroll, *Secrets of Electronic Espionage,* (New York: E. P. Dutton, 1966), 168,170.
6. Research Publications, CIA 758, 1985.
7. Bailey, *We See All,* 119,120.
8. "Martin/General Dynamics RB-57F", *Aerophile,* (Vol. 2 #3), 22.
9. Robert C. Mikesh, *B-57 Canberra at War 1964–1972,* (New York: Scribners, 1980), 135.
10. Lt. Col. Richard E. Fitts, *The Strategy of Electromagnetic Conflict,* (Los Altos: Peninsula Publishing, 1980), 58–60.
11. Carroll, *Secrets of Electronic Espionage,* 129–134.
12. David A. Anderton, *Strategic Air Command,* (New York: Scribners, 1975), 164.
13. Charles A. Mendenhall, *Delta Wings,* (Osceola: Motorbooks International, 1983), 114,115.
14. Anderton, *Strategic Air Command,* 235.
15. Bailey, *We See All.*
16. "Soviets May Have Lured RB-47 to Bolster Confidence in Defenses", *Aviation Week & Space Technology,* (18 July 1960): 30.
17. Bailey, *We See All.*
18. Andrew Brookes, *V-Force: The History of Britain's Airborne Deterrent,* (London: Jane's, 1982), 106.
19. Aileen Clayton, *The Enemy Is Listening,* (New York: Ballantine Books, 1982).
20. Glenn T. Seaborg, *Kennedy, Khrushchev and the Test Ban,* (Berkeley: University of California Press, 1981), 77.
21. James Bamford, *The Puzzle Palace,* (New York: Penguin Books, 1983), 218,219. Also see *Newsweek* 24 July 1961 and 30 July 1962.
22. Carrollton Press DoD 101B 1977.
23. Carrollton Press DoD 288A 1977.
24. Philip J. Klass, "Military Satellites Gain Vital Data," *Aviation Week & Space Technology,* (15 September 1969): 58,59.
25. Philip J. Klass, *Secret Sentries in Space,* (New York: Random House, 1971), 190–194.
26. Correspondence, *Journal of the British Interplanetary Society,* (April 1983): 191.
27. Anthony Kenden, "U.S. Reconnaissance Satellite Programmes," *Spaceflight,* (July 1978): 256,257.
28. Correspondence, *Journal of the British Interplanetary Society,* (April 1983): 191.
29. Anthony Kenden, "U.S. Reconnaissance Satellite Programmes," *Spaceflight,* (July 1978): 256,257.
30. Brookes, *V-Force,* 128,129,134,135.
31. "Satellite Needs Push Reconnaissance Gains," *Aviation Week & Space Technology,* (22 May 1961): 27.
32. Anthony Kenden, "U.S. Reconnaissance Satellite Programmes," *Spaceflight,* (July 1978): 256,257.

33. Klass, *Secret Sentries in Space,* 195.
34. Anthony Kenden, "Recent Developments in U.S. Reconnaissance Satellite Programmes," *Journal of the British Interplanetary Society,* (January 1982): 38–41.
35. Col. N. Gavrilov (Reserve) "Military Journal on U.S. Intelligence Satellites," *Zarubezhnoye Voyennoye Obozreniye,* (November 1984): 54–59.
36. Satellite Digest 158, *Spaceflight,* (November 1982): 417.
37. "Soviets Charge U.S. Aligned 747 With Satellite, Aircraft," *Aviation Week & Space Technology,* (26 September 1983): 42,43.
38. NASA Satellite Situation Report, 31 December 1978.
39. James E. Oberg, "Sakhalin: Sense and Nonsense," *Defense Attache Magazine,* (January/February 1985): 37–47.
40. Letter from James E. Oberg to Rupert Pengelley, Editor, *Defense Attache Magazine,* 6 July 1984.
41. James E. Oberg, "Sakhalin: Sense and Nonsense," *Defense Attache Magazine,* (January/February 1985): 37–47.
42. Satellite Digest 170, *Spaceflight,* (February 1984): 94.
43. *Daily Telegraph,* 6 August 1984.
44. John Prados, *The Soviet Estimate,* (New York: The Dial Press, 1982), 57,58.
45. Prados, *The Soviet Estimate,* 35.
46. Klass, *Secret Sentries in Space,* 26.
47. Francis Gary Powers and Curt Gentry, *Operation Overflight,* (New York: Holt Rinehart Winston, 1970), 47.
48. Prados, *The Soviet Estimate,* 36.
49. Eloise Engle and Kenneth H. Drummond, *Sky Rangers,* (New York: John Day, 1965), 144–149.
50. Prados, *The Soviet Estimate,* 178.
51. Bamford, *The Puzzle Palace,* 209.
52. Prados, *The Soviet Estimate,* 275.
53. Prados, *The Soviet Estimate,* 177.
54. Bamford, *The Puzzle Palace,* 256.
55. Philip J. Klass, "U.S. Monitoring Capability Impaired," *Aviation Week & Space Technology,* (14 May 1979): 18.
56. Robert Lindsey, *The Falcon and the Snowman,* (New York: Pocket Books, 1980), 61,62,126,127.
57. Anthony Kenden, "Recent Developments in U.S. Reconnaissance Satellite Programmes," *Journal of the British Interplanetary Society,* (February 1982): 38–41.
58. Space Shuttle Navy Space Systems Activities, October 1977, 55.
59. Lindsey, *The Falcon and the Snowman,* 24,25.
60. Lindsey, *The Falcon and the Snowman,* 35.
61. Robert Lindsey, *The Flight of the Falcon,* (New York: Pocket Books, 1985), 247.
62. Lindsey, *The Falcon and the Snowman,* 51,55,57.
63. Lindsey, *The Falcon and the Snowman,* 93,209,371.
64. Philip J. Klass, "U.S. Monitoring Capability Impaired," *Aviation Week & Space Technology,* (14 May 1979), 18.
65. Strobe Talbott, *Endgame,* (New York: Harper Colophon Books, 1980), 200,201.
66. Lindsey, *The Falcon and the Snowman,* 420.
67. *Webster's New Twentieth Century Dictionary,* (New York: The World Publishing Co., 1966), 100.

68. Philip J. Klass, "U.S. Monitoring Capability Impaired," *Aviation Week & Space Technology,* (14 May 1979), 18.
69. Lindsey, *The Falcon and the Snowman,* 63,420.
70. Bamford, *The Puzzle Palace,* 209,210.
71. Bamford, *The Puzzle Palace,* 257–259.
72. Talbott, *Endgame,* 312,313.
73. Talbott, *Endgame,* 258,259.
74. Bamford, *The Puzzle Palace,* 258.
75. Prados, *The Soviet Estimate,* 275,276.
76. Clarence A. Robinson, Jr., "Soviets Push Telemetry Bypass," *Aviation Week & Space Technology,* (16 April 1979): 14–16.
77. Lindsey, *The Falcon and the Snowman,* 417,418.
78. Talbott, *Endgame,* 285,288–290.
79. *Soviet Military Power 1984,* (Department of Defense, 1984), 132.
80. Lindsey, *The Flight of the Falcon,* 322,339.
81. *New York Times,* 18 June 1981, A-1.
82. "Filter Center," *Aviation Week & Space Technology,* (12 July 1976): 53.
83. Col. William V. Kennedy, *Intelligence Warfare,* (New York: Crescent, 1983), 174.
84. William Stockton and John Noble Wilford, *Spaceliner,* (New York: Times Books, 1981), 50,51,58.
85. "Washington Roundup," *Aviation Week & Space Technology,* (4 June 1979): 11.
86. Stockton and Wilford, *Spaceliner,* 62,63.
87. Lindsey, *The Flight of the Falcon,* 289–292,303,333,337–339.
88. Alcestis R. Oberg, *Spacefarers of the '80s and '90s,* (New York: Columbia University Press, 1985), Chapter 5.
89. Edgar Ulsamer, "In Focus . . . ," *Air Force Magazine,* (July 1984): 21.
90. "Washington Roundup," *Aviation Week & Space Technology,* (30 January 1984): 15.
91. Jack Anderson and Dale van Atta, "Poison and Plague: Russia's Secret Terror," *Reader's Digest,* (September 1984): 58.
92. Brian Ford, *German Secret Weapons,* (New York: Ballantine Books, 1969), 108–110.
    Sarin is the code name of a nerve gas developed by the Germans in World War II. The chemical name is Fluoroisopropoxymethylphosphine Oxide.
93. Edgar Ulsamer, "In Focus . . . ," *Air Force Magazine,* (August 1984): 24.
94. Richard G. O'Lone, "Improvements in Inertial Upper Stage Renew Confidence in System Viability," *Aviation Week & Space Technology,* (11 February 1985): 67,68,71.
95. "Shuttle Test for Tiles and Upper Stage," *Flight International,* (19 January 1985): 12.
96. *San Diego Union,* 18 December 1984, A-1.
97. *San Diego Union,* 20 December 1984, A-1.
98. *San Diego Union,* 22 December 1984, A-33.
99. NBC News Report, 24 January 1985, (launch coverage).
100. NBC Nightly News, 24 January 1985.
101. "IUS Meets Mission Objectives on Defense Department Shuttle Flight," *Aviation Week & Space Technology,* (4 February 1985): 20.
102. "Defense Department IUS Uses Liquid Thruster Aid," *Aviation Week & Space Technology,* (18 February 1985): 22,23.

103. *San Diego Union,* 29 January 1985, A-2.
104. William H. Gregory, "A Media Event," *Aviation Week & Space Technology,* (14 January 1985): 11.
105. *San Diego Union,* 23 January 1985, A-2.
106. CBS Evening News, 24 January 1985.
107. "British Designing Independent Sigint Satellite System," *Aviation Week & Space Technology,* (2 February 1987): 25.
108. *San Diego Tribune,* 23 January 1987, A-23.
109. *San Diego Tribune,* 4 February 1987, A-18.
110. "Shuttle Launch Schedule 1986," *Spaceflight* (February 1986): 57.
111. "Ariane Launch Manifest," *Spaceflight* (March 1986): 119.
112. "Shuttle Launch Manifest," *Spaceflight* (November 1986): 378.
113. "Skynet Switched from Shuttle," *Spaceflight* (July/August 1986): 295.

CHAPTER 12

Soviet ELINT

The development of Soviet ELINT was affected by geographic and technical factors. In the years following World War II, it was relatively easy for the Soviets to monitor NATO signals from Eastern Europe. They began building ground ELINT stations called Box Brick. Although they were believed to be radar receivers and direction finders, western intelligence was not certain of their function. The British, therefore, mounted a major effort against one station in Austria. ELINT monitoring, coordinated with ground teams, determined the station did not transmit. Next, the British directed ground and airborne radar signals towards it and noted the reaction. They were able to determine not only that Box Brick was a receiver/direction finder but also the range over which it could pick up signals.[1]

Such facilities were suitable for Europe but were of no use against the U.S. during the 1940s and 1950s. The reason was that until the Cuban revolution, there were no friendly countries near the U.S. where the Soviets could build a station. To cover the U.S. and other areas of interest, the Soviets used submarines, which could provide both mobility and a minimal risk of detection—particularly valuable if the submarine had to enter U.S. territorial waters. One example of the use of Soviet submarines to monitor U.S. activities occurred during the 1958 nuclear test series in the Pacific. The Soviet announcements of the U.S. tests caused some in the Pentagon and the Atomic Energy Commission to worry that there was a leak. An examination of the circumstances, however, clearly indicated the Soviets had been monitoring the test countdown from submarines in or near the safety zone. Several factors supported this. The countdown was broadcast to all the ships and installations involved in the tests, which were scattered across a good part of the Pacific. The transmission would be relatively easy for the Soviets to pick up. More important was that two of the Soviet test announcements were false. In the first case, the detonation was stopped at the last moment. In the other, the explosion was not of a nuclear weapon but a 112-pound (51-kilogram) high-explosive charge. In both cases, however, a normal countdown was made. Clearly, the Soviets had monitored these and assumed they were actual tests. The standard search of the safety zone was meant to chase off any wayward freighters, not to find a submarine.[2]

At the end of the 1950s, another type of ELINT ship was put into service, one that would become a symbol for ELINT activities—the "Russian trawler"—ships that carried a large set of antennas and never seemed to fish. The first Russian trawlers were converted Okean-class fishing boats built in East

A Russian Whiskey-class submarine cruising on the surface. The Soviets have made extensive use of submarines for ELINT probing missions. A diesel submarine, like this one, is quiet and able to operate in shallow coastal waters. Although long used in the ELINT role, it did not come to public attention until a Whiskey-class submarine ran aground in Swedish coastal waters. *U.S. Navy Photo.*

German shipyards starting in 1959. The ELINT equipment carried would vary over time and even from one voyage to the next. A typical trawler may carry direction finders, high-frequency antennas, rod and dipole antennas, as well as multiple radar receivers.[3] For monitoring rocket tests, they carry multiple helical antennas. Trawler operations were underway by 1960. The trawler *Vega* operated off the Virginia coast in April 1960 and entered Long Island Sound to watch the launching of a dummy Polaris missile from the USS *George Washington*. Other trawlers conducted missions off Newfoundland and in the Mediterranean during September and November 1960.[4] From only four ships in 1963, the ELINT trawler fleet grew during the 1960s, undertaking various types of missions. Some

A Soviet ''trawler'' shadowing a U.S. naval exercise. Various antennas crowd the two masts and bridge. The only ''fishing'' these boats do is for radar, radio, and telemetry transmission. Russian trawlers are stationed off western ports and tests areas. *U.S. Navy Photo.*

shadowed naval task forces. Others loitered off port, particularly those used by U.S. missile submarines. Areas of interest included northern Scotland, the English Channel, Gibraltar, southern France, Hawaii, and both U.S. coasts.[5] Soviet trawlers also frequently appear off the Kennedy Space Center to monitor satellite launches and missile tests. They often have to be chased out of danger zones and have even interfered with submarine missile tests. Using trawlers has certain advantages. Unlike the Iranian ground station, they are not several hundred miles from the launch pad but as close as the 3-mile limit. This puts the shipboard technicians in a much better position to receive both telemetry and ground communications.

Soviet Airborne ELINT

In the early 1950s, the Soviets were making probing flights over Western Europe. MiG 15s, the Soviets' highest performance fighters, were initially used. The pilots were East German, Czech, or Bulgarian (presumably, to

A TU-16 Badger D photographed during the 1976 NATO exercise Operation Teamwork. The TU-16, a medium bomber, has been extensively used for ELINT missions off Alaska and Japan. The aircraft first flew in 1952; the Soviet Air Force operates about 90 ELINT Badgers while the navy has about 40. *U.S. Navy Photo.*

avoid the political consequences of a Russian pilot being shot down over Western Europe). One or more MiG 15s would fly along the West German border, just inside communist airspace, then suddenly turn west. The penetrations went as far as Munich. As the now-alerted Western air defenses tracked the MiGs, Soviet ground stations would monitor reaction time, radar, and radio transmissions and other ELINT data. Since some MiG 15s were equipped with cameras, the flights could also provide aerial coverage of NATO bases. The MiGs would then run for home. If they were intercepted, the pilots would use the MiG 15's superior rate of climb to escape.

There was also an element of harassment. Often the sweeps were timed to interrupt U.S. fighter squadrons' beer call. These probing missions occurred on a weekly basis during the 1950s and continued into the mid-

A TU-95 Bear D photographed while flying off the Virginia coast. The Bear D is the Soviets' main long-range ELINT aircraft. The prototype TU-95 first flew in mid-1954 with deliveries beginning in 1956. Although slower than the Mya-4 jet bomber, the TU-95's superior range caused it to be selected as the Russians' primary strategic bomber. *U.S. Navy Photo.*

1960s. (Between 1962 and 1964, penetrations of Western air space occurred ninety-five times.) [6,7]

Such flights were not devoid of "incidents." On 20 January 1962, a Bulgarian MiG 17, equipped with cameras, crash-landed near a Jupiter IRBM base in Italy. It had made two low passes over the base before clipping a tree. The pilot, Sub-Lt. Milusc Solakov, survived and was taken to a hospital, where he told several conflicting stories. On 1 February 1962, he was charged with espionage. The case did not become another "U-2 affair," however, since a year later he was returned to Bulgaria. [8]

The Soviets were slow to fly long-range ELINT missions. It was not until the end of the 1950s that the first Soviet medium and heavy bombers were converted for this purpose. They were assigned to the Soviet naval air force for both long-range ELINT and ocean-surveillance flights. One series of flights began in early 1959, when a single Soviet aircraft would fly along the Japanese coast. In early 1960, this was expanded to two or three aircraft formations, which would go as far south as Tokyo Bay before turning back. The flights were made on a bimonthly schedule, so regular that they were dubbed "the Tokyo express." [9]

Operations against Alaska were also underway. The aircraft would take off from the Kamchatka Peninsula and fly north along the international date line, passing east of St. Lawrence Island. On 7 September 1960, the U.S. protested these flights. The missions were to probe the coverage of the Distant Early Warning network of air defense radars. By the early 1960s, these patrols off the Alaskan coast became routine. They were not without mishaps, however. On 15 May 1963, two wayward Soviet aircraft penetrated 30 miles (48 kilometers) into Alaskan airspace over Kuskokwim Bay and Etolin Strait; they spent 25 and 19 minutes, respectively, inside U.S. airspace. Although tracked, neither was fired upon.[10] The mainstay of these flights was the TU-16 Badger medium bomber. Specific ELINT versions, the Badger D and F, were soon identified. To operate the equipment, an extra ELINT crewman was carried in the tail. The TU-16's short range limited the area it could cover. For longer range flights, the Soviet navy began to use TU-95 Bears and Mya-4 Bison bombers. The specific ELINT aircraft were the Bear C and D and the Bison B and C. Being large aircraft, they could carry heavy loads of equipment. The Bear D, for instance, has a huge belly radar and up to forty aerial blisters or fairings.[11] The tapes are analyzed by the GRU's Electronic Intelligence Directorate. Commanded by a lieutenant general, it has overall control of all ELINT analysts.

The Soviet Need for ELINT Satellites

This buildup of Soviet ELINT activity also had a space dimension. ELINT satellites met several Russian needs, specifically geographic. Soviet aircraft could not reach the continental U.S. during the 1960s. Ground stations were limited to Cuba. The trawlers and submarines could be substituted for aircraft and even be sent to cover important targets. They could not, however, "see" very far inland (the trawler antennas were only 20 or 30 feet (6 or 9 meters) above the water). U.S. radar and radio signals from the interior were out of reach. The emphasis in the Soviet program may also have been different from that of the U.S. One main target of U.S. ELINT satellites was the Soviet air defense network. The reverse may not have been true. Throughout the 1960s, U.S. air defenses were in decline. Radar and missile sites were being closed and interceptor squadrons phased out. With the increasing number of Soviet ICBMs, the U.S. felt air defenses would be destroyed in an attack. The funding would be better spent on U.S. offensive nuclear forces to deter the Soviets in the first place.

China's air defenses were even more limited; they were based on Soviet equipment provided in the 1950s. Moreover, effectiveness was disrupted, first by the cutoff of Soviet aid in the early 1960s and then by the cultural revolution. Aircraft were grounded and personnel spent most of their time in political lectures. A more productive area would seem to have been monitoring communications.

It is possible the Soviets flew experimental ELINT packages aboard reconnaissance satellites. This would test the equipment and allow them to gain experience.

The first specialized ELINT test satellite was apparently *Cosmos 103*, launched on 28 December 1965, from Tyuratam. The C-1 booster placed it into a 373-mile (600-kilometer) circular orbit inclined 56 degrees. A second satellite with similar orbital characteristics followed a year and a half later. *Cosmos 151* was launched on 24 March 1967, also from Tyuratam. Its 391-mile (630-kilometer) circular orbit was slightly higher than the earlier one. The two satellites were apparently successful, because they were followed on 30 October by the first satellite to use the operational orbit and launch site. *Cosmos 189* went into a 373- by 332-mile (600- by 535-kilometer) orbit, which was more elliptical than the test satellites' orbits. The big difference was the inclination—74 degrees rather than 56 degrees. Also, the launch site was Plesetsk. These differences marked a shift towards operational status. The satellite was a cylinder 6.6 feet (2 meters) long and 3.3 feet (1 meter) in diameter, with paddle-shaped solar panels. It weighed about 1,929 pounds (875 kilograms)—more than U.S. subsatellites—possibly because it carried more extensive equipment than U.S. satellites. Another possibility is Soviet ELINT equipment is not as advanced as that of the U.S.; so for a given role, the Soviet system would be heavier.

The launch of *Cosmos 189* was one part of a major expansion of Soviet military space activity. Up to this point, it had been limited to reconnaissance satellites. Not only did the debut of the ELINT satellite take place in 1967, so did that of the nuclear-powered ocean surveillance and the antisatellite interceptor. There was also an intensive series of test flights of the FOBS (Fractional Orbit Bombardment System).

Operations

Although the ELINT satellite was flying, full operational status took several years to achieve. Three launches were made in 1968— *Cosmos 200* (20 February), *Cosmos 236* (27 August), and *Cosmos 250* (30 October). *Cosmos 236* had orbital characteristics like those of the two test satellites. Possibly, additional work was needed to sort out technical problems (the short life of U.S. tape recorders comes to mind).

Two launches took place in 1969 and another three satellites went up in 1970. All used the operational-type orbit. In 1971, the program underwent an explosive growth. Five of the satellites were put into orbit, which presumably marked the shift from limited to full operational status. It was not without mishap—eighteen fragments were detected in connection with the *Cosmos 436* launch on 7 September 1971. Apparently, an explosion disabled it. The replacement, *Cosmos 437,* was launched only 3 days later on 10 September. Subsequently, three or occasionally four launches per year were made. The standard orbit was 345 by 321 miles (555 by 516 kilometers) inclined 74 degrees, the orbital period was 95.3 minutes, and the ground track repeated every fifteen orbits. Like other Soviet space systems, they have a short operating lifetime. The interval between launches indicates an average of 280 days.

For the first few years, the satellites were not arranged in any type of orbital pattern.[12] Starting in 1972, this changed. Four satellites were placed 45 degrees apart, forming an orbital network. This would allow the maximum coverage, which would be particularly important if the satellites' mission was monitoring communications. Radio messages are transmitted only once. Thus, multiple satellites are needed if the messages are to be picked up on a systematic basis. Radars, on the other hand, transmit constantly; thus, multiple satellites would not be as necessary.[13,14]

At this same time, the Soviets were also flying a heavy ELINT satellite. The launch vehicle was the A-1 booster, which allows a much greater payload than the C-1 booster. Weight estimates of the Soviet heavy ELINT

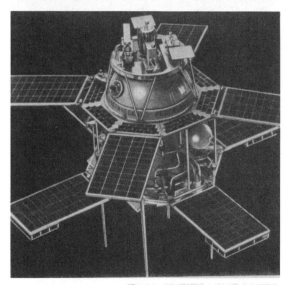

A variety of Cosmos satellites. Presumably these are comparable to the C–1–launched small ELINT satellites. The exact configuration of the satellite is unknown. *Soviet Photo via James E. Oberg.*

are 4,410 pounds (2,000 kilograms) or more. It has been suggested that they use the Meteor weather satellite body and that the ELINT equipment and tape recorders replace the television camera. If this is correct, the satellite is a cylinder about 16.4 feet (5 meters) long and 4.9 feet (1.5 meters) in diameter, with two solar panels.[15]

The first launch was *Cosmos 389* on 18 December 1970. It went into a 434- by 407-mile (699- by 655-kilometer) orbit inclined 81 degrees. Western space analysts did not immediately realize it was an ELINT satellite, because its orbit was identical to those used by the Meteor weather satellites. *Cosmos 389* was, therefore, thought to be a failed weather satellite, the Cosmos cover name being used to hide the malfunction. As time passed and one or two such launches per year were made, this explanation became untenable. Also starting in 1972, Meteor satellites started to go into roughly circular 559-mile- (900-kilometer-) high orbits. The Cosmos satellites continued to use the old lower orbit. Eventually, it was realized they were heavy ELINT satellites.[16] The flights continued sporadically through the mid-1970s—one in 1971, two in 1972, one each in 1973 and 1974, two in 1975, and none in 1976. One, and later two, of these satellites would be operating at a time.

It is tempting to draw parallels between the Soviet ELINT program in the 1970s and the U.S. effort during the 1960s. Both used two separate types of satellites, one smaller than the other. Presumably, there was some division of operations between them. With a larger payload, the heavy ELINT satellites could undertake detailed surveys. The small number implies their object was monitoring radars rather than communications.

This activity in space was part of a wide-ranging buildup in Soviet ELINT activities. On 18 April 1970, the year of the first heavy

Photo of a Meteor 2 weather satellite. It is believed the Soviet heavy ELINT satellite uses a similar design outfitted with radio and radar receivers. Given the satellite's large size, it can carry an extensive range of equipment and make a wide variety of measurements for each signal. *Soviet Photos via James E. Oberg.*

ELINT launch, the Soviets made their first ELINT flights off the U.S. East Coast to Cuba. This is the longest of any such Soviet mission. The TU-95 takes off from northwest Russia and flies around the North Cape of Norway. The aircraft then heads out over the Atlantic, passing between Iceland and Ireland. It can monitor NATO air defense radars, military and civilian radio communications, and ship positions on the crowded North Atlantic sea lanes. Several hours later, the TU-95 flies along the East Coast of the U.S., repeating its ELINT activities, and finally, lands in Cuba. The flight is about 5,000 miles (8,047 kilometers) long and takes about 13 hours. For such long flights, the TU-95 can carry two complete eight-man crews. TU-95s were permanently stationed in Cuba starting in September 1972.

The Soviet trawler fleet was also expanding, reaching fifty-four ships by the mid-1970s. No longer were these small, converted fishing boats. Some were large, specially designed ships specifically tailored for ELINT.[17]

The Soviet ELINT ground station near Havana, Cuba, was also enlarged by 60 percent, making it the most extensive facility outside Russia. The Cuban station, along with others inside Russia, allowed complete coverage of the transmissions of all U.S. geosynchronous communications satellites.[18]

A Comic Interlude

ELINT is a very serious subject. At times, however, even it has its lighter, even bizarre, aspects. After a 1-year layoff, three heavy ELINT satellites were launched in 1977— *Cosmos 895* (27 February), *Cosmos 925* (7 July), and *Cosmos 955* (20 September). The last of these caused the Soviets several years of acute embarrassment. *Cosmos 955* lifted off from Plesetsk at 3:58 A.M. local time. As the A-1 booster climbed, it left the earth's shadow. The exhaust plume from the twenty engines was backlit by the sun. From the ground, which was still in darkness, the now-illuminated exhaust plume looked like a huge, jellyfish-shaped apparition slowly moving through the night sky. The weather over northern Russia was clear, so early risers in Leningrad, Estonia, and as far west as Finland saw the plume (Moscow was clouded over). Nikolay Milov, a local Tass correspondent in the city of Petrozavodsk, wrote up a brief account of an "unusual natural phenomenon observed in Karelia." Two days later, the report was sent out over the international wires. Soon, American newspapers were filled with accounts of the "jellyfish UFO" seen over Russia. To Soviet space program expert James Oberg, however, it sounded like a rocket launch. A check with the Goddard Spaceflight Center confirmed *Cosmos 955* was launched at the same time the jellyfish UFO was seen.

Oberg's solution was widely publicized in such diverse sources as the *International UFO Reporter,* the *Christian Science Monitor, Omni* magazine, and the *Reader's Digest.* However, this information was not available inside Russia, where UFOs and other "occult" subjects have wide popular interest. Russian UFO buffs began collecting second- and third-hand accounts of the events of that exciting morning. Such accounts being notoriously unreliable, it is not surprising that they soon had a fine collection of wild stories, rumors, and fantasies—all wrong. One account stated that rays from the UFO had drilled holes in pavement, stones, and windows. The holes were not seen, of course; the KGB had taken them away. Some people, other stories went, seemed to become ill and mentally confused because of the "UFO." As the rumors were

passed from one UFO buff to the next, the events became so distorted as to be unrecognizable. The mundane rocket launching soon became a huge object seen for 2 hours at an altitude of 60 miles (97 kilometers). Small UFOs were seen to dart in and out of the huge mother ship.

As the stories spread, the Soviet government became concerned. It was hostile to UFOs, calling them "inventions" of Western "yellow journalism." The irony was that the Soviet government had the explanation but could not use it; Plesetsk and its military launches were state secrets. Despite Plesetsk's discovery in 1966 by the Kettering group, the Soviets had never publicly acknowledged its existence. To both cover up Plesetsk and explain away the UFO, the Soviets trotted out a number of lame explanations. In October 1977, the sighting was explained as a reentering satellite. When this didn't work, in August 1978, M. Dmitriyev, a doctor of chemical sciences, called it a "chemoluminescence zone"—industrial smog, which glowed due to solar radiation. Oberg dubbed it "Swampsky gas" after an Air Force explanation of a series of UFO sightings during 1966 in Michigan. In the spirit of try, try again, Dr. V. V. Migulin in January 1979 referred to it as "physical changes in the upper atmosphere" similar to the northern lights due to solar radiation.

The Soviet dilemma was made worse on the night of 14 June 1980, when the launch of *Cosmos 1188* (an early-warning satellite) was seen from central Russia, including Moscow, Gorkiy, and Kazan. On 15 May 1981, it happened again when a *Meteor 2* weather satellite, launched from Plesetsk, was seen from Moscow. The Soviet government was between a rock and a hard place. Their bogus explanations could not stop the UFO stories, while in the West, where Plesetsk was common knowledge, space analysts were enjoying their discomfort. Finally, the Soviets surrendered to the UFOs. The 20 June 1983 issue of *Pravda* carried an article entitled "Place of lift off—Plesetsk." It began, "To the legend of 'flying saucers' (UFO) the Plesetsk cosmodrome makes a great contribution." The article contained several accounts of how rocket launches had sparked UFO reports. The article did not mention that most of Plesetsk's launches were military.[19,20] Because of his efforts to publicize the incident and expose the Soviets' attempts to cover it up, James Oberg deserves full credit for forcing the Soviets to (finally) acknowledge Plesetsk. Looking back on the entire incident, one can't help wondering if the ELINT data from *Cosmos 955* was really worth all the trouble it caused.

Changes in the ELINT Program

The embarrassment of *Cosmos 955* coincided with several changes in the Soviet ELINT effort. In 1977, three of the C-1–launched ELINT satellites went into orbit, which was the pattern of the past 5 years. In 1978, however, only two went up—*Cosmos 1008* (17 May) and *Cosmos 1062* (15 December). The next year only one launch was made—*Cosmos 1114* (11 July); it was the last such mission. As the C-1–launched satellites were phased out, the launch rate of the heavy ELINT satellites went up—four in 1978, five in 1979, and four in 1980.[21] They were placed into an orbital network—first, three satellites 60 degrees apart and then, starting in 1977, six satellites positioned 60 degrees apart. Use of a network arrangement implied their role had expanded to include both radar and communications. The use of a network allowed the Soviets to cover their targets and also helped Western analysts to determine the lifetimes of the satellites. When one satellite failed or became degraded, a replacement would be launched into the same orbital plane. This could be easily determined and is a reliable indicator. Satellites launched since 1974 showed an average lifetime of 506 days,[22] which implies the full network is operational at one time. (The same probably could not be said of the C-1 ELINT satellites; their lifetimes were too short.) By the early 1980s, the launch rate of the A-1 heavy ELINT satellites had stabilized at three or four per year.

The A-1 booster is basically the SS-6 ICBM designed in the mid-1950s, with an upper stage developed in the late 1950s. Its technology is old and it uses cryogenic fuel. Clearly, a more modern booster would have better reliability, a greater payload, and more flexibility. The debut of just such a new Soviet booster—the F-2—came in the late 1970s. It was to take over some of the missions of the A-1, including launch of the Meteor weather satellite and the heavy ELINT payload. The beginnings of the F-2 ELINT program are confusing. The first candidate was *Cosmos 1025* on 28 June 1978, from Plesetsk. Its orbital characteristics, 423 by 403 miles (680 by 649 kilometers), inclined 82.5 degrees, were very close to those of the A-1 ELINT satellites (typically 410 by 390 miles (660 by 627 kilometers), inclined 81 degrees). The next F-2 launch with similar characteristics was *Cosmos 1076* (12 February 1979), which went into a 421- by 402-mile (678- by 647-kilometer) orbit, inclined 82.5 degrees. The Soviets, however, called it an oceanographic satellite carrying visible light sensors. A year later, *Cosmos 1151* went into a virtually identical orbit. The satellite was described as an improved version that covered a spectrum wider than that of *Cosmos 1076.*[23]

Despite the Soviets' description, these may have been ELINT test satellites. In any event, the operational ELINT satellite began flights in 1981. *Cosmos 1300* (24 August) went into a 413- by 395-mile (664- by 636-kilometer) orbit. On 3 December, it was followed by *Cosmos 1328,* which went into an identical orbit. It was positioned 44 degrees east of *Cosmos 1300,* implying the Soviets were starting to build a network. Two more F-2 ELINT satellites followed in 1982—*Cosmos 1378* (10 June) and *Cosmos 1408* (16 September). Both satellites were positioned 90 degrees ahead of *Cosmos 1328. Cosmos 1378* and *Cosmos 1408* were either working together, or *Cosmos 1408* was a replacement for the failure of the earlier satellite.[24]

The variation in orbital characteristics between the different satellites is very narrow. The high points of the first six operational satellites differ by only 4 miles (7 kilometers), the low point by only 1 mile (2 kilometers)—a testimony to the F-2's guidance system.

The launch rate increased to four F-2 satellites in 1983. One of these, *Cosmos 1500* (28 September), was described by the Soviets as an oceanographic satellite. Its orbit, however, was identical to those of the earlier unidentified satellites. Western analysts, despite the Soviet claims, have decided that these are really heavy ELINT satellites, which is backed up by the fact that the F-2 satellites are positioned in a pattern with the A-1 ELINT satellites.[25]

Far below, on the seas and in the skies of Earth, the Soviets have also been busy. On 27 October 1981, the world got a close look at Soviet ELINT activities when a Whiskey-class submarine ran aground inside Swedish territorial waters. The Soviets blamed "navigation error due to faulty radar and bad weather."[26] Since the submarine was only 9 miles from the Karlskrona Naval Base, nobody believed this. To get to the point where it ran aground would require careful navigation through a maze of small islands and narrow channels. The incident took a sinister turn when radiation was detected coming from the torpedo tubes. The Swedes believed it was coming from uranium 238 shielding material inside nuclear warheads. The news the submarine was carrying nuclear torpedoes further soured Swedish-Soviet relations and made a mockery of Soviet calls for making the Baltic a nuclear-free zone—a "sea of peace." After a 10-day standoff, the submarine was released to sail towards home and the questions that awaited its commander and crew.[27,28]

That year, the nature of the ELINT flights to and from Cuba changed. Until this point, the effort was sporadic. In 1981, the Soviets began to permanently base the aircraft at the San Antonio de los Banos airfield.[29] A squadron of up to twelve TU-95 aircraft could be in Cuba at any one time.[30]

In the Far East, Soviet airborne ELINT activities had also increased. In 1979, Vietnam, in exchange for increased aid needed to support its war in Cambodia and along the Chinese border, granted the Soviets access to the former U.S. naval base at Cam Ranh Bay.[31] TU-16 and TU-95 aircraft, flying from the Soviet Far East to Cam Ranh Bay, can collect ELINT data from Japan, South Korea, China, and Taiwan. These flights can also be used to intimidate. Between 1967 and 1983, Soviet aircraft violated Japanese airspace about fifteen times. In November 1984, they made an unprecedented show of force. On 12 November, seven TU-16s and two TU-95s were detected flying southward. When the formation reached Kyushu, the southernmost main island, four of the TU-16s turned back northward. One of them entered Japan's 12-mile territorial airspace. Despite signals from Japanese interceptors, the aircraft remained inside for about 2 minutes. Since the Japanese pilots were under standing orders not to fire unless attacked first, it escaped unharmed. The other planes continued southward.[32] Eleven days later, the incident was repeated. Two TU-95s entered Japanese airspace, again off Kyushu. Despite repeated warnings from both interceptors and ground stations, the aircraft spent about 3 minutes inside Japanese airspace, finally leaving on a course that took them to Vietnam. Nearby were five TU-16s, which remained outside. Both incidents occurred in daylight.[33] Such unusually large formations were sure to generate considerable radar and communications activity, which would increase when the intruder aircraft violated Japan's airspace.

Coinciding with the upgrading of Soviet ELINT satellites was similar upgrading to their ELINT aircraft. A new version of the TU-16—the Badger K—made its debut.[34] More important were two entirely new ELINT

The IL-20 Coot A, one of the Soviets' latest generation of strategic ELINT aircraft. Converted from the IL-18 airliner, it has a crew of five plus 20 systems operators. The large cylinder under the fuselage is a side-looking radar. Two other cylinders are mounted on either side of the forward fuselage. The IL-20 first appeared in 1978 and is believed to be operated by both the Soviet Air Force and Navy. Some 20 to 25 have been built. It has an endurance of about 12 hours and has been observed off Great Britain and over the Baltic. *U.S. Navy Photo.*

aircraft. One was the IL-20 Coot-A, a turbo-prop airliner fitted with two large antennas atop the fuselage; a long blister on either side; and a large, canoe-shaped, side-looking radar underneath the aircraft. The other was the An-12 Cub-D, a turboprop military cargo airplane. It carried five large blisters and various other antennas. The large interiors and cargo capacities of both aircraft would allow them to carry a heavy load of receivers, monitoring equipment, and crewman. As a sour postscript to the shooting down of Korean Airlines' flight 007 in August 1983, several An-12s have been photographed painted with civilian registration numbers (CCCP-11417 and 11916).[35]

The Saga of Cosmos 1603

The most important shift in Soviet ELINT activities came in space. On 28 September 1984, the Soviets launched *Cosmos 1603*. This was the first flight of a new, very heavy ELINT satellite, one more capable than any previous Soviet design. It is the largest single military satellite the Soviets have put up. The launch sparked an intensive effort on the part of NORAD due to the satellite's peculiar behavior. In the words of Col. William E. McGarrity, NORAD director of space operations, "it did not fit anything that had happened before."

The first unusual aspect was the launch vehicle—the D-1 Proton booster. *Cosmos 1603* was the first military satellite it had launched (other than the military Salyut space station in the 1970s). The D-1 can put 50,000 pounds (22,680 kilograms) into low earth orbit. *Cosmos 1603* was placed initially into a 118- by 112-mile (190- by 180-kilometer) orbit, inclined 51.6 degrees. The satellite and attached upper stage then maneuvered, chang-

ing not only their altitude but also their inclination. *Cosmos 1603* went into a nearly circular 528-mile (850-kilometer) orbit, inclined 66.6 degrees. Soon after, it maneuvered again, raising the orbit only slightly to 530 miles (853 kilometers) but changing its inclination to 71 degrees. This was a 19.4 degree change in all. The maneuvers were difficult for the U.S. tracking system to follow. When each orbital change was made, the U.S. had no information about the intended orbit. *Cosmos 1603* was "lost" two or three times. As it maneuvered, a special team at NORAD tried to determine what it would do next, as well as to determine where the tracking systems should look for it. After 1 or 2 hours, the satellite would be picked up again. The main problem was the seeming irrationality of its behavior. Again quoting Colonel McGarrity, "The thing that mystified us most about it was there is no logical reason for putting something into orbit the way they did in this particular case. There is a great deal of speculation on why it was done that way." This is understandable. A change in inclination requires a lot of fuel. Because of the large amount of energy needed, about two-thirds of the payload would have to be fuel (the exact amount would depend upon the satellite's weight, efficiency of the engine(s), weight of the empty upper stage, etc.). An altitude change requires only that the satellite be speeded up or slowed down, which uses only a small amount of fuel. Changing inclination means not only a change in speed but also a change in the direction of the velocity. This involves adding a significant portion of the original orbital velocity to the satellite. When the upper stage fired, it was at an angle to the orbital path. The profile also has the disadvantage of added complexity. The multiple burns require precision, and upper-stage failures are a dismally regular feature of the Soviet space effort. A launch directly into a 71 degree orbit would seem to be more efficient all around.

Cosmos 1603's unusual profile raised questions about its mission. At first, it was suspected it might be an antisatellite interceptor, but this possibility was quickly eliminated. After reaching its final orbit, *Cosmos 1603*'s ground track was stabilized. The amount of drift was limited to only 0.1 degree per day, which allowed coverage of the same area at the same time each day. *Cosmos 1603* makes four passes over the U.S. each day. Because of its 530-mile (853-kilometer) altitude, it can pick up signals from the same targets on consecutive passes.[36]

The very heavy ELINT satellite program underwent expansion during 1985 with three launches. The first was *Cosmos 1656* (30 May). After completing the two-step maneuver, the satellite was placed into an orbit 45 degrees away from *Cosmos 1603*.[37] Next was *Cosmos 1697* (22 October). Although its characteristics were identical to the first two satellites, ground observers found it visually fainter.[38]

The years' operations were concluded with *Cosmos 1714* (28 December). This time, the maneuver failed (it was alleged the satellite remained attached to the upper stage). Air drag lowered its orbit until it reentered on 27 February 1986. *Cosmos 1714*'s path carried it across Australia, over the North Pole, and across the northeastern corner of the U.S., ending over the Atlantic Ocean.[39,40]

The large size of *Cosmos 1603* and those that followed it indicates they carried a more comprehensive package of ELINT equipment than the older A-1– and F-2–launched satellites. It also indicates that the Soviets, despite the expansion in their ELINT capabilities in the past quarter century, still feel the need to collect even more communications and radar data.

NOTES Chapter 12

1. Lt. Col. Richard E. Fitts, *The Strategy of Electromagnetic Conflict,* (Los Altos: Peninsula Publishing, 1980), 60.
2. Research Publications, Department of State 2726, 1983.
3. Col. William V. Kennedy, *Intelligence Warfare,* (New York: Crescent Books, 1983), 173.
4. John M. Carroll, *Secrets of Electronic Espionage,* (New York: E. P. Dutton, 1966), 139–141.
5. P. H. Vigor, et al, *The Soviet War Machine,* (London: Salamander Books, 1976), 48.
6. General Chuck Yeager and Leo Janos, *Yeager,* (New York: Bantam, 1985), 235,240.
7. "The 120 Mile Error", *Time,* (20 March 1964): 32.
8. *New York Times,* 21 January 1962, 1. 23 January 1962, 7. 4 January 1963, 2.
9. "Soviet Bombers Reconnoiter Japan", *Aviation Week & Space Technology,* (9 May 1960): 36.
10. Carroll, *Secrets of Electronic Espionage,* 139.
11. Bill Gunston, *An Illustrated Guide to the Modern Soviet Air Force,* (New York: Arco, 1982), 116–119, 120–123.
12. G. E. Perry, "Cosmos at 74°," *Flight International,* (30 November 1972): 788–790.
13. Phillip S. Clark, "The Skean Programme," *Spaceflight,* (August 1978): 298–304.
14. *Soviet Space Programs 1971–1975,* (Library of Congress, 1976), 406.
15. Satellite Digest 165, *Spaceflight,* (July/August 1983): 334.
16. *Soviet Space Programs 1971–1975,* 434.
17. P. H. Vigor et al, *The Soviet War Machine.*
18. *Soviet Military Power 1984,* (Department of Defense, 1984), 126.
19. "Counterfeit UFO Flap Forces Moscow to Reveal Space Secret," (News Release), 13 July 1983. Committee for the Scientific Investigation of Claims of the Paranormal.
20. James E. Oberg, *UFOs & Outer Space Mysteries,* (Norfolk: Donning, 1982), Chapter 9.
21. *Soviet Space Programs 1976–1980 Part 1,* (Library of Congress, 1982), 436–445.
22. Nicholas L. Johnson, "Estimated Operational Lifetimes of Several Soviet Satellite Classes," *Journal of the British Interplanetary Society,* (July 1981): 282, 283.
23. *Soviet Space Programs 1976–1980 Part 1,* 439, 442.
24. Phillip S. Clark, "The Soviet Space Year of 1982," *Journal of the British Interplanetary Society,* (June 1983): 259.
25. Nicholas L. Johnson, "The Soviet Year in Space: 1983 Part 3," *Space World,* (November 1984): 29, 30.
26. *Los Angeles Times,* 28 October 1981, 1.
27. *Los Angeles Times,* 5 November 1981, 1.
28. *Los Angeles Times,* 8 November 1981, 1.
29. *Soviet Military Power 1984,* (Department of Defense, 1984), 126.
30. Edgar Ulsamer, "In Focus . . .", *Air Force Magazine,* (November 1984): 32.
31. *Soviet Military Power 1983,* (Department of Defense, 1983), 97.
32. *San Diego Union,* 13 November 1984, A-9.
33. *San Diego Union,* 23 November 1984, A-3.
34. Gunston, *An Illustrated Guide to the Modern Soviet Air Force,* 118.
35. "SovAir: The New Wave," *Air International,* (September 1984): 140.

36. ''Soviets Orbit Large New Military Electronic Intelligence Satellite,'' *Aviation Week & Space Technology,* (14 January 1985): 19, 20.
37. Satellite Digest 186, *Spaceflight,* (November 1985): 428.
38. ''News Digest,'' *Aviation Week & Space Technology,* (28 October 1985): 28.
39. *Los Angeles Times,* 24 February 1986, 2.
40. *San Diego Union,* 28 February 1986, A-17.

CHAPTER 13

Manned Orbiting Laboratory

A human being has certain unique advantages not found in automated systems. A human is intelligent, he can reason, he can improvise to meet an unexpected situation. Separate factors can be combined to find patterns. With a crew on board, a military spacecraft would gain operational flexibility. No longer would the satellite be limited to its preprogrammed instructions—blindly photographing a cloud-covered landscape.

The idea of manned military spacecraft is an old one. In the early 1920s, Hermann Oberth suggested that a space station would be valuable in monitoring a war in areas of low population. It is also worth remembering that early space theorists, such as Oberth, thought almost exclusively in terms of manned spacecraft. The electronics needed for an automated satellite simply did not exist until after World War II. Oberth believed the station would communicate with the earth via flashes of light. Radio was too crude in the 1920s to be used.[1]

It was not until the late 1940s and early 1950s that spaceflight became a real possibility. In the early 1950s, *Collier's* magazine ran a series of articles on the coming space age. They depicted a large, wheel-shaped space station. One justification for the station, the magazine offered, was its ability to undertake military reconnaissance. The station would control a telescope with a 100-inch (254-centimeter) mirror, orbiting nearby. From 1,000 miles (1,609 kilometers) up, the telescope would have a resolution of 16 inches (41 centimeters). The reconnaissance officer in the station would use pictures from a television sighting scope to pick out a target. The telescope would then automatically lock on and begin taking pictures. The television image would allow the reconnaissance officer to keep the telescope centered on the area of interest. During each 2-hour orbit, 100 photos could be taken. The editor of the series, Cornelius Ryan, called the station "man's guardian in the sky." He wrote, "No nation could undertake preparations for war without the certain knowledge that its massing troops were being observed by the ever watchful eyes aboard the 'sentinel in space'. It would be the end of Iron Curtains wherever they might be."[2]

The political support was lacking, however; nobody was willing to put up the billions needed to build the station. Even such limited programs as Vanguard had to get by on minimal funding. Then, literally in one night, all that changed. With the launch of *Sputnik I*, the political priorities and funding were now within reach. Since then, reality has surpassed the hopes of the early visionaries—with one exception. After 25 years of manned spaceflight, the U.S. has yet to undertake manned photo reconnaissance in space. The reasons are tied up in a project few have heard of—the Manned Orbiting Laboratory (MOL) * — an attempt to build a military space station. The obscurity surrounding MOL comes from two factors—first, the secrecy surrounding its mission; and second, the fact that it was cancelled before the first flight. To this extent, MOL highlights the political factors that shape the fate of military space activities.

Blue Gemini

The history of MOL spans nearly a decade. The first studies began in early 1960. One of these studies, called SMART (Space Maintenance And Resupply Techniques), envisioned manned spacecraft regularly servicing unmanned satellites. This would extend their operational lifetime. Small contracts were issued. An emphasis developed around orbital rendezvous. This emerged in early 1961 as the MMSCV (Manned Military Space Capability Vehicle). These studies centered on use of a winged spacecraft for reentry but, more importantly, on justifying a manned as opposed to an automated system.[3]

At the same time, NASA was making studies of an advanced version of the Mercury spacecraft. It would have a crew of two and could maneuver in orbit as well as rendezvous and dock with another spacecraft. It could also stay in orbit for a prolonged period—up to 2 weeks. On 7 December 1961, approval of development was announced in Houston. This new spacecraft was subsequently named Gemini. Its advanced capabilities matched those envisioned by the earlier studies. The idea of using it as a military spacecraft first surfaced in February 1962. In June, with studies by the Air Force Space Systems Division, the concept became firmer of using Gemini hardware to test rendezvous, docking, and transfer. This was to be the first step in a program called MODS (Manned Orbital Development System). The name was carefully selected; it was to *demonstrate* the capabilities of men in space rather than being an operational system. MODS itself was a space station with a crew of four or more plus equipment, a reentry capsule based on either the Gemini or Apollo, and a supply module that would carry cargo and provide propulsion. In August 1962, the idea was further expanded with a proposal to fly six Gemini missions with Air Force pilots for preliminary orientation and training. The program was given the name "Blue Gemini."

The idea raised political difficulties. Some in the Air Force wanted a larger role in the Gemini program. Others feared that any involvement with Gemini would endanger plans to build a one-man space glider called Dyna-Soar. In the background were civilian officials who doubted man had any role in military space activities. NASA, for its part, supported Air Force involvement in Gemini. It had been designed as an operational system and the Air Force was the obvious user. Also, the program could use additional funding.

NASA administrator James Webb and associate administrator Robert Seamans, Jr., went to the Pentagon to talk with Deputy Secretary of Defense Roswell L. Gilpatric to argue for a bigger Air Force role. Defense Sec-

---

* Pronounced as either "mole" or "mall."

retary McNamara happened to drop in. He liked the presentation—so much, in fact, that he suggested the Air Force and NASA efforts be combined and the entire package be moved to the Department of Defense. Both NASA and the Air Force were taken aback. NASA opposed the idea because of the interference with Apollo; the Air Force didn't want the competition with DynaSoar, and felt the costs were too high for the limited data provided. By January 1963, the idea was dropped. One fallout, though, was an agreement to fly Air Force experiments aboard NASA Gemini missions.[4]

While this was going on, the MODS efforts were broadened and detailed studies were made to provide the cost and technical information needed to begin full-scale development.[5]

MOL

Even as MODS was expanding, the DynaSoar was in its final stages. Over the course of the program, it had suffered several reorientations, delays, changes in launch vehicles, and cost increases. Moreover, it now appeared that subscale models could provide the same reentry data as DynaSoar but at lower costs. The kind of advanced missions Gemini could do would require major modifications to DynaSoar. Finally, Gemini would be flying about 2 years before DynaSoar ever could. Although a winged spacecraft was seen as an important future activity, ballistic capsules such as Gemini were satisfactory for the present.

Accordingly, nobody held much confidence in DynaSoar's future. It, therefore, came as no surprise when on 10 December 1963, Defense Secretary McNamara announced DynaSoar's cancellation. In its place, the Air Force was to begin work on the Manned Orbiting Laboratory. In making the announcement, McNamara also set the configuration. The MOL vehicle was to be a modified NASA Gemini (called a Gemini B) attached to a trailer-sized laboratory module. It was to be launched as a single unit by a Titan III C to allow the maximum use of existing hardware. Only the laboratory was new. This would also reduce the technical risks. An example of this was launching the Gemini and Laboratory as a single unit. If they were launched separately, the Gemini would have to rendezvous. In late 1963, this was an unknown and was seen as best avoided.

In this way, man's military usefulness in space could be determined in the most cost-effective manner possible.[6,7] The words "cost effectiveness" were the key ones. The announcement seemed to indicate the Air Force was to begin building MOL. In practical terms, it simply meant that the Air Force could continue its studies. The Air Force would still have to justify its development, and that was going to be a difficult task. McNamara and those around him had a skeptical "show me" attitude about military space activity.[8] The Air Force realized early that unless they could put together a program that would show the worth of man in space, MOL would not be approved.

One very basic problem was the lack of knowledge about man's ability to function in space. At the time of the December 1963 press conference, the U.S. had only 54 hours of orbital experience. There were some indications, however, on the *Mercury 9* flight. Astronaut Gordon Cooper had seen houses, trucks, ships, and other small objects from orbit.[9] Balancing this were doubts that men could survive prolonged weightlessness. It was feared the bones of the crewmen would turn soft.

From another point of view, there was also the matter of cost. Man could think and improvise, but was his presence in space worth the added expense? MOL would cost $1.5 billion, of which $500 million was needed to give it a human crew. The Air Force had to justify a 30 percent cost increase over an equivalent unmanned system. The skills of the crew would have to provide a 30 percent or greater increase in capability. If they didn't, the MOL would not be cost effective and it would never see the light of day. This may have been "good management," but it put the Air Force in a bizarre position. It had to prove man had a military role in space before it could test man's military role in space.

In its justification, the Air Force said that before man's contribution could be determined, it was necessary to determine what military missions could best be done from space and if this contributed to the security of the U.S. The MOL could also show if So-

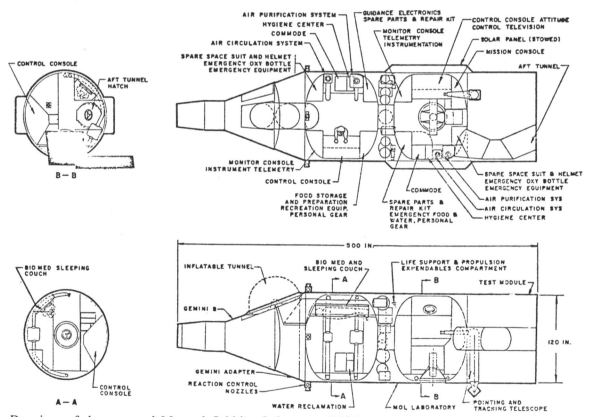

Drawings of the proposed Manned Orbiting Laboratory (MOL) configurations: *Above:* This used two separate pressurized compartments and an inflatable tunnel to connect the lab with the Gemini B. *Opposite page:* This has a single laboratory module.

The final design used a single pressurized compartment divided into two areas by a wall. The rear tunnel was not used. The MOL grew considerably in size and weight between these 1964 drawings and the final design in 1967–68. *U.S. Air Force Diagrams.*

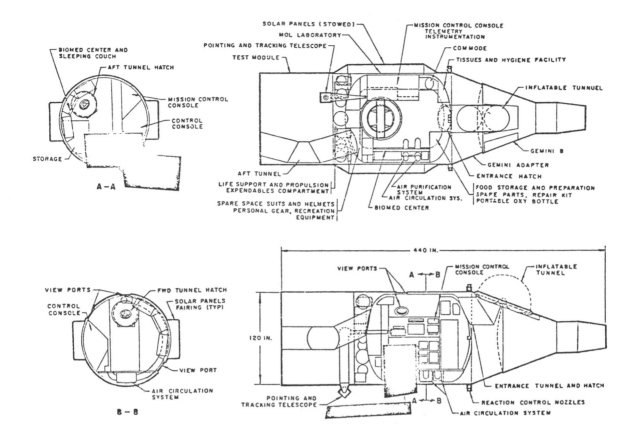

viet manned activities would pose a threat. The specific tests undertaken would be aimed at measuring the value of a crew as part of the system ("in the loop" is how engineering vernacular described it). MOL, in short, was to be an experiment, which was reflected in the flight plan; MOL would be launched from Cape Kennedy. Its orbit would range from 125 to 250 nautical miles (232 to 463 kilometers). Its orbital inclination would be less than 36 degrees and so would never pass over Russia. In fact, there were no plans to take photos of reconnaissance quality.[10]

The Air Force's initial proposals in January 1964 did not go over well, and revisions were quickly begun. Such paper studies were not the only activity, however. The Air Force looked at various design questions during 1964. One, in particular, was the best way for the crew to transfer from Gemini back to the space lab. The early choice was a hatch cut into the Gemini's heat shield. In fact, this was one of the first features selected. There were concerns, however, about how well this arrangement could survive reentry. The hatch would have to take the full brunt of reentry. Any flaw could be fatal. Accordingly, the Air Force looked at three alternatives—an inflatable tunnel from one of the Gemini's crew hatches back to the lab, having the crew space walk, and rotating the entire Gemini on a large hinge to make a direct connection.[11] In the end, the heat-shield hatch remained the preferred method.

Most importantly, the Air Force began looking for a contractor in February 1964. Douglas Aircraft, Martin Marietta, and General Electric were awarded study contracts. This was followed in May by several small studies. Martin looked at what role the Titan III C would play, while Hamilton Standard and Garrett Corporation studied the life-support system. Finally, on 24 July, McDonnell Aircraft was awarded a $1,189,500 fixed-price contract for Gemini B studies.[12]

Politics was also making itself felt: 1964 was a presidential election year. When the contractors objected to the slow pace, they were told not to expect any change until after the election. The Johnson administration was attempting to project an image of economy to the voters.[13] Another potential reason was the Republican platform. It called for a reorganization of the U.S. manned effort in space, away from the moon towards military activities in low earth orbit. If Johnson pressed for MOL, it might be seen as confirming Goldwater's position.

A more serious political objection was one of duplication. At this time, NASA was looking beyond Apollo to proposals for advanced earth orbital missions, called Apollo X. This raised the possibility the U.S. would have two separate space station programs—one civilian, the other military—a seeming duplication. This concern had appeared early. On 19 December 1963, NASA headquarters stated that MOL was designed to fulfill short-term military goals rather than the long-term objectives of a national space station program. Ten months later, NASA associate administrator Seamans stated that Apollo X would not receive hardware funding until the 1970s. This was much later than MOL. Also, MOL would

not meet the NASA requirements for scientific data.[14]

Others were not yet convinced. On 9 November 1964, Senator Clinton P. Anderson (D-New Mexico), the chairman of the Committee on Aeronautics and Space Sciences, sent a letter to President Johnson. He recommended that MOL and Apollo X be merged into one program, which he said would save $1 billion over 5 years. He noted that MOL could not grow beyond a 30-day mission without development of a resupply system.[15] To ease such concerns, NASA and Air Force representatives met and agreed to coordinate their activities and share hardware and equipment. Senator Anderson conceded that this went a long way toward meeting his objections.[16]

Approval

As 1964 was ending, the future status of MOL was unclear. For the past year, the Air Force had presented various proposals. Each time, McNamara had rejected them as not proving the need for military men in space. Gradually, the Air Force reoriented MOL, going from test project towards one more akin to an operational system.

Two experiments were added—P-14, the in-orbit assembly of a large radar antenna; and P-15, large orbital optics. The latter would be a reconnaissance camera operated by the crew.[17] MOL had become an intelligence system. This change in emphasis was the turning point. Though there still remained hardware and other questions, the fiscal year 1966 budget included $150 million for MOL. This would be withheld until the remaining questions were settled.[18]

In January 1965, the White House ordered reports from both the Air Force and NASA; also, contractors were requested to submit proposals. These would be used to determine whether MOL or Apollo hardware would best meet the military requirement. There were three options for the president—approving MOL, combining MOL and Apollo X, or cancelling MOL. The decision on whether Apollo or MOL hardware would be selected was not expected until May or June 1965.[19] On 13 and 14 May, the four possible contractors—Douglas, Boeing, General Electric, and Lockheed (North American had been eliminated earlier)—briefed the Air Force selection committee on their proposals.[20]

Congress was becoming more supportive. On 2 June, the military operations subcommittee of the House Committee on Military Operations urged MOL to begin full-scale development without further delay. This was coupled with a request that duplication be avoided whenever possible. A few weeks later, Representative John W. Wydler (R-New York) went further. In a letter to the *New York Times,* he urged that MOL be speeded up by a year, the Gemini Bs be ordered at once, DOD be given responsibility for all low earth orbit missions, and the U.S. Air Force be reorganized into the U.S. Aerospace Force.[21]

Attitudes within the administration were changing. The reasons were two-fold. The first was MOL's usefulness for intelligence gathering, which was described as convincing a large segment of the administration. MOL's ability to photograph and provide ELINT data on Soviet ABM defenses, smaller, solid-fuel ICBMs, and Chinese efforts to develop a ballistic missile force was considered critical. Such data would enable the U.S. to make more rational decisions on defense policies just as early reconnaissance satellites had removed concern over the missile gap.[22]

The clincher, however, was the beneficial effects MOL would have on arms control. One

Artist's conception of a MOL/Titan III C launch. The two solid-fuel strap-ons have separated, while the first-stage engines continue to burn. The original development plan envisioned the heat-shield test flight, two unmanned MOL orbital flights, and five manned missions. Only the 1966 heat-shield flight was made. The MOL would have made a number of firsts—the first manned military mission, first manned booster to use solid fuel, first manned polar orbit mission, and the first manned launch from the West Coast. *U.S. Air Force Photo.*

official was quoted as saying, "MOL will be the greatest boon to arms control yet." The union of human judgement and reconnaissance satellite technology could result in true open skies, which in time could lead to a formal agreement for inspection from space. Even without such a treaty, MOL could strengthen the U.S. deterrent by allowing a better understanding of Soviet nuclear forces. One who was swayed by this possibility was Vice President Hubert H. Humphrey—long an advocate of disarmament. On 9 July, as chairman of the National Aeronautics and Space Council, he presided over a meeting on MOL. Humphrey was concerned about the implications of manned military missions on the world's perception of the U.S. space program. He asked about twenty-five questions. The answers convinced him that MOL's value for arms control outweighed any political drawbacks.[23] At the same meeting, both McNamara and NASA administrator Webb appeared—again stressing that MOL and Apollo X were not duplications.

The meeting was not the final step, however. According to one report, a final delay was caused by DOD not briefing the State Department. As a result, State began to express concern about MOL's international implications. Because of this, the final decision was not expected until mid-August.[24]

Finally, on 25 August 1965, President Johnson announced the formal go-ahead for MOL. Douglas was to build the laboratory, McDonnell the Gemini B, and General Electric was to manage the experiment package. MOL was to be 54 feet (16.8 meters) long, of which the laboratory would make up 41 feet (12.8 meters). The total weight was to be 25,000 pounds (11,340 kilograms), which included a 6,000-pound (2,727-kilogram) Gemini B (slightly less than a NASA Gemini)

and a 5,000-pound (2,272-kilogram) reconnaissance payload. This was less than the payload carried by unmanned reconnaissance satellites. However, with a human crew, it was hoped much of the automated equipment necessary for these satellites could be eliminated.

In the president's announcement, "reconnaissance" was, of course, not mentioned—only that MOL would perform "very new and rewarding experiments." The first of two unmanned flights would be made in early 1968. The five manned flights would start in late 1968. All would be made from Vandenberg Air Force Base into a polar orbit. The launch vehicle was to be a Titan III C. MOL was to cost $1.5 billion.[25–27]

Johnson's announcement generated various reactions. Some bemoaned the "militarization" of space. The Russians responded in a predictable way. Col. Gen. Vladimir Tolubko, deputy commander of Soviet Strategic Rocket Forces, raved that MOL was for "the practical testing of orbital nuclear weapons, not scientific space laboratories," a claim the U.S. was quick to deny.[28] (The Russians often translated MOL as standing for military orbiting laboratory.)

Another source of criticism was from "Florida Nationalists," led by Martin Anderson, publisher of the *Orlando Sentinel*. The objections were not to MOL but to its use of Vandenberg Air Force Base (and the loss of Florida jobs and payroll). The newspaper accused the Air Force of empire building and wasting money, as well as claiming that the payload penalty of launching a satellite into a polar orbit from Cape Kennedy was minor.

The DOD's response was to note that a due south launch from the Cape would overfly Miami. The likelihood of an impact and fatalities in a failed launch from the Cape was 100 times greater than that from Vandenberg. If the launch were made to the east with the rocket then turning south in a dogleg maneuver into polar orbit, the payload penalty would be 10 to 15 percent over that of a Vandenberg launch. With this large a penalty, it would be impossible for MOL to perform an intelligence mission.[29] MOL stayed at Vandenberg.

Development and Management

With the president's approval, the Air Force could begin serious development of MOL. The first step was the phase 1 contract definition, which refined the equipment requirements and lasted into the fall of 1966.

As with NASA missions, the MOL crewmen would participate in spacecraft design. The Air Force planned to select twenty pilots in three groups.* The first eight names were announced on 12 November 1965: Maj. Michael J. Adams (Air Force), Maj. Albert H. Crews, Jr. (Air Force), Lt. John L. Finley (USN), Capt. Richard E. Lawyer (Air Force), Capt. Lachlan Macleay (Air Force), Capt. F. Gregory Neubeck (Air Force), Capt. James M. Taylor (Air Force), and Lt. Richard H. Truly (USN). Their selection followed the basic pattern used by NASA (review of records, medical tests, and an interview). All were graduates of the Air Force's Aerospace Research Pilot School (ARPS) at Edwards Air Force Base. Their backgrounds ranged from recent graduates to pilots with several years of experience. Major Crews had originally been selected for the DynaSoar program.[30]

As the phase 1 effort continued in 1966,

MOL ran into increasing technical difficulties, which translated into weight increases. The weight problem was aggravated in the spring of 1966 when the Air Force requested an automatic operating mode, which added about 2,000 pounds (907 kilograms). This and other problems pushed the MOL's weight to more than 30,000 pounds (13,608 kilograms), too much for the Titan III C. Accordingly, a new version of the booster rocket was designed—the Titan III M. It had new seven-segment, solid-fuel strap-ons instead of the Titan III C's five-segment units. The first-stage engines had new 15:1 expansion nozzles. The first stage itself was stretched 68 inches (173 centimeters). Takeoff thrust was 3 million pounds (1,363,636 kilograms). The Titan III M had no transtage.[31]

The technical difficulties also resulted in a 1-year slippage in the launch schedule. The first unmanned launch was to be made on 15 April 1969; the first manned launch was scheduled for 15 December. Despite the problems, support from congress and the administration remained strong; $150 million was included in the budget for 1966. Congress added $50 million and DOD reprogrammed an additional $28.4 million.

In June 1966, a second group of MOL crewmen was selected. They were Capt. Karol J. Bobko, Capt. Charles G. Fullerton, Capt. Henry W. Hartsfield (all Air Force), Lt. Robert L. Crippen (USN), and Capt. Robert F. Overmyer (USMC). All were younger than the men who made up the first group. Several months later, Adams, dissatisfied with MOL's pace, left the program to fly the X-15.[32]

---

*Initially, the Air Force was reluctant to call them astronauts, believing that that should be reserved for those who had actually flown in space.

In September 1966, the MOL program office began negotiations with the prime contractors. This brought phase 1 to an end and was the beginning of phase 2 hardware development contracts—a rather complex process. The Douglas contract specification ran 4,480 pages. About sixty program exhibits, each running fifteen to sixty pages, were also prepared.

Space exploration is not a triumph of technology as much as it is of management. The top level of MOL management used a complex arrangement. The initial management team was announced in August 1965. The program director was Gen. Bernard A. Schriever, who was also commander of the Air Force Systems Command. He had responsibility for the broad overview. General Schriever's selection was an excellent choice. During the time preceding MOL's approval, he had worked diligently to convince key congressmen of the project's importance. The actual engineering details of MOL were the responsibility of the deputy program director. The first to hold this position was Brig. Gen. Russell Berg, who was also the chairman of the selection committee that interviewed the prospective MOL crewmen. The political aspects of MOL, such as relations with Congress, preparing briefing materials, and giving testimony at budget hearings, were handled by the vice program director, Brig. Gen. Harry Evans. The MOL science adviser was Dr. Peter Leonard of the Aerospace Corporation. To coordinate activities, the group would communicate daily and meet several times per month. They also held regular monthly meetings with a high-level group at the Pentagon, composed of the secretary and under secretary of the Air Force, the assistant secretary for research and development, and the Air Force chief of staff. General Berg later called his time as deputy program director the most rewarding and difficult part of a 30-year Air Force career. He found the key to success was having the best, most talented people at all levels and providing them with the necessary authority and responsibility.[33]

To keep track of each part of the MOL and its place in the large picture, a control center was established. It was a huge room, at least 75 feet (22.8 meters) on a side, containing status boards covering each segment of the project. The boards were updated every 4 hours. Each MOL contractor had a miniature board covering its part. There was a constant exchange of information back and forth between the contractors and the control center; thus, everybody knew what everybody else was doing.[34,35]

The year's activities ended on a spectacular note—the sixth Titan III C launch, to test the Gemini B heat shield and hatch. The Air Force had equipped the old NASA *Gemini 2* capsule with the MOL hatch. It had been used in an unmanned suborbital test flight in January 1965. The *Gemini 2* would be attached to a simulated MOL canister built from a Titan II first stage. As a secondary goal, the canister carried nine on-board experiments. Although only 34 feet (10.6 meters) in length—shorter than a real MOL—the canister would provide information on the launch stresses of the MOL/Titan III M. On 3 November 1966, the Titan III C was launched from Cape Kennedy. When the transtage and payload reached 125 miles (210 kilometers), the transtage pitched down towards the earth and separated the *Gemini 2*. It made a severe reentry that put maximum heating on the hatch. The *Gemini 2* landed 5,500 miles (8,871 kilometers) downrange, only 7 miles (11.3 kilometers) from the recovery ship USS *LaSalle*. Meanwhile, after separating the *Gemini 2,* the tran-

Pre-dawn photo of the MOL heat-shield test. The spacecraft, carried aboard the Titan III C, was the Gemini 2 modified with a heat-shield hatch. It was separated before the booster reached orbit, subjecting the spacecraft to a more severe reentry than a normal MOL/Gemini B profile. This was the first time a spacecraft had flown in space twice (although both were sub-orbital and unmanned flights). The cannister, built from a converted Titan II tank, continued to orbit, where it deployed several sub-satellites. *U.S. Air Force Photo.*

stage and canister were in a near-circular 184- by 183-mile (296- by 295-kilometer) orbit. Three subsatellites were then ejected, two of which were the OV4-1R and OV4-1T communications satellites. They would test satellite-to-satellite communications. The purpose of the other satellite, OV-1-6, was not disclosed, but it was rumored to be an inflatable decoy. Postflight analysis of the *Gemini 2* indicated that the hatch had survived the 17,500-mile-per-hour (28,255-kilometer-per-hour) reentry intact. There had been no damage or loss of strength. The hatch had welded shut, sealing it.

The heat shield qualification flight was the high point of the MOL program so far. A decline would soon begin.[36]

A Casualty of War

In the larger world, events were taking place that would have a critical effect on MOL. In 1961–62, when the MODS/Blue Gemini studies were being made, U.S. involvement in Vietnam was limited. Advisers were sent to train the South Vietnamese army and U.S. personnel flew air support missions against the Vietcong in the south. By the time President Johnson approved MOL, U.S. involvement had grown into a full-scale war. Air strikes were being flown against the north and increasing numbers of ground troops were pouring into the south. As early as December 1965, it had been foreseen that the Vietnam war costs would put a crimp on MOL.[37]

To make up for the time lost because of the technical problems, the Air Force requested $510 million. They were allowed only $480 million due to war costs. Later, this was cut further to $430 million. The best DOD could offer was that if development problems were solved, money could be reprogrammed from other sources. The promise was academic, since there was no money to spare.[38] The Air Force estimated the reduced budget cut would result in a 15-month slip. DOD allowed only a 12-month delay. The first unmanned launch would come in late 1970.

In May 1967, formal contracts were issued. Douglas Aircraft received a $674,703,744 fixed-price contract for the laboratory. McDonnell Aircraft received a $180,469,000 fixed-price contract for the Gemini B.[39] (Actually, it was two contracts

to one company, since McDonnell and Douglas had formally merged on 28 April.)[40] The use of fixed-price contracts was Defense Secretary McNamara's decision. MOL management had opposed the move on grounds of the many technical unknowns of MOL. The supposedly fixed price would last only until the first problem appeared that required a design change.[41]

Management became more and more involved with budgetary problems, which by the fall of 1967 were too great to cope with by simple adjustments. The Aerospace Corporation began a program review called Project Upgrade to cut costs and preserve the schedule. The portion dealing with the program development effort was Project Emily. It involved program deletions, elimination of certain tests, changes in assembly requirements, and uses of hardware components in contractor tests.[42]

On 30 June 1967, the last group of MOL crewmen was selected from sixteen candidates—all Air Force. The four selected were: Maj. James A. Abrahamson, Lt. Col. Robert T. Herres, Maj. Donald H. Peterson, and Maj. Robert H. Lawrence, Jr. All were students at ARPS. Major Lawrence was the first black selected for astronaut training. He had a multifaceted background. In addition to being a test pilot, he also had a doctorate in nuclear chemistry and had worked as a research scientist.[43]

In 1966 and 1967, there were also changes in the management team. In September 1966, Gen. James Ferguson became director of the MOL program and commander of the Air Force Systems Command, replacing General Schriever. In March 1967, vice director Major General Evans was replaced by Maj. Gen. James T. Stewart, and deputy director Brigadier General Berg was succeeded by Brig. Gen. Joseph S. Bleymaier.[44]

There were also tragedies for MOL. On 15 November 1967, Major Adams made his seventh X-15 flight. As the X-15 neared its peak altitude of 266,400 feet (81,199 meters), Adams apparently misread an instrument and turned the X-15 until it was sideways to its normal flight path. As it came down, the X-15 went into a supersonic spin, then began to oscillate uncontrollably. The aerodynamic forces began to build up and the X-15 broke apart. Major Adams did not eject and was killed on impact.[45] A month later, there was a second loss. After each group of MOL crewmen was selected, they returned to ARPS for 6 months of additional training. Part of this was simulated lifting body landings in an F-104. On 8 December, Major Lawrence was making such a flight. With him was Maj. Harvey J. Royer, chief of operations at ARPS. Witnesses later reported that their F-104 did not slow its descent before hitting the runway. It skidded down the runway and into the brush. Both pilots ejected; Major Royer was injured, Major Lawrence was killed.[46]

MOL Described

The new year of 1968 brought both good and bad news. On the positive side, the last of the technical problems had been solved. One major difficulty was stabilizing the MOL during the photo run. The large optics magnified any vibration, which degraded photo resolution.[47]

One irony about all of this was despite the secrecy surrounding MOL, many technical details are known—many more, in fact, than about contemporary unmanned reconnaissance satellites.

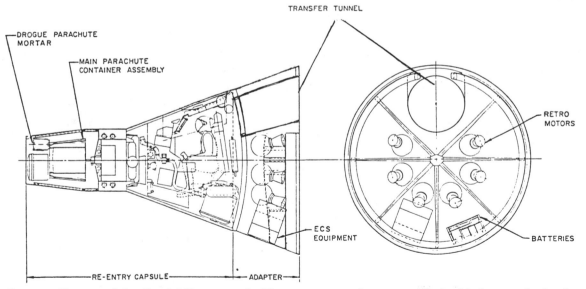

Cutaway diagram of the Gemini B spacecraft. The reentry capsule was modified with the transfer hatch, and controls for the laboratory module was added. Its systems were also changed to allow a 30-day flight (twice that of NASA Gemini); the spacecraft could be powered down when the crew entered the laboratory and then powered up for reentry at the end of the mission. The adapter is a new design. The NASA Gemini, for example, had only four retro-rockets. *U.S. Air Force Diagram.*

The Gemini B itself was very similar to the NASA version. Some systems would have to be changed because of the satellite's prolonged stay in space and since it would be powered up only for launch and recovery. One change was in the atmosphere—the NASA Gemini used pure oxygen during launch. The Gemini B would use an oxygen/helium mix at the request of the crewmen, on the grounds of fire safety. The change was made in late 1966, before the Apollo fire. The major change, of course, was the hatch cut in the heat shield and the transfer tunnel which led back to the laboratory. Once the MOL was

A model of the Manned Orbiting Laboratory in the collection of the San Diego Aerospace Museum. The black Gemini B is at the forward end; the laboratory is the black and white segment in the middle. The section between the two holds the transfer tunnel, fuel cells, liquid oxygen, liquid hydrogen, and attitude-control rocket tanks. The long section at the rear contains the oxygen and helium tanks used to pressurize the lab. This is the final configuration and color scheme. MOL was made in 1967 by McDonnell Douglas. *Photo by Curtis Peebles.*

in orbit, the crew would begin to activate the laboratory systems and prepare the Gemini B for semidormant storage. The crew would then open the 26-inch- (66-centimeter-) in diameter hatch and float through the tunnel into the laboratory. When the life-support system was checked out, the crew would remove their space suits and begin work.

The laboratory was 10 feet (3 meters) wide and 14 feet (4 meters) long—roomy by contemporary spacecraft standards. It was divided into two parts: the forward section held the experiments and work area; the rear was the living area and contained the crew's sleeping bags, an exercise bicycle, and the toilet. A wall divided the two areas. A floor ran the length of the laboratory; this up-and-down reference was another request of the crew. Velcro carpets were tested as a way to hold the crew in place. The laboratory used a mixture of 31 percent helium and 69 percent oxygen at 5 psi (351.5 grams/cm^2). Use of a two-gas atmosphere avoided medical problems caused by breathing pure oxygen for a prolonged time; it also had fire safety advantages. One disadvantage was the mixture's effect on the crews' vocal chords; it made their voices high and squeaky like Donald Duck. The gas was kept in tanks located in the equipment module behind the laboratory. The outer skin of the laboratory bulged slightly, apparently as a meteorite shield. The laboratory had no windows. The crew could not sit and watch the world go by.[48]

The purpose of MOL was, of course, to test man's military usefulness in space. A total of twenty-five experiments were selected—fifteen primary and ten secondary. The primary experiments were large optics in space; the assembly, erection, and alignment of large structures such as radar antennas; tracking ground and space targets; spotting targets of opportunity; ELINT surveys; ocean surveillance; multispectral photos; poststrike target assessment; extravehicular activity (EVA) tests; in-space maintenance; navigation tests; biological and medical experiments; and general performance in the military area.[49] The advantage that man brought to these activities was that of on-the-spot analysis. The crew could spot a new target visually and photograph it with either the high-resolution optics or the multispectral system. There would not be the several-month delay before a high-resolution satellite could be orbited. In effect, the crew would be their own photo interpreters, determining on the spot what required a closer examination.

ELINT, it was hoped, would similarly benefit. ELINT satellites were essentially orbiting tape recorders—they just took it all in. Analysis of the tapes would have to come later. With MOL, the crew would determine if a particular radar signal was a new site or type, or one previously cataloged. They could then use the proper equipment for the best analysis. In effect, MOL would be the equivalent of a specialized ELINT aircraft with its skilled crew.[50] The film and tape would be returned to earth via several reentry capsules.

The Navy also hoped that MOL would provide a space-borne means of ocean surveillance. Each day, the satellite would pass over all the world's oceans. In clear weather, the crew could visually spot the wakes of ships. Early astronauts had proven this was possible. For bad-weather surveillance, MOL would use radar. The crew would detect shipping, locate it, then identify and classify the ships according to type and, finally, track their course.[51] It would be necessary both to separate warships from commercial vessels and Allied from Soviet-bloc shipping. Some in the Navy hoped the fourth MOL would be specifically equipped for ocean surveillance and antisubmarine warfare as well as having an all-Navy crew.[52]

Not surprisingly, a major emphasis was on medical tests. MOL would be in orbit for 30 days—twice as long as any previous U.S. space mission. The tests were to both determine the physical effects of prolonged zero g and the effects on the crews' ability to perform their mission. Should there be an adverse reaction or if there was a deterioration in the crews' performance over time, the utility of an operational military space station would be in question.

The secondary experiments were more broadly based and less military oriented. They were communications propagation, recovery of space objects, material degradation, passive propellent settling, air glow photography and analysis, electron density, plasma, ultraviolet photos, and multispectral observations of the planets. The communications tests would show how well an orbiting command post could transmit and receive messages from ground stations and aircraft. Material degradation would examine the effects of the space environment on various components. It was also hoped these tests would show if the stringent quality requirements were actually needed. If they could be eased without compromising safety or reliability, then significant cost savings could be made. The scientific experiments were similar to those flown aboard NASA Gemini missions. The multispectral observations of the planets, for instance, would determine the composition of the atmospheres of Venus and Mars. This data, in turn, would be used to support NASA's unmanned planetary missions.

Each of the five manned MOL missions would have its individual experiment package. The first mission would probably have been an engineering flight to see how well both men and machine held together. Based on its results, subsequent missions would move either quickly to fly military experiments or if there had been problems, move more slowly.[53]

The MOL's orbit would be similar to that of NASA's Gemini, although an ELINT MOL might go as high as 300 miles (483 kilometers). Inclination would be 85 to 92 degrees.[54,55]

After the crew's 30-day mission was completed, they would shut down the laboratory systems, transfer back to the Gemini B, then separate. The retrofire and splashdown would be as on a NASA mission. The laboratory would be left to burn up on reentry.

A shadow on this bright vision was the continued problem of funding. In 1968, MOL reached peak funding. The year's budget included structural testing and fabrication of the first three flight MOLs (the two unmanned test vehicles and the first manned spacecraft), development of the seven-segment, solid-fuel strap-ons, and installation of equipment at the SLC-6 pad at Vandenberg.[56] To cover the cost and keep the program on schedule, the Air Force believed that $600 million was needed.

The request was made against an expanding war in Vietnam. In the early months of 1968 were the Tet offensive, Lyndon Johnson's withdrawal from the presidential campaign, the never-ending air strikes in North Vietnam, and increasing domestic opposition. The war was devouring an increasing amount of defense funding. Programs not related to the war effort were cut back or cancelled outright to release additional money. Traditionally, the first casualty of such conditions is military research and development funding. This meant MOL, since it was the largest such item in the Air Force budget.

Congress expressed no enthusiasm for MOL and directed that a $6 billion cut be made in Defense Department funding. Fears were expressed that MOL's share of the cuts could reach $100 million, causing a major delay. Congress finally directed an $85 million cut, leaving $515 million. Once more the schedule slipped. The first unmanned launch in late 1970 was not affected; however, the first manned MOL, scheduled for the summer of 1971, was pushed back 3 months.[57] Adding to the problems, the total cost of MOL, because of the technical changes and budget "savings," had increased from $1.5 to $2.2 billion.[58]

In November, Richard M. Nixon was narrowly elected president. Among the questions he faced when he took office in January 1969 was the direction of the U.S. space program in the post-Apollo era. For a time, things seemed to improve. The budget included $576 million for MOL. Although below the $700 million some believed necessary, it was better than the underfunding of the previous years. The amount would cover thermal and dynamic testing, delivery of the first flight MOL and the beginning of construction on those to follow, completion of the launch pad, pilot training, and work on booster components.[59] The needed political support was also available. Dr. Robert C. Seamans publicly called MOL "important, even urgent." [60]

But the problems of Vietnam continued. Defense Secretary Melvin R. Laird found the outgoing Johnson administration had been too optimistic. War costs would continue to be high for at least 2 more years. Additionally, several defense projects, including the C-5A transport aircraft, had major cost overruns. As before, the money to cover such costs would come, in part, from MOL; $20 million was cut. This was done by cancelling the fifth

manned mission. Seamans agreed, believing this was the least damaging alternative. It was argued that because of improvements in technology, the test goals could be accomplished in four manned flights. The Air Force was given an "out," however. If funding restraints had eased by 1971, then money for a fifth mission could be requested. The production line would still be open.[61]

This proved a rather academic possibility, since soon after, a further $31 million was cut. This meant the first unmanned MOL flight would be pushed back to early 1971. The first manned launch would not be until mid-1972. Costs had gone up to $3 billion—twice the original estimate. Despite such setbacks, management was still confident that MOL would succeed.

When the end came, it struck out of the blue. It is understood that the actual decision was made at a White House meeting between President Nixon, national security adviser Dr. Henry Kissinger, and Budget Bureau director Robert Mayo. Mayo opposed continued funding for MOL. After the decision to cancel MOL was made, Mayo called Defense Secretary Laird. When told, Laird, normally a calm gentleman, was reported to have become very upset and objected strongly.[62] The public announcement was made on 10 June 1969. A total of $1.4 billion had been spent. MOL was dead.

The Reasons Why

At the time of MOL's cancellation, the delays it had suffered were a source of problems. Originally, MOL would have flown in the late 1960s. NASA's *Skylab* would not have been launched until the early 1970s— several years later. The repeated slips in MOL's schedule meant the two programs would probably overlap. This only served to underline the seeming duplication of separate military and civilian manned programs. The Air Force and NASA assurances never seemed to stick.

The technology of unmanned satellites had also been progressing. By the early 1970s, they would have capabilities not possible when MOL began. Also, while such unmanned satellites would lack the flexibilities of MOL, they would provide an operational system. An operational MOL would require a separate, additional development effort after the initial flight tests had been made and the results examined. An unmanned system like Big Bird would also avoid the nagging questions of man's usefulness in space—something that had been continuously studied right up to MOL's last day.

These were, however, just the symptoms of a much more severe difficulty—MOL's continuous financial problems. Like a living thing, a space project must be provided with steady, reliable funding. Without it, a project's growth is stunted and the inevitable technical problems cannot be dealt with. Delay after delay is the inevitable result. With the lost time comes increased costs, the evaporation of political support, and changing circumstances. All put the effort at risk. MOL's funding became just another source to be tapped when more money was needed for Vietnam.

The most important legacy of MOL is as an object lesson for future space efforts—to underline the need for commitment to a goal. That is the key to success.

NOTES Chapter 13

1. Willy Ley, *Rockets, Missiles and Men in Space,* (New York: Signet, 1969), 364,365.
2. *Across the Space Frontier,* Cornelius Ryan, Editor, (New York: Viking Press, 1952), xiv,44,45.
3. *The Aerospace Corporation: Its Work 1960–1980,* (Los Angeles Times: Times Mirror Press, 1980), 86.
4. Barton C. Hacker and James M. Grimwood, *On the Shoulders of Titans,* (NASA SP-4203, 1977), 117–122.
5. Fact Sheet—Manned Orbiting Laboratory Program Chronology, (USAF), November 1965.
6. "Air Force to Develop Manned Orbiting Laboratory," (USAF Press Release), 10 December 1963.
7. Remarks by Brig. Gen. Joseph S. Bleymaier Before the AIAA Space Station Symposium, 15 April 1964.
8. Michael Getler, "Accuracy, Payload, Cost Favor Ground Based Anti-Satellite Systems," *Missiles and Rockets,* (28 October 1963): 19.
9. Richard S. Lewis, *Appointment on the Moon,* (New York: Ballantine Books, 1969), Chapter 7.
10. Preliminary Technical Development Plan for the Manned Orbiting Laboratory Status as of 30 June 1964, 1-3,6-3.
11. "Single Contractor Choice Planned for Integration of Gemini, MOL," *Aviation Week & Space Technology,* (20 April 1964): 39.
12. *Astronautics and Aeronautics 1964,* (NASA SP-4005).
13. "Washington Roundup," *Aviation Week & Space Technology,* (23 March 1964): 15.
14. Roland W. Newkirk, Ivan D. Ertel and Courtney G. Brooks, *Skylab a Chronology,* (NASA SP-4011, 1977), 28,29,36.
15. *Astronautics and Aeronautics 1964,* 408.
16. *Astronautics and Aeronautics 1964,* 425.
17. Donald E. Fink, "Defense Department Expand Capability of MOL," *Aviation Week & Space Technology,* (15 February 1965): 16,17.
18. *Astronautics and Aeronautics 1964,* 428.
19. Donald E. Fink, "First Manned MOL Mission Slips to 1968," *Aviation Week & Space Technology,* (5 April 1965): 26,27.
20. "MOL Gemini," *Aviation Week & Space Technology,* (24 May 1965): 16.
21. *Astronautics and Aeronautics 1965,* (NASA SP-4006), 267,290,291.
22. "Detection of ICBMs Key in MOL Approval," *Aviation Week & Space Technology,* (27 September 1965): 26,27.
23. *Washington Post,* (5 September 1965), Collection: NSF Intelligence File Box 10, Folder: Manned Orbiting Laboratory, Lyndon Baines Johnson Library, Austin, Texas.
24. Donald E. Fink, "CIA Control Bid Slowed Decision on MOL," *Aviation Week & Space Technology,* (20 September 1965): 26,27.
25. William J. Normyle, "Air Force Given Manned Space Role," *Aviation Week & Space Technology,* (30 August 1965): 23.
26. Ken Gatland, *Manned Spacecraft Second Revision,* (New York: Macmillan, 1976).
27. Philip J. Klass, *Secret Sentries in Space,* (New York: Random House, 1971), Chapter 17.
28. *Astronautics and Aeronautics 1965,* 400,401,423,424,494.

29. Memo from Vance to Johnson, 12 February 1966, Collection CF OS-3 Launching Site, and Letter from Johnson to Anderson, 18 February 1966, Collection CF OS-2 Box 75, Folder OS-3 Launching Site, Lyndon Baines Johnson Library, Austin, Texas.
30. *USAF Test Pilot School 1944–1979,* (American Year Book Company, 1979).
31. Frank A. Burnham, ''Manned Orbiting Laboratory,'' *Above and Beyond Vol. 8,* (Kingsport: New Horizon Publishing Inc., 1968), 1452–1457.
32. Walter Cunningham, *The All American Boys,* (New York: Macmillan, 1977), 22.
33. Curtis L. Peebles, ''The Manned Orbiting Laboratory: Part 3,'' *Spaceflight,* (June 1982): 274–277.
34. Carrollton Press Retrospective Collection DoD 109 B.
35. Curtis L. Peebles, ''The Manned Orbiting Laboratory: Part 3,'' *Spaceflight,* (June 1982): 274–277.
36. ''Titan 3C Passes 6th Test, Furnishes MOL Support,'' *Aviation Week & Space Technology,* (14 November 1966): 30.
37. *Astronautics and Aeronautics 1965,* 567.
38. Cecil Brownlow, ''Budget Cuts Threaten MOL Project,'' *Aviation Week & Space Technology,* (5 May 1969): 22,23.
39. *Astronautics and Aeronautics 1967,* (NASA SP-4008).
40. Douglas J. Ingells, *The McDonnell Douglas Story,* (Fallbrook: Aero Publishers, 1979), 159.
41. Curtis L. Peebles, ''The Manned Orbiting Laboratory: Part 3,'' *Spaceflight,* (June 1982): 274–277.
42. *The Aerospace Corporation: Its Work, 1960–1980,* 88,89.
43. Biography—Major Robert Henry Lawrence, Jr., Air Force Press Release.
44. *MOL Program Information Available to the Public,* 2-2.
45. ''Heading Shift Cited in X-15 Loss,'' *Aviation Week & Space Technology,* (12 August 1968): 104, 105, 107, 108, 110–113, 115–117.
46. Message from Lt. Col. Maston O'Neal, Edwards AFB History Office—Lawrence File.
47. ''Big Bird: America's Spy in Space,'' *Flight International,* (22 January 1977): 165.
48. Curtis L. Peebles, ''The Manned Orbiting Laboratory: Part 3,'' *Spaceflight,* (June 1982): 274–277.
49. ''MOL Experiment Areas,'' *Aerospace Technology,* (22 April 1968): 30.
50. Klass, *Secret Sentries in Space,* Chapter 17.
51. *Astronautics and Aeronautics 1964,* 434.
52. ''Industry Observer,'' *Aviation Week & Space Technology,* (18 October 1965): 13.
53. Curtis L. Peebles, ''The Manned Orbiting Laboratory: Part 3,'' *Spaceflight,* (June 1982): 274–277.
54. MOL Program Information Available to the Public, 12-7.
55. Donald E. Fink, ''CIA Control Bid Slowed Decision on MOL,'' *Aviation Week & Space Technology,* (20 September 1965).
56. ''DOD Details Funding Requests,'' *Aviation Week & Space Technology,* (12 February 1968): 20.
57. *The Aerospace Corporation: Its Work 1960–1980,* 89.
58. MOL Program Information Available to the Public, 3-1.
59. ''MOL Delayed by Funding Cut,'' *Aviation Week & Space Technology,* (21 April 1969): 17.
60. ''Washington Roundup,'' *Aviation Week & Space Technology,* (27 January 1969): 15.
61. Cecil Brownlow, ''Budget Cuts Threaten MOL Project,'' *Aviation Week & Space Technology,* (5 May 1969): 22,23.
62. Curtis L. Peebles, ''The Manned Orbiting Laboratory: Part 3,'' *Spaceflight,* (June 1982): 274–277.

CHAPTER 14

The Military Salyut Space Station

For both the U.S. MOL and the Soviet military Salyut, only half the story can be told. With MOL, the history of its development and the reasons for its approval and cancellation are known. Yet, because MOL was cut short, it will never be known how the missions would have gone. With the military Salyut, the reverse is true. The missions are known, but the story of its development is limited to a few fragmentary reports and logical suppositions. It is not known, for example, if the Soviets ever agonized over man's military usefulness in space. The official approval date of the Salyut program was 1 January 1969.[1] It came at a turning point in the history of the Soviet space effort. From the October 1957 launching of *Sputnik I* until mid-1965, the Soviets had led in space. With the start of the Gemini program, superior U.S. technology and management skills began to overtake the Russians. Over the next 3 years, the U.S. demonstrated the skills and hardware needed to go to the moon. The Soviet lunar effort went into eclipse. A week before Salyut's approval, *Apollo 8* had orbited the moon and brought the space race to a close. The Soviets were still a year away from being able to land a man on the moon.[2]

Although unmanned tests would continue, no Russian cosmonaut would walk the dusty, gray plains of the moon. The propaganda humiliation of being second in a race the Russians said they would win was too great. The moon-landing program was quietly abandoned. Salyut was one part of the Soviet effort to establish a post-Apollo program.

Civilian and Military Salyuts

Rather than a single Salyut program, one must speak of the Salyut programs—one civilian, the other military in its goals. Both drew on the same technology. The Soviets were able to do what the Air Force and NASA could not—combine both efforts. Doing so, however, required major modifications to the civilian station. The civilian Salyut is made up of three cylinders. In order of size, they are the transfer module at the front (where the Soyuz spacecraft docks), the midsection, and the large-diameter work module. The three cylinders make the Salyut look like an old-fashioned telescope. Attached to the rear is the propulsion module. It is the same as the service module on the Soyuz.

It took a decade before Western space analysts fully understood the modifications needed to perform military reconnaissance from space. It was originally thought that the military Salyut was simply a civilian station equipped with a reconnaissance camera. Only after the program was completed was it apparent just how great the modifications were. The Soyuz transport spacecraft docked not at the front but at the rear, which required the propulsion system's rocket engines to be moved to the rim of the Salyut to make room for the docking collar and hatch. This was necessary because the forward transfer module, where the Soyuz docks on a civilian station, was removed to make room for a large reentry capsule, used to return film and tapes after the mission is completed. The capacity may be approximately 1,103 pounds (500 kilograms).

The military Salyut is powered by two large solar panels placed farther back on the work module. They can be rotated to face the sun even when the station is being oriented to photograph the earth. Solar cells are placed on both sides of the panels rather than just on the sun-facing side. The reflected sunlight from earth provides about 20 percent efficiency. The military Salyut is 47.25 feet (14.4 meters) long and 13.1 feet (4 meters) at its widest point. The launch vehicle for both civilian and military Salyuts is the D-1. The total weight is 41,675 pounds (18,900 kilograms).[3,4]

The interior of the military Salyut is also highly modified. On the civilian Salyut, the interior resembles a long hallway. The arrangement of the military stations is more complex. The midsection is the living area, with a special sofa for medical tests and two bunks, one of which folds into a bulkhead to save room. There are also hot- and cold-water taps; a shower; toilet; dining table; and storage space for clothes, linen, and recreation equipment (a tape recorder for playing music, a chess set, and a small library).

The work module has a corridor down the left side and is divided into several areas.

The first is the control area, where the station instrument panel is located. Behind this is the work area and the reconnaissance camera. The floor is a dark color and the ceiling is light to allow the cosmonauts to orient themselves in zero g.[5]

Access to the reentry capsule, which is forward of the living area, is through a tunnel. If the hatch is cut into its heat shield, the military Salyut uses MOL's most distinctive technical feature.

The most important part of the military Salyut—indeed, the whole reason for its existence—is the reconnaissance camera. It uses a 33-foot (10-meter) focal length Cassegrain telescope. With the best quality Soviet film, the telescope's camera would have a resolution of 12 to 18 inches (30 to 46 centimeters) from an altitude of 150 miles (240 kilometers). The camera housing stretches from the floor almost to the ceiling. It was designed to operate both with and without a crew.[6]

The high resolution requires the station to be held steady, because the long focal length would magnify any vibration. To control the station's attitude, the Soviets used an "inertial sphere"—a spherical gyroscope spun within a magnetic chamber. As it spun, the station would move in the opposite direction without using any rocket fuel.[7] The civilian station had no need for such accurate position control and did not use it.

Overall, the design of the military Salyut gave it a greater degree of flexibility than MOL. Because the crew was launched separately, a station could be used for several missions (MOL, it will be remembered, was launched as a single unit. Once the crew left, the canister could no longer be used). With the military Salyut, the limiting factors were its 6-month lifetime and the large reentry capsule. As with MOL, once it separated, the

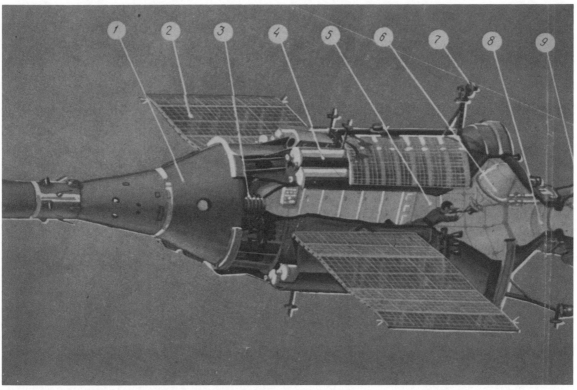

Soviet diagram of the Cosmos 1443 space station module. Some Western space analysts believe it is a modified military Salyut due to a number of similarities—both use similiar radio frequencies, return capsule, gyroscope for altitude control, and rear docking unit. Also the V. N. Chelomey Bureau designed both the military Salyut and the space station module (as well as the D-1 booster). The S. P. Korolev Bureau is believed to have designed the civilian Salyuts. If the space station module and military Salyut are related, it is not clear what specific modifications are made.

1. return module
2. solar panels
3. transfer tunnel

4. fuel tanks
5. crewmen
6. rear hatch

7. docking radar antenna
8. rear docking tunnel
9. *Salyut 7* space station

Photo via James E. Oberg.

station apparently lacked any way to return data to earth. This suggests the military Salyut program was a test rather than an operational system.

Only 2 years after the go-ahead, the first civilian Salyut station was being readied for launch. *Salyut 1,* the world's first true space station, was launched on 19 April 1971—a few days after the tenth anniversary of the first spaceflight by Yuriy Gagarian. The program's name, "Salute," is meant as commemoration. Three days later, *Soyuz 10,* with its three-man crew, was launched. They docked with *Salyut 1* but were unable to enter. They returned safely to earth after only 2 days.

Six weeks later, on 6 June 1971, *Soyuz 11* was launched. It docked and the crew—Lt. Col. Georgiy Dobrovolskiy, Vladislav Volkov, and Viktor Patsayev—began working. Each night the Soviets saw television views of their activities. The coverage excited the public's imagination and filled them with pride. The old days of Soviet space leadership were back. After 24 days, the crew packed up to head home. They oriented the *Soyuz 11* spacecraft and fired the retros. They were now only 40 minutes from touchdown. The retro burn complete, the orbital and service modules separated from the command modules. As the orbital module was released, it jarred open a vent and the atmosphere in the capsule was lost in 40 to 50 seconds. The capsule made a normal landing, but the crew was dead.[8]

In a moment, a great human achievement had turned to bitter tragedy. For the Soviet people, the loss was as deeply felt as the assassination of President Kennedy had been for Americans. The crew was given a hero's funeral and their ashes were interred in the Kremlin Wall.

To prevent a recurrence, the Soyuz underwent major changes. The *Soyuz 11* crew died because they were not wearing space suits. These were not even carried. On future missions, the crew would wear space suits during launch and reentry. To make room for the added equipment, one of the seats was removed. For the next decade, all Soviet spaceflights had two-man crews.[9]

As the modifications were underway, the cosmonauts for the military Salyut missions began their training. The Soviet practice is to pick a group of cosmonauts from the available personnel. The group would undergo intensive training to prepare them for a particular type of mission. Once it was completed, the cosmonauts would be assigned to specific crews. The training group for the military Salyut was organized in 1972. Two veteran cosmonauts were the training managers—Maj. Gen. Gherman Titov (*Vostok 2* in 1961 and the second Russian in space) and Col. Yevgeniy Khrunov (*Soyuz 5* in 1969). The group was divided between mission commanders and flight engineers. The mission commanders were Col. Pavel Popovich (Air Force), Col. Boris Volynov (Air Force), Col. Viktor Gorbatko (Air Force), Lt. Col. Gannadiy Sarafanov (Air Force), and Lt. Col. Vyacheslav Zudov (Air Force). The flight engineers were Engineer Col. Lev Demin (Air Force), Lt. Col. Yuriy Artyukhin (Air Force), Lt. Col. Vitaliy Zholobov (Army), Lt. Col. Yuriy Glazkov (Air Force), and Engineer Lt. Col. Valeriy Rozhdestvenskiy (Navy). Of the ten, only Popovich (*Vostok 4* in 1962), Volynov (*Soyuz 5* in 1969), and Gorbatko (*Soyuz 7* in 1969) had flown in space. There were also two other training group members—Lt. Col. Anatoliy Berezovoy (Air Force) and a flight engineer, possibly Valeriy Illarionov. It is not clear whether they joined in 1972 or sometime later.[10,11] When the group was organized, the first military Salyut mission was still a year away.

Salyut 2

A distinctive feature of the early Soviet space program was its use by Khrushchev to impress the West. The schedule and goals of flights were not based on technical or scientific needs but were to upstage the American program. This constant seeking of "firsts" disrupted the Soviet effort, making it impossible for it to develop in a logical manner. Although Khrushchev had been deposed in October 1964, his mismanagement practices continued to hamper Soviet space activities. They reached their final climax in the aborted flight of *Salyut 2*—the first military space station.

When *Salyut 1* was launched, the U.S. equivalent, *Skylab,* was still 2 years away. It had taken time to correct the *Soyuz 11* problems. More time was lost when, on 29 July 1972, a Salyut was lost due to the failure of its D-1 booster to reach orbit. By late 1972 and early 1973, the *Skylab* launch was fast approaching. The Soviets began to feel the old urge to once more "beat the Americans." The plan they developed would involve flying two Salyuts simultaneously—one civilian, the other military. Such an ambitious plan greatly increased the difficulty the Soviets faced—and the risks.[12]

The first part of the effort began on 3 April 1973, with the launch of *Salyut 2*. It was placed into a 162- by 133-mile (260- by 214-kilometer) orbit, inclined 51.6 degrees. This put *Salyut 2* into position for a manned Soyuz launch during the space station's sixteenth orbit. A mission as long as 28 days was believed planned.[13] No Soyuz, however, was launched on 4 April. One possible reason was a series of major solar flares that occurred on 4 and 5 April. These flares could increase radiation levels in low earth orbit and possibly endanger the crew.

It is known that the station's orbit was raised slightly to 162 by 148 miles (260 by 239 kilometers). This changed the launch window for the manned Soyuz to the morning of 10 April. On 8 April, however, *Salyut 2* was put into a parking orbit high above that

normally used for rendezvous. Apparently, whatever had delayed the manned launch was proving more troublesome than first believed. Still, the Soviets probably intended to launch a manned mission.[14]

On 14 April, *Salyut 2*'s orbit was raised again. It then tore apart. Western radar tracked twenty-three pieces of debris—pieces of the solar panels and various antennas. The main body was spinning at 30 revolutions per minute—a complete revolution every 2 seconds. It is believed that a thruster began firing uncontrollably, throwing *Salyut 2* into a spin.[15,16] Four days later, the Soviets, trying to put the best face on the disaster, claimed they had never intended to man the space station.[17]

Still, the impending launch of the U.S. *Skylab* bore down on the Soviets. They continued their preparations for the launch of the second Salyut. Through it, the Soviets could salvage some prestige. Its launch came on 11 May 1973, a bare 4 days before *Skylab*. The space station went into a 151- by 133-mile (243- by 214-kilometer) orbit. Having reached orbit, it "died." It suffered a failure in either the propulsion system or the unit that received radio commands from earth. Whatever the reason, although it continued to transmit signals, the station made no orbital maneuvers. The malfunction disabled it so quickly that the Soviets did not even have time to announce it was a Salyut. Instead, to cover up the failure, it was called *Cosmos 557*.

In just over a month, the Soviets had launched two Salyut space stations. All they had to show for their efforts were two derelict pieces of high-priced space junk. Their performance was all the more humiliating when compared to the triumph of the U.S. *Skylab*. It was rescued from the brink of a disaster that was due to the loss of the thermal shield and one solar panel during launch. As *Salyut 2* and *"Cosmos 557"* decayed toward reentry and disintegration, Western space watchers were beginning to realize the two spacecraft were different.

Cosmos 557's telemetry was transmitted at 15.008 MHz and 922.7 MHz—the same as *Salyut 1*. The radio signals from *Salyut 2*, however, were on 19.944 MHz.[18] This frequency had been previously used by Soviet reconnaissance satellites. The two missions were the first solid evidence of military and civilian halves to the Salyut program, *Salyut 2* being the military station and the equally ill-fated *Cosmos 557* being its civilian counterpart.

The Soviets had tried to "beat the Americans at any price." *Salyut 2* and *Cosmos 557* were the price. We do not know what arguments raged in the Kremlin in the aftermath of the twin failures. The results, however, were obvious. The attempts to score meaningless propaganda points were dropped. The Soviet space program virtually started over. Mission planning was to be step-by-step, a process that had been lacking. The rebuilding effort began cautiously. *Soyuz 12* was launched on 27 September 1973. It was a 2-day manned test flight of the modified Soyuz. *Soyuz 13* (18 December 1973) flew for 8 days, carrying an ultraviolet telescope for astrophysical observations. The success of these two manned flights opened the way for a Salyut mission.

Salyut 3

It took 14 months for the Soviets to sort out the problems that had destroyed the two Salyuts, and prepare the next satellite. *Salyut*

3 was launched on 24 June 1974, into a 162-by 131-mile (260- by 210-kilometer) orbit, inclined 51.57 degrees. It was maneuvered several times. On 3 July, it was in a 168-by 154-mile (270- by 247-kilometer) circular orbit. Western observers were quick to note *Salyut 3* was transmitting on 19.946 MHz—a reconnaissance satellite frequency.[19,20]

On 3 July 1974, *Soyuz 14* was launched. Its commander was Col. Pavel Popovich and the flight engineer was Lt. Col. Yuriy Artyukhin. After a 24-hour chase, *Soyuz 14* had closed to within 330 feet (100 meters) of *Salyut 3*. At this point, Popovich took over manual control and docked. Artyukhin entered the *Salyut 3* and, after checking out the life-support system, invited Popovich aboard. The crew busied itself with medical tests, atmospheric studies, solar measurements, hardware testing, housekeeping, and what the Soviets officially described as "earth resources studies." On 9 July, the crew set up the earth resources equipment. The next day, they began earth photography. On 14 July, they photographed Soviet central Asia, looking for ore deposits. They also charted the movements of glaciers. The photo sessions were continued over the next 2 days.[21]

The crew had a normal work schedule—8 hours of experiments, 8 hours of housekeeping, and 8 hours of sleep. The crew's activities were followed by two engineers in a ground-based mock-up of the Salyut. If a change in plan was requested, it would be tested by the ground crew. At one point, Popovich's sleep was disturbed by a noisy ventilator. Tests in the mock-up indicated the ventilator could be shut down without interfering with the life-support system.[22,23] The crew also learned to cope with life in zero g. Popovich engaged in a wrestling match with a vacuum cleaner

Pavel Popovich and Yuriy Artyukhin at a public appearance after the *Soyuz 14/Salyut 3* flight, July 1974. Popovich (*left*), the commander, flew the Vostok 4 mission but had to wait over a decade for his next flight—*Soyuz 14*. Artyukhin (*right*) served as *Soyuz 14* flight engineer. This was the first military Salyut mission and the first manned military spaceflight. *Soviet Photo via James E. Oberg.*

hose. Artyukhin was blown across the station by the draft from the air-conditioning system—almost colliding with a bulkhead.[24]

Yet, amid the experiments they performed (measuring blood flow, monitoring polarized light reflected from the atmosphere, tests of water-recycling equipment, determining the amount of flex of the solar panels due to vibration, etc.), it was apparent the crew was performing other tasks. The cosmonauts used code words to hide their activities. More importantly, objects had been laid out near Tyuratam as photo-resolution targets, which the crew could observe and photograph. The results would indicate Salyut's usefulness as a reconnaissance platform.[25]

On 17 July, the crew began preparing to return to earth. They packed research results

and documents into *Soyuz 14*, checked out its systems, and shut down *Salyut 3*. *Soyuz 14* separated from *Salyut 3* on 19 July and flew alongside for an orbit before retrofire and reentry. Touchdown on the plains of Kazakhstan came after a flight of 377.5 hours. Popovich and Artyukhin were able to leave the capsule without the aid of the recovery crew. It was the first successful Salyut mission.

Six weeks after their return, the Soviets launched a second mission to *Salyut 3*. *Soyuz 15* was launched on 26 August 1974. Its crew was Lt. Col. Gennadiy Sarafanov (commander) and Col. Lev Demin (flight engineer). Again, both were military officers. Demin, at 48, was then the oldest man to go into space and the first grandfather. *Soyuz 15* went into a 144- by 112-mile (230- by 180-

The crew of *Soyuz 15/Salyut 3* pose in the Soyuz simulator. Lev Demin (*left*), the flight engineer, was at the time of the mission the oldest man to go into space. On the right is Gennadiy Sarafanov, the *Soyuz 15* commander. This was intended to be the second flight to *Salyut 3,* but a docking problem forced a return to earth. *Soviet Photo via James E. Oberg.*

kilometer) orbit. Over the next sixteen orbits, it maneuvered to close the distance to *Salyut 3*. As *Soyuz 15* approached to within 100 to 165 feet (30 to 50 meters) of the space station, the automatic rendezvous system began firing the engines too long. This caused the approach speed to be too high. The crew made several attempts to dock but failed each time. They were now in a critical situation. Part of the modifications after the *Soyuz 11* disaster was the removal of the spacecraft's solar panels. This made it lighter and more maneuverable but also meant the spacecraft had to use batteries, which could supply power for only 2.5

days. There was not time for additional attempts at troubleshooting. To conserve power, the crew stopped voice and telemetry transmissions. On 29 August, the spacecraft reentered, and landed at 11:10 P.M. Moscow time.[26,27]

The Soviets tried to cover up the failure, saying it was a completely successful test of a new docking system. They claimed the flight had been planned for only a 2-day stay in space. Nobody in the West believed them. The launch time was such that a normal midafternoon landing could be made after about 30 days (roughly double that of *Soyuz*

14). More importantly, a story in a weekly Moscow newspaper, written on the day of launch but not published until after the failure, explicitly said, "Like its predecessor, [*Soyuz 15*] is a ferry, intended as a transport rather than a research vehicle." [28]

The results of the two flights were mixed—one was a success, the other a failure. On the positive side, the *Soyuz 15* crew survived the mishap. However, the failure had wrecked Soviet plans for back-to-back missions. The next launch attempt probably could not have been made until October. The *Salyut 3* had an operating lifetime of only 90 days, which would have been exceeded by the time the next manned mission could be launched. The Soviets felt this would be too risky.

On 23 September, the reentry capsule, carrying exposed film and tape, separated from *Salyut 3* and was successfully recovered. After this, the station continued to orbit while engineering tests were made. Finally, on 24 January 1975, *Salyut 3*'s engines were fired for the last time, sending it to reenter and burn up over the Pacific.

Salyut 5

It would be 22 months after the failure of *Soyuz 15* before military Salyut activities resumed. This provided time to analyze the results of *Salyut 3*. The Soviet space program had not been static, however; *Salyut 4* had been launched a month before *Salyut 3*'s reentry. It was a scientific station and underlined the differences between the two programs. *Salyut 4* went into a 218-mile- (350-kilometer-) high orbit—higher than the 155-mile (250-kilometer) orbit of the military station. The crew makeup was also different. The commander was always a military officer; the flight engineer, however, was a civilian. On *Salyut 3,* the crew had been all military. The final difference was the all-important telemetry frequencies. Two successful manned missions were flown—*Soyuz 17* (30 days) and *Soyuz 18* (63 days). Another launch was aborted when the booster malfunctioned. The crew made a rough but survivable landing uncomfortably close to the Chinese border. [29]

Salyut 5 was launched on 22 June 1976. It went into an initial orbit of 161 by 135 miles (259 by 217 kilometers). After 2 days, it maneuvered into a circular 144- by 139-mile (232- by 219-kilometer) orbit. This low orbit and the telemetry frequency of 19.944 MHz identified it as a military station. Several more maneuvers followed over the next few days. *Salyut 5* ended up in a 170-mile (274-kilometer) circular orbit in preparation for its first crew. On 6 July 1976, *Soyuz 21* was launched. Its crew was Col. Boris Volynov (commander) and Army Lt. Col. Vitaliy Zholobov (flight engineer). The next day they docked with the *Salyut 5* and began work. According to Soviet press accounts, the crew busied itself with medical tests. Exercise and special clothing that put stress on the muscles were used to prevent deterioration in zero g. They had a zero g scale to measure weight loss and instruments to measure lung volume, balance, and brain activity. They also made space processing tests. The first was the melting of a sample of bismuth, lead, tin, and cadmium the size of a match head. It was then allowed to cool and solidify into spheres. Another test involved growing crystals of potash-soda and alum. A final series of processing experiments was making an alloy of dibenzyl and tolane, as well as the soldering of small-diameter, stainless-steel tubes. Volynov and Zholobov also scanned the earth's atmosphere for smoke, dust, and other pollutants

The *Soyuz 21/Salyut 5* crew of Boris Volynov (*left*) and Vitaliy Zholobov (*right*). This mission was apparently cut short due to hardware and/or medical problems. *Soviet Photo via James E. Oberg.*

using a hand-held spectrograph. They monitored the behavior of guppies and danio fish, as well as the incubation of fish eggs for information on their adaptation to weightlessness.

They tested a pump that used capillary and surface tension forces to transfer liquids. The results could be used to design fuel systems that had no moving parts. The crew performed time and motion studies to improve work efficiency. Volynov and Zholobov also did earth resources photography.[30,31]

This was all very scientific, but it was only window dressing for *Salyut 5*'s military activities. One potential target of interest was Operation Sevier—a large air and naval maneuver east of Siberia. Monitoring the activities would allow the Soviets to determine the station's usefulness. Photography of both the U.S. and China was also possible. The level of military activities was shown by several things. First, the announced activities were substantially less than those on the *Salyut 4* scientific space station. Soviet press coverage was also muted; 2 or 3 days would pass with no mention of the mission. In reports that were published, few details were given about crew activities. During the *Salyut 4* missions, long and detailed accounts were published.[32] Most important, voice and telemetry were scrambled to prevent eavesdropping. The Kettering group found that voice transmission would begin when Salyut was in range of a ground station. Once contact was established, voice signals would stop and be replaced with highly irregular telemetry, presumably encrypted voice signals.[33] Since Zholobov was an army officer, some of the military tests probably were aimed at the needs of ground forces.

In the first 3 weeks of August, the scientific experiments continued. The crew photographed the moon as it set through the earth's night horizon. This was to determine the real level of the horizon to develop scanners for navigation systems. They also made infrared studies of the atmosphere. The scanner viewed the sun as it set through a 621-mile (1,000-kilometer) horizontal length of the earth's atmosphere. The reading gave a better idea of the water vapor content of the upper atmosphere. The difference in water vapor between two leading atmospheric models was 200 to 300 percent. Earth resources studies also continued. Black and white, color, and spectrozonal films were used to photograph various areas. The targets included suspected mineral deposits, areas of seismic activity, areas endangered by mud slides, and proposed dam locations and railroad lines. Volynov and Zholobov also took time out for public relations duties. They answered questions from a group of children from the U.S. and other countries who were touring the Yevpatoriya mission control center in the Crimea. Despite the attention given to the scientific experiments, these made up only a small part of the crew's activities. Most of their time was spent on tasks the Soviets were unwilling to talk about.[34]

Before the launch, the Soviets had talked about attempting a 90-day mission. This would have fitted the pattern of the previous flights (16 days, 30 days, 64 days, then 90 days)—an orderly, step-by-step pattern. As the crew began their eighth week in space, there was no indication it would end soon. Normally, the crew changes its schedule about 10 days before returning. The amount of exercise is increased from the standard 2.5 hours per day and the crew's diet is changed. Also, about 2 days are spent loading the Soyuz, and a day or two beforehand, its engine is test-fired. More importantly, it would be another 3 weeks before a landing could be made

in the recovery zone during the midafternoon. Indeed, on the morning of 24 August 1976, the Soviets issued a statement that implied the flight was to continue.[35] Then, at noon Moscow time the same day, the Soviets suddenly announced *Soyuz 21* was coming down. Volynov and Zholobov quickly loaded the Soyuz with film, experiment samples, and their log and set the Salyut's equipment to operate unmanned. That evening, they separated from *Salyut 5* and retrofired, landing at 9:33 P.M. Moscow time after 48 days in space.[36]

Their "general condition" was called "satisfactory" rather than the "excellent" normally used in such announcements. Clearly, the crew was called down early. The night landing was evidence enough. The question was why?

Initial speculation centered on a medical reason. It was noted that shortly before the cosmonauts' return, the Soviet press had talked about the difficulty of the isolation and sensory deprivation. Perhaps these were signs the crew was suffering from just such problems. Indeed, when medical data from the flight was published a year later, it showed the crew's condition had deteriorated during the last 10 days of the mission.[37]

Another possibility, a hardware failure, was suggested by *Aviation Week & Space Technology* magazine. According to its account, an acrid odor began to come from the life-support system. The crew was unable to find the cause or fix the problem.* Finally, the smell became unbearable and the crew had to leave.[38] Possibly, the two were linked—the crew's worsening condition was

---

*Apparently, this included venting part of the station's atmosphere. In July, the Soviets announced the crew had "depressurized" *Salyut 5*. No further information was made public.

due to the stress caused by enduring the odor.

More Missions to Salyut 5

Whatever the cause of *Soyuz 21*'s emergency descent, the Soviets apparently believed it could be fixed, since 2 months later, the next manned attempt was made. *Soyuz 23* was launched on 14 October. The crew was Lt. Col. Vyacheslav Zudov (commander) and Engineer Lt. Col. Valeriy Rozhdestvenskiy (flight engineer). Rozhdestvenskiy was a navy engineer and a former commander of a unit of the Baltic Fleet Rescue divers.[39] Possibly, he was to undertake ocean surveillance as one of the mission goals. After *Soyuz 23* went into orbit, everything went well until the automatic rendezvous system was activated. Its radar detected the Salyut at a range of several miles. It then fired the thrusters to slow the rate of approach and closed the distance until the crew took over manually at 330 feet (100 meters). Soon after the automatic system was activated, it suffered an electronic failure. The crew was unable to get close enough to manually dock. The mission was immediately aborted and the crew prepared to return to earth. Once more there would be a race between the time remaining to landing and the battery power. The crew began radio silence to conserve battery power. Retrofire came after nearly 48 hours in orbit. Conditions in

The *Soyuz 23/Salyut 5* crew of Vyacheslav Zudov (*left*), commander, and Valeriy Rozhdestvenskiy (*right*), flight engineer, in the Soyuz simulator. Docking with *Salyut 5* failed, forcing an early return; the capsule made the Soviets' only splashdown, in a salt lake. *Soviet Photo via James E. Oberg.*

the recovery zone were poor. In addition to darkness, there were high winds, falling snow, near-zero visibility, and below-freezing temperatures. Yet, as they descended beneath the capsule's parachute, Zudov and Rozhdestvenskiy could well have thought their troubles were over. They were wrong. Their problems were just starting.

The recovery zone is a flat prairie much like the American Midwest. The only bodies of water are a few small salt lakes and, at the northern end, the 20-mile- (32-kilometer-) long Lake Tengiz. *Soyuz 23* splashed down in Lake Tengiz about 1 mile (1.6 kilometers) from the shore. While the crew remained inside the floating capsule, Soviet divers struggled in the cold water and darkness to secure a line. The first attempt failed. Amphibious vehicles were unable to reach the capsule. Finally, a line was put on the capsule, but the helicopter was unable to lift it. The capsule had to be dragged towards the shore and then, crew still inside, across a frozen swamp. It finally reached an area firm enough for the other helicopters to land. It was 6 hours after the crew had splashed down that they were finally able to leave the spacecraft. The recovery crew truly performed their duties with "high courage and heroism" under the harshest conditions.[40]

The failure of *Soyuz 23* showed that the Soviet space program still lacked consistency. One mission might be a success with the next becoming a cliff-hanging adventure. *Salyut 1, 3, 4,* and now *5* all had failures that either prevented docking or cut missions short; but the Soviets were learning. The failure of *Soyuz 23* also raised questions about the future of *Salyut 5*. It was nearing the end of its 6-month life and time would be needed to correct systems that failed on *Soyuz 23*. Months passed, the New Year came and still no flight.

Finally, 4 months after the abort, on 7 February 1977, *Soyuz 24* was launched. Its commander was Col. Viktor Gorbatko; his flight engineer was Lt. Col. Yuriy Glazkov. This time no problems appeared and *Soyuz 24* docked successfully. Rather than entering the station immediately, as is usual practice, the crew slept in the Soyuz. It was not until 11 hours after the docking that they opened the hatch. No crew before had done this. If the atmosphere was bad, the Soviets would want the crew to be fully prepared when they entered.[41] In any event, the air was at least tolerable and the crew began activating *Salyut 5*'s systems.

The crew performed a number of experiments (infrared observations of the earth's atmosphere as well as biological, medical, zero g metal crystallization, and earth resources photography). The crew also replaced several electronic modules in *Salyut 5*'s navigational computer. The main activity remained military reconnaissance.

On 16 February 1977, the Soviets announced the mission had reached its halfway point. Five days later, an unusual test was made. Soviet television audiences saw Gorbatko and Glazkov vent part of the station's atmosphere into space while simultaneously releasing compressed air from storage bottles. The crew said that "a light breeze blew through the compartment of *Salyut 5*." The Soviets denied there was anything wrong with the air, but the chain of events tend to indicate there was a life-support problem.

On 23 February, the crew began packing up for the return to earth. On the morning of 25 February, *Soyuz 24* separated from *Salyut 5*. For the next orbit, the crew ran through the landing checklist, oriented the spacecraft, and finally fired the retro-rocket. Weather in the recovery zone was poor, with high winds

The crew of the *Soyuz 24/Salyut 5* mission undergoing splashdown training in the Black Sea: Yuriy Glazkov (*left*), flight engineer, and Viktor Gorbatko (*right*), commander. This is contingency training that all Soviet space crews undergo. The mission was of relatively short duration to conduct a few important experiments before the military Salyut program was concluded. *Soviet Photo via James E. Oberg.*

and blowing snow. The Soviets shifted the landing site to a relatively clear area. Touch-down came at noon Moscow time after 18 days in orbit.[42]

It appeared the mission had always been planned as a short one. The landing time was the standard one for a successful mission and the flight's midpoint had been announced well beforehand. Salyut was 8 months old—be-yond its planned lifetime. Both systems and food would be going bad. If this view was correct, *Soyuz 24* was to wind things up and salvage some critical experiments from what must have been a rather disappointing series of flights. The day after the crew landed, the

recovery capsule, with its accumulated film and tape, was deorbited and recovered.

Salyut 5 continued to orbit the earth until 8 August 1977, when the engines fired and it burned up on reentry. With this, the military Salyut program came to a close.[43]

Later Military Activities

No operational system was developed from the military Salyut. Indeed, no military stations have been flown since *Salyut 5*. This would tend to imply that the Soviets found that manned military reconnaissance was too costly or inefficient compared to unmanned satellites, at least with available technology.

At the same time, the military Salyut seems to have indicated humans have some military role in space. The crews on both *Salyut 6* and *7* have performed military experiments as part of Soviet ground, sea, and air exercises. A number of specific categories have been noted. The crews have observed the release of smoke, gas, or mists of air fuel explosives over wide areas. They could either monitor the spread or actually indicate what areas needed additional coverage. The cosmonauts have observed the interception of reentering dummy warheads by ABM missiles. They have played a key role in naval exercises. During these maneuvers, the activities have been coordinated with passes of the space station. These observations could be for either ocean surveillance or to act as an orbiting command post to direct the ships.[44] Salyut crews have made measurements of the infrared plumes from Soviet ICBM launches.[45] Most significant in the long term is involvement with the Soviet laser development effort. The *Salyut 7* station carried sensors to measure laser light striking the station from the ground.

Tests of laser pointing and tracking experiments have also been made. They involve acquiring a target, then following it.[46] Crew conversations relating to these activities have been scrambled. Voice transmission relating to scientific experiments and normal operations are not so encrypted.

Western analysts see the basic thrust as gaining experience in manned military activities. These include using Soviet crews in space to direct ground forces. The flights of the military Salyuts and the subsequent experiments on later missions could ultimately serve as the foundation for operational military space stations.[47] The Soviets are believed to be planning the launch of a twenty-man space station in the late 1980s. Specialized modules, which could be added for specific experiments, could allow the Soviets to expand the station indefinitely. The station could undertake military research, repair of satellites, reconnaissance and interpretation, as well as antisatellite (ASAT) and ABM support. Personnel and supplies would be brought up by the Soviet heavy shuttle. It is believed the station would greatly enhance Soviet military abilities both on earth and in space.[48]

NOTES Chapter 14

1. James E. Oberg, *Red Star in Orbit,* (New York: Random House, 1981), 126.
2. Memorandum for the President, Collection: EX CO 303 11/22,66 Box 73, Folder: CO 303 7/16/68, Lyndon Baines Johnson Library, Austin, Texas.
3. ''Soviet Launches of More Military Salyuts Expected,'' *Aviation Week & Space Technology,* (4 December 1978): 17.
4. *Salyut Soviet Steps Towards Permanent Human Presence in Space—A Technical Memorandum,* (Office of Technology Assessment, December 1983).
5. *Soviet Space Programs 1976–1980 Part 2,* (Library of Congress, 1984), 542.
6. ''Design Advances Boost Salyut Flexibility,'' *Aviation Week & Space Technology,* (25 February 1974): 36–37.
7. Oberg, *Red Star in Orbit,* 132.
8. Nicholas L. Johnson, *Handbook of Soviet Manned Space Flight,* (San Diego: American Astronautical Society, 1980), 297–299.
9. Oberg, *Red Star in Orbit,* 101–105.
10. Rex Hall, ''The Soviet Cosmonaut Team, 1971–1983,'' *Journal of the British Interplanetary Society,* (October 1983): 475.
11. *Astronauts and Cosmonauts Biographical and Statistical Data Revised March 31, 1983,* (Library of Congress, 1983).
12. Oberg, *Red Star in Orbit,* 106.
13. ''Salyut-2 Placed in Parking Orbit; Manned Launch Abortive,'' *Aviation Week & Space Technology,* (16 April 1973): 18,19.
14. Johnson, *Handbook of Soviet Manned Space Flight,* 235.
15. ''Soviet Salyut 2 Total Failure; Debris Decays,'' *Aviation Week & Space Technology,* (7 May 1973): 22,23.
16. ''Reaction Controls Cited in Salyut Breakup,'' *Aviation Week & Space Technology,* (14 May 1973): 12.
17. Johnson, *Handbook of Soviet Manned Space Flight,* 236,237.
18. Oberg, *Red Star in Orbit,* 129.
19. ''Soviets Circularize Salyut-3 Orbit,'' *Aviation Week & Space Technology,* (1 July 1974): 21.
20. ''Soviets Prepare to Dock with Salyut,'' *Aviation Week & Space Technology,* (8 July 1974): 13.
21. Johnson, *Handbook of Soviet Manned Space Flight,* 303.
22. ''Soviet Salyut-3 Crew Focuses on Medical-Biological Tests,'' *Aviation Week & Space Technology,* (22 July 1974): 16.
23. Correspondence, *Spaceflight,* (April 1979 and November/December 1980): 186 and 361.
24. ''Soyuz 14 Crew Back,'' *Flight International,* (25 July 1974): 95.
25. ''Washington Roundup,'' *Aviation Week & Space Technology,* (29 July 1974): 11.
26. ''Soyuz Ends Flight, Salyut Linkup Fails,'' *Aviation Week & Space Technology,* (2 September 1974): 23.
27. Johnson, *Handbook of Soviet Manned Space Flight,* 305–307.
28. Oberg, *Red Star in Orbit,* 133.
29. Oberg, *Red Star in Orbit,* 134–136,252.

30. "Cosmonauts Experiment With Space Processing," *Aviation Week & Space Technology*, (19 July 1976): 243.
31. Johnson, *Handbook of Soviet Manned Space Flight*, 325–327.
32. "Salyut Communications Indicate Efforts to Conceal Discussions," *Aviation Week & Space Technology*, (9 August 1976): 20,21.
33. Harry L. Helms, *How to Tune the Secret Shortwave Spectrum*, (Blue Ridge Summit: Tab Books, 1981), 77,78.
34. "Salyut Cosmonauts Begin Eighth Week of Mission," *Aviation Week & Space Technology*, (23 August 1976): 19.
35. Johnson, *Handbook of Soviet Manned Space Flight*, 324–327.
36. "Soyuz-21 Cosmonauts Return to Earth," *Aviation Week & Space Technology*, (30 August 1976): 23.
37. Oberg, *Red Star in Orbit*, 145,146.
38. "Industry Observer," *Aviation Week & Space Technology*, (18 October 1976): 13.
39. *Astronauts and Cosmonauts Biographical and Statistical Data Revised March 31 1983*, 247.
40. Oberg, *Red Star in Orbit*, 148.
41. Oberg, *Red Star in Orbit*, 148.
42. "Military Salyut Returns Data Quickly," *Aviation Week & Space Technology*, (7 March 1977): 20.
43. Johnson, *Handbook of Soviet Manned Space Flight*, 251,332,333.
44. "Salyut Cosmonauts Support Military Exercises," *Aviation Week & Space Technology*, (28 January 1985): 22.
45. "Salyut Could Have Tracked Columbia," *Aviation Week & Space Technology*, (22 November 1982): 24.
46. "Salyut Cosmonauts Support Military Exercises," *Aviation Week & Space Technology*, (28 January 1985): 22.
47. "Soviets Scramble Salyut 7 Radio Signals," *Aviation Week & Space Technology*, (31 October 1983): 20.
48. *Soviet Military Power 1985*, (Department of Defense, 1985), 56.

CHAPTER 15

Soviet Ocean Surveillance Satellites

In all the fields covered so far, it is the U.S. that has taken the lead. In the area of ocean surveillance from orbit, however, the Soviets made the first move. The reasons have to do with history, the respective nations' naval policy, and even geography. The early years of the space age coincided with fundamental shifts in Soviet naval policy. During World War II, the Soviet navy played a very minimal role. After the war, Stalin ordered the building of a massive surface and submarine fleet. Although fitted with equipment like radar and sonar, these ships were still prewar designs meant to refight World War II. These ambitious plans were handicapped by the destruction of Soviet shipyards during the war. So it was not until the last years of Stalin's life that the naval buildup began in earnest.

With the death of Stalin and the rise of Khrushchev, these plans were fundamentally altered. Khrushchev, perhaps alone among the Soviet hierarchy, realized that nuclear weapons had profoundly changed warfare. Accordingly, Khrushchev scrapped plans for a large surface fleet. He decided to concentrate on a force built around submarines and smaller ships like destroyers and patrol boats, a fleet more suitable for a nuclear age.[1] These ships would not be armed with the traditional gun turrets but with long-range cruise missiles. The SS-N-3 Shaddock, the largest of these missiles, had a range of 280 miles (450 kilometers) or more and carried a 350- or 800-kiloton nuclear warhead. The Shaddock was carried by both submarines and surface ships.[2] Medium- and intermediate-range ballistic missiles, based in southwestern Russia, could also be fired against carrier task forces in the Mediterranean. In the longer term, the Soviets were developing the SS-11 Sego as a long-range antishipping ballistic missile (later converted to an ICBM). These raised a targeting problem. The position of the target ships would have to be found before they fired. The Shaddock could hit targets far out of range of the launching ship's own radar. The ballistic missiles would be fired at a point in the ocean. Some outside source was needed to provide this location data. One possibility was to have submarines shadow the target ships and radio their position. TU-95 patrol aircraft could do the same. Both of these, however, are vulnerable to the target ship's own defenses. Space, on the other hand, offered the means of locating Western shipping on a worldwide basis without such vulnerability. With the buildup of the Soviet navy in the early and mid-1960s, the stage was set for development of the Soviet's most notorious space system.

Ocean Surveillance from Space

Photo reconnaissance from space is primarily a strategic activity. As such, it is not usually time-critical. With ocean surveillance, on the other hand, time is everything. Ships can move at 30 knots (56 kilometers per hour) or more. The position report must be relayed quickly to the attacking ships. A long delay would allow the targets to change course and move off, and the Shaddock or ballistic missile would be fired at an empty ocean. Another requirement for an ocean surveillance system is that it could detect ships in any weather. There are many examples of ships using fog or rainsqualls to evade pursuit.[3] To meet this requirement, the Soviets selected a radar-equipped satellite. From U.S. examples, it is possible to get an idea of this radar's capacity. The antenna itself is a flat, panellike device between 23 and 28 feet (7.1 and 8.5 meters) long and 4.6 feet (1.4 meters) wide. It puts out 3 to 5 kilowatts of power. The radar's coverage extends perhaps 746 miles (1,200 kilometers) on either side of the satellite's orbital path. Because of this wide-area scan in northern and southern latitudes, there will be overlaps in coverage between orbits. The resolution of the radar is estimated to be 65 to 98 feet (20 to 30 meters).[4] The echo of a ship would appear as a "blob." With suitable analysis, the types of ships could be separated (for example, aircraft carriers from destroyers). The arrangement of the blobs could also give a clue as to their function. A supertanker, although it may give a radar return like a carrier, travels alone. A carrier, on the other hand, would be surrounded by several protective rings of cruisers, destroyers, and frigates.

The satellite is estimated from radar and optical tracking measurements to be 48 feet (14.58 meters) long and 8 feet (2.4 meters) in diameter. Total weight is estimated to be about 10,551 pounds (4,785 kilograms). The launch vehicle is the F-1-m, a three-stage adaptation of the SS-9 Scarp ICBM. The Soviet radar ocean surveillance satellites are launched from Tyuratam into a 165- by 155-mile (265- by 250-kilometer) orbit. The satellites have an unusual operating profile. Two satellites work as a team; they are placed into the same orbit so that one follows behind the other by 20 to 30 minutes.[5] When the location of the ships from each satellite is compared, their speed and course can be determined. The overlap in coverage between ground tracks in northern and southern areas can also be used for the same purpose. A second pass would be made about 90 minutes later. The two sets of positions can show if a course change has been made. These areas of overlap are vital ones, because they cover the north Atlantic shipping lanes between the U.S. and Europe and the routes taken by oil tankers going around the southern tip of South America and Africa.

In addition to surface ships, the Soviet satellites may also be able to track submerged submarines. The U.S. Seasat's radar detected the waves generated by submarines as deep as several hundred feet. Given the threat to the Soviet Union posed by U.S. missile submarines, the ability to detect them would seem to be a desirable goal. The satellites are operated by the GRU's Fleet Cosmic Intelligence Directorate, a part of Soviet naval intelligence. The information is sent to the intelligence directorates of the Northern, Pacific, Black Sea, and Baltic fleets.

Because of the low orbit, the relationship between the two satellites is not a stable one. Air drag will change the orbit and cause the satellites to drift apart. To maintain the spacing, each satellite carries ion thrusters, which provide a low but constant thrust to counteract the drag. The low orbit is necessary so the radar can have the best resolution. If the orbit were higher, the radar signals would be spread out over a larger area and the echo would be correspondingly weaker. This low orbit and the quick reentry that follows cause a far more serious problem. To operate, the radar needs electrical power in prodigious quantities. The only way to provide it is with a nuclear reactor.

Topaz

Topaz is the name the Soviets have given to the reactor that powers the radar. It converts the heat generated by the nuclear material into electrical current. The process, called the thermionic principle, works in the following way. Certain materials like molybdenum, if heated to 1,000–2,500 degrees Kelvin, will easily lose electrons. These electrons will flow to a cooler collector a few microns away. This flow generates an electrical current. To generate the necessary heat, a Topaz reactor uses 110 pounds (50 kilograms) of 93 percent enriched uranium 235, which is packed into seventy-nine multilayered tubes called "electrogenerating channels." In the center of each are the uranium pellets. Wrapped around this is the beryllium moderator; surrounding this is the molybdenum emitter, which is at a temperature of 1,500 degrees centigrade. The collector tube is made of niobium. The small gap between the two is filled with cesium vapor, which assists the electrons in jumping the gap. The niobium collector is kept at 500 degrees centigrade (one-third the temperature of the emitter) by sodium potassium reactor coolant, which flows through the circular

channel between the back of the collector and the stainless steel outer casing. The arrangement is, essentially, a series of tubes fitted one inside another. The seventy-nine electro-generating channels form the 11- by 10-inch (28- by 26-centimeter) reactor core. The amount of energy the reactor turns out is regulated by twelve control drums, which resemble kitchen rolling pins. One side of each drum is made of beryllium, which reflects neutrons back into the reactor. This causes it to generate more energy. The other side of the control drum is a boron neutron absorber. The drums can be rotated just like a rolling pin on its shaft. The amount of absorbing material facing the core controls the reactor. The complete assembly weighs 232 pounds (105 kilograms). The core produces 85,000 watts of heat energy, about 12 percent of which is converted into electricity (5–10 kilowatts).

The design has one serious drawback. In a commercial reactor, as more heat is generated, the reaction moderates. In effect, it is self-regulated. The Topaz does not have this feature. The hotter it gets, the more reactive it becomes. This generates more heat, which makes it still more reactive. The control rods are the only thing stopping it from melting down. Thus, the ground controllers must perform a delicate balancing act to keep it within the narrow limits of safety.

The Topaz reactor is apparently started up on the pad before launch. After going into orbit, the satellite operates until a system failure occurs and it can no longer track ships. As noted previously, the satellite is in a low orbit and would reenter and scatter radioactive debris. To prevent this, the Soviets command the satellite to separate into three parts—the reactor unit, the F-1-m's third stage, and the radar antenna. The reactor unit has its own booster rocket, which places the reactor into a 559-mile- (900-kilometer-) high circular orbit. More than 600 years will pass before atmospheric drag brings the reactor down. This is a sufficient period of time for the highly radioactive waste products generated by the reactor—including strontium-90, cesium-137, and xenon-135—to decay. The debris that reenters will pose only a minimal danger to twenty-sixth or twenty-seventh century Earth.[6]

Operations

A generic weakness of Soviet space systems is their poor reliability and short lifetimes. These shortcomings are highlighted by the nuclear-powered, ocean-surveillance satellites. From the program's start, there have been repeated examples of missions abruptly cut short. The first of the nuclear-powered satellites was orbited on 27 December 1967. *Cosmos 198* went into a 175- by 165-mile (281- by 265-kilometer) orbit. Then, 2 days after launch, it split and the reactor was boosted into a 592- by 556-mile (952- by 894-kilometer) final orbit.

The second of this type, *Cosmos 209,* followed 3 months later on 22 March 1968. It remained in the lower orbit for 6 days, after which the flights were halted. It now appears this hiatus was because of a launch failure on 25 or 26 January 1969, when an F booster is understood to have malfunctioned. Nuclear material was detected in the upper atmosphere. A year and a half would pass before the Soviet's next attempt. It was only a slight improvement. *Cosmos 367* (3 October 1970) failed almost immediately after going into orbit. So sudden was the failure that the Russians announced only the reactor's final 640- by 579-mile (1,030- by 932-kilometer) orbit.

Despite this, the Soviets not only persisted but expanded their effort. Two nuclear-powered satellites were launched in 1971—*Cosmos 402* on April Fool's Day and *Cosmos 469* on Christmas. They operated for 8 and 10 days, respectively.

By this time, private Western space analysts had taken note of this unusual type of satellite but were unable to determine its purpose.[7] The orbit of a satellite is one of the indications of its function. The nuclear-powered satellites had two orbits—the low initial one and a final high orbit. Was the first just a parking orbit and the satellite's real mission began only when it reached the higher orbit? Or did it operate only during the low orbit phase? But if this were true, why the high orbit? Further confusing the matter was the brief time the satellite spent in the low orbit. Without knowledge of the use of radar and the nuclear reactor that powered it, Western space analysts could not solve the puzzle. It would not be until 1974 that the satellite's function would be made public.[8]

Between 1967 and 1971, the Soviets had poor luck with the satellites. In 1972, however, their persistence paid off with their first success. *Cosmos 516,* launched from Tyuratam on 21 August 1972, continued to operate until 21 September—a record 31 days. The next attempt, on 25 April 1973, was a launch failure. U.S. fallout sampling aircraft picked up traces of radioactive material like those produced by a reactor.[9] The failure was, however, only a minor setback. On 29 December 1973, *Cosmos 626* went up and did even better than its predecessor, lasting 44 days. With this success, the Russians embarked on the next step—the operational two-satellite profile. It began on 15 May 1974, when *Cosmos 651* went into a 172- by 159-mile (276- by 256-kilometer) orbit. Two days later, *Cosmos 654* followed it into a nearly identical orbit.

They continued to operate for 71 and 74 days, respectively. Not only had the operational network been established, but the satellites had lasted nearly the full 75-day design lifetime. It was a level of success that the Soviets would not achieve again.

The ELINT Ocean Surveillance Satellite

The debut of a second type of ocean surveillance satellite also came in 1974. The first was *Cosmos 699,* launched on 24 December 1974 (7 months after the dual flight). It went into a nearly circular 282- by 271-mile (454- by 436-kilometer) orbit. As with the nuclear satellites, the launch vehicle was an F-1-m. Unlike the nuclear-powered satellites, however, *Cosmos 699* was a passive satellite that listened for transmissions from the target warships, which put out a wide variety of signals—surface search and air defense radars, as well as messages to other ships, aircraft, and shore bases. These transmissions, although necessary, are also dangerous. During World War II, the Allies had radio direction finders on both sides of the Atlantic monitoring the radio frequencies used by German U-boats. Within a few seconds of a U-boat beginning to transmit, three or four stations would have a bearing. Often, a few hours later, the U-boat would be on the bottom, killed by its own message.[10]

A satellite in low orbit can perform the same function as a ground-based direction-finding network, though with changes in procedure. For one thing, a single fixed station cannot locate a source's position. It can show only that it is somewhere along a particular line—close or quite distant. One or more additional bearings from other stations are needed to pin it down (the intersection of the bearings

shows the location). In contrast, a single satellite can be used to locate the target, because the satellite is not fixed but moving. Moreover, the satellite is traveling along a known orbit, which can provide the necessary reference point. A ship, from the satellite's viewpoint, would appear on the horizon. As the satellite moves past the ship, it could take repeated bearings on the transmissions. These would vary from moment to moment and would provide the intersections to locate the ship. The amount of time the satellite had to pick up the ship would vary according to the distance between the two. A ship at extreme range may be visible for only a brief time. This procedure is the simplest; a more sophisticated approach is to use the Doppler shift. As the satellite moves toward the ship, the frequency of its transmissions will be higher than normal. At the satellite's closest approach, it will be normal. As the satellite moves away, the frequency will be lower. By examining how the frequency changes, the position of the source can be determined. Presumably, the Soviet ELINT ocean-surveillance satellites use one or both of these methods.

In addition to simply locating the ships, the ELINT satellites might also be able to determine the type of ship. The transmissions from an aircraft carrier would be different from those of a frigate. Analysis of the signals picked up could show which was which. It might also be possible to identify individual ships within a class by their individual peculiarities.

The obvious countermeasure to this is radio silence. However, it is probably not a realistic option for a task force. If the ships shut off all of their radar when a satellite passed overhead, they would be left unable to detect any threats.

Cosmos 699 continued to operate until 18 April 1975, when it was blown up. This is a standard procedure to dispose of any equipment the Russians consider "sensitive." Since *Cosmos 699* did not carry a nuclear reactor, there was no need to boost it.[11]

About 2 weeks earlier, the second set of nuclear-powered satellites—*Cosmos 723* and *Cosmos 724* (2 and 7 April 1975, respectively) went up. They used a different orbital procedure than the first dual set. Rather than having one satellite follow behind the other, *Cosmos 723* and *724* were positioned one revolution apart. This was the only pair to use this arrangement. The outcome of the mission was something of a letdown. The satellites lasted only 43 and 65 days, respectively.[12] Their flight coincided with phase 2 of the Soviets OKEAN 75 naval war games (15–17 April), which covered Soviet naval activities in the Atlantic, Mediterranean, Pacific, and Indian oceans. Both air and space-borne reconnaissance were coordinated to hunt down "enemy" naval forces.[13] It is worth remembering that despite the short operating lifetimes of the nuclear-powered satellites, they could still support a coordinated worldwide attack, such as that demonstrated in OKEAN 75. The satellites may not have much staying power, but they could support the opening phases of such a war.

Four months after *Cosmos 724* was boosted into its higher orbit, the second ELINT ocean-surveillance satellite was launched. *Cosmos 777* (29 October 1975) went into the standard 283- by 272-mile (456- by 437-kilometer) orbit. Sven Grahn picked up radio signals from *Cosmos 777* on 166 MHz—the same frequency used by the nuclear-powered satellites. This was one of the factors that led to its identification as an ocean-surveillance satellite. *Cosmos 777* also estab-

lished a pattern of operation: a solo ELINT satellite would be launched between pairs of nuclear-powered satellites. The Soviets deteriorating fortunes with their nuclear-powered satellites continued with *Cosmos 785*. Only 16 hours after its launch on 12 December 1975, the reactor was boosted into the higher orbit; clearly, the malfunction struck very quickly. Presumably, a second launching was planned to complete the pair. If so, it was cancelled.

The pattern of previous years was continued in 1976—*Cosmos 838* (2 July 1976), an ELINT satellite; *Cosmos 860* and *861* (17 and 21 October, respectively), the nuclear pair; and *Cosmos 868* (26 November 1976), the year's second ELINT ocean-surveillance satellite—were the year's activities. The two nuclear satellites continued the type's mediocre record (24 and 60 days of operation). The periods of operation of *Cosmos 861* and *868* overlapped. ELINT satellite operations were continued with *Cosmos 937* (24 August 1977).

The next pair of nuclear-powered satellites, *Cosmos 952* and *Cosmos 954,* went up on 16 and 18 September 1977, respectively. *Cosmos 952* failed after 22 days. *Cosmos 954* was left to carry on alone. One, 2, then 3 months passed, longer than any satellite had operated without the reactor being boosted into the final orbit. By mid-December, the situation was clear—*Cosmos 954* was not going up; it was coming down.

Cosmos 954

With the news that the satellite and its radioactive cargo were going to reenter, the U.S. government began to prepare. One early decision would later spark controversy: no public announcement would be made about the danger overhead. The U.S., however, quietly notified the governments of Great Britain, Australia, New Zealand, France, Italy, Japan, and Canada. The Nuclear Emergency Search Team (NEST) was also alerted in mid-December. This is a group of about 100 personnel equipped with airborne and hand-held radiation detectors, computers to process the data, and a navigation system to guide aircraft in their search for contamination. Earlier in the year, the group had conducted a practice exercise called NEST 77. If radioactive debris survived reentry and came down on land, the NEST team would handle the cleanup. Until now, *Cosmos 954* had remained stable, which meant atmospheric drag was minimized and reentry would not occur until March or April 1978. Then, in early January 1978, the situation suddenly deteriorated. The satellite began to tumble, increasing the drag. It would come down before the end of the month.[14]

On 12 January 1978, National Security Adviser Zbigniew Brzezinski met with Soviet Ambassador Anatoly Dobrynin. Brzezinski spoke of the danger if the reactor core landed in a city. The core could injure somebody who came within 1,000 feet (305 meters) of it. He asked the Soviets to provide any information that could reduce the hazard. Dobrynin called back the next day with a minimum of information. Brzezinski and Dobrynin had several subsequent phone conversations and on 17 January, a second meeting. Dobrynin assured him that there was no danger the uranium could explode in a nuclear blast (something a high school physics student would know). Other than general information about the reactor and an assurance that it would burn up on reentry, the Soviets provided almost no information.[15]

As reentry neared, NEST members Milo Bell and Ira Morrison began to calculate the reentry footprint—the size of the area over which the satellite's debris would be scattered. This involved such variables as the ability of the satellite to withstand the aerodynamic forces and heating of reentry. Lacking official Soviet information, Bell and Morrison used design characteristics from U.S. satellites and fragmentary data about the construction of Soviet spacecraft. Still, unknown variables exceeded the known ones. Their task required use of a Control Data Corporation 7600 computer.[16]

On 22 and 23 January, as Bell and Morrison were putting in 20-hour days, C-141 Starlifter jet transport aircraft at Las Vegas, Travis Air Force Base, and Andrews Air Force Base were loaded with all of NEST's equipment. The personnel were ordered to be ready to move with 2 hours' warning. In the early morning hours of 24 January 1978, the North American Air Defense Command warned that *Cosmos 954* was on its final orbit. During its 2,060th revolution, at 6:53 A.M. EST, *Cosmos 954* hit the atmosphere. High above the frigid terrain of the Canadian Northwest Territories, air friction began to heat it. In the early morning hours, a few people in the city of Yellowknife saw its fiery demise. Marie Ruman, a cleaning woman arriving home from work, saw it streaking across the star-filled sky. Peter Pagonis, a truck driver in Yellowknife, saw three brilliant objects with fiery tails.[17] High above the astonished spectators, *Cosmos 954* was being torn apart. At an altitude of 37 nautical miles (69 kilometers), several large beryllium cylinders separated from the reactor. At the same time, the core began to disintegrate into bits from fingernail to microscopic in size. As the core fragmented, a set of smaller beryllium rods broke free. Thirty nautical miles (56 kilometers) above the western end of Great Slave Lake, the reactor completely disintegrated. The breakup had taken about 70 seconds. The surviving fragmentary debris continued its long fall through the night until it hit the ice- and snow-covered landscape.[18] One piece of debris had a more spectacular landing; it contained a lithium radiation shield. When the hot lithium hit the ice-covered Thelon River, it ignited and burned a shallow, 8-foot- (2.4-meter-) wide crater around it. Other bits of lithium dug small pits into the snow. Had anybody been nearby to see it, the display would have looked like a huge sparkler burning in the night.[19] As dawn brightened the eastern sky, the debris was scattered along an area about 500 miles (805 kilometers) long in some of the coldest, most inhospitable terrain found in the Western Hemisphere. In this cruel land, the NEST team would have to track down the wreckage. At 7:15 A.M. EST, barely 20 minutes after reentry, President Jimmy Carter phoned Prime Minister Pierre Trudeau to advise him the satellite had landed in Canada and to offer U.S. assistance. The Canadians accepted.

Morning Light

The search effort was code named Operation Morning Light. In the first few hours, several things occurred. A U-2 and a KC-135 checked for any radiation at high altitude. Also, computer projections, along with eyewitness accounts, were used to locate the footprint. At 10:30 A.M. EST, the commander of Canadian Forces Base (CFB) Edmonton was alerted that he was to assume control of the search effort. At 7:30 P.M. EST, the first two C-141s arrived at CFB Edmonton. The NEST team immediately began installing the airborne search equipment in a Canadian Air

Force C-130 transport. The first search mission took off at 3:15 A.M. the next morning.[20]

The C-130s put in 12- to 14-hour missions. They flew in a grid pattern as the onboard gamma ray detectors looked for abnormal radiation. After completing their mission, the aircraft would return to CFB Edmonton, be refueled, receive a new crew, and be sent out again.[21] The data tapes from each mission were processed in the NEST computer van. Four hours of computer time were needed for each flight hour. The NEST team's task was complicated by natural uranium deposits in the rocks of northern Canada: analysis was necessary to separate these from the Cosmos debris. Within several days of the satellite's landing, manmade radiation sources had been detected and confirmed.[22]

The search area between Yellowknife and Baker Lake covered about 15,000 square miles (38,849 square kilometers). In this whole vast wilderness, there lived only thirty-two or thirty-three people, most of them in the settlement of Fort Reliance. Another six or seven were native trappers and the remaining six were a group of adventurers on a 15-month trip through the Northwest Territories. They were wintering over at an area called Wardens Grove along the Thelon River.[23] On 26 January (2 days after the satellite came down), they were awakened at 3 A.M. by the noise of the low-flying C-130s. A radio message to the weather center at Yellowknife brought the first word of the satellite crash. Two of the adventurers, John Mordhorst and Mike Mobley, had left on a 2-day trip and were heading back to camp at about 3 P.M. on 28 January. As they rounded a bend in the frozen Thelon River, they noticed a crater in the ice. Sitting in it was an assembly of rods attached to a crumpled plate—in the words of Mobley, "metal that looked kind of gray—copperish, kind of burnt." Around the crater were the small pits in the snow caused by the burning lithium. They didn't know what it was. Mordhorst touched it with his gloved hand. After spending about 20 minutes at the site, they resumed their dogsled ride home. Arriving at about 5:30 P.M., they told the others of their find. They, in turn, were told about the satellite. Convinced that they had found it, they passed the news by radio on to Yellowknife and were warned not to approach within 1,000 feet (305 meters). As one of them later recalled, "When we told them two of us had approached it and one had handled pieces, they got very excited." Two hours later, they were told to be ready to evacuate.[24]

At 2 P.M. the next day, a Twin Otter and a Huey helicopter landed at Wardens Grove. The Twin Otter flew out the four men who had remained in camp; Mordhorst and Mobley flew to the crash site, dubbed "Satellite One." The helicopter landed on a small rise 1,640 feet (500 meters) from the debris, and a radiation survey team gingerly approached it. They found it was giving off only 10 to 100 milliroentgens per hour. Mordhorst and Mobley were then sent to Edmonton and tested for contamination. They were found to be in perfect health.

The finding of the first actual debris from the satellite generated great press interest. The six adventurers were barraged with interview requests. Four aircraft were also chartered by the press to fly out to the Wardens Grove area. Because of the known radiation, and to prevent an influx of the curious, the Canadians decided to close the area. Four paratroopers were dropped by flare light before dawn on 30 January. They secured the campsite and took care of the dogs that had been left behind. Charter operators were reminded that permission was needed to fly into the area. It was well they were warned, for in the western section of the search area, far more dangerous debris awaited the NEST team.

On 1 February, the crew of a Twin Otter saw an object lying on the snow about 20 miles (32 kilometers) northwest of Fort Reliance. A helicopter was diverted to check it out. The crew found a large tube about the size of a trash can; various smaller pieces were scattered around it. When fitted together, they formed a base plate on one end. None of the pieces was radioactive. After picking it up, the team continued on to a hit site a few miles away. There they discovered a metal slab about 3 by 10 inches (7.6 by 25.4 centimeters). It was emitting 200 roentgens per hour—an extremely dangerous level of radioactivity.[25] A few minutes' exposure would exceed the 1-day limit of 3 rads. (U.S. radiation standards set a 5-rad limit in 1 year.)[26] If a person was exposed to the slab for about 2 hours, he would suffer vomiting and nausea within a day, followed by fever, hemorrhaging, and diarrhea. Half the people exposed to this level of radiation are dead in a month; the survivors require 6 months of hospital care.[27]

The NEST team had to leave the slab on the ice until a specially shielded container was prepared. Red marker ribbons were laid in circles around the slab; the outermost was 2,000 feet (610 meters) away. The lead and steel box needed to safely hold the slab was built overnight at the University of Alberta; it weighed 1,600 pounds (726 kilograms). The lid alone weighed 240 pounds (109 kilograms). The box was flown to the site; then, four of the cleanup crew dragged it on a sled to within 50 feet (15 meters) of the slab. On 4 February, two Canadian officials, Tom Robertson and Wick Courneya, made the pickup. While the news media watched (from a safe distance), Courneya scooped up the slab in a snow shovel and trotted over to Robertson and the box. Robertson picked up the slab with a pair of 5-foot- (1.5-meter-) long tongs, dropped it in, and slammed the lid.[28]

By this time, the aerial and ground surveys indicated the debris was concentrated in three areas. The first was 20 miles (32 kilometers) northwest of Fort Reliance, where the 200-roentgen-per-hour piece was found along with other metal plates, disks, rods, and the nonradioactive pipe. Another area was 20 miles (32 kilometers) northeast of Fort Reliance; here were found several metal cylinders with radioactive levels ranging from 25 up to a dangerous 100 roentgens per hour. The last area was the Thelon River site, where the first debris had been discovered. The recovery was made extremely difficult by the weather conditions. On the day the 200-roentgen-per-hour slab was found, gale-force winds at Baker Lake lowered the temperature to −105 degrees Fahrenheit (−76 degrees centigrade).

Despite the debris already recovered, the main concern was still the reactor core and the danger it posed. By late February, however, it was apparent the core had not survived reentry as had been feared but had broken up. The particles, because of their small size, were rapidly slowed down and driven by a north wind in the western end of the footprint. They came down on the ice-covered Great Slave Lake and the area south. Ground surveys of towns, camps, and roads were made. Because of the danger of contamination, the first priority was to clean up all inhabited areas. In Fort Resolution, six particles were

found, about 200 feet (61 meters) apart. Cleanup was accomplished simply by shoveling the contaminated snow into garbage cans. All detected particles found in towns were picked up. Outside towns, only the larger particles were collected; the remaining ones were deemed too small to add a significant amount to the natural background radiation. It was estimated that the reactor debris added about 1 percent to the amount of strontium 90 and cesium 137 already in Canada from nuclear fallout. These were the isotopes causing the most concern.[29]

The long, slow cleanup of the small particles marked the beginning of the end for Operation Morning Light. By early March, Canadian personnel and equipment began to replace the U.S. NEST team. On 22 March 1978, the last U.S. equipment left Canada, and on 18 April, the last U.S. representative departed Edmonton.

Aftermath

Even though the cleanup was finished, the political repercussions were continuing. Under international agreement, the Soviets were liable for damages caused by their satellite. The Canadians presented them with a bill for the cleanup, totalling $6,041,174.70. So began several years of negotiations. Also, at the UN, the U.S., Canada, and Sweden pressed for strict safeguards to prevent a recurrence. President Carter called for a complete ban on nuclear reactors in space. The Russians opposed these efforts, and Western desires for a complete ban or meaningful controls collapsed in the face of Russian intransigence. In the end, the UN passed a resolution permitting the launch of a reactor, providing safety guidelines were followed.[30] Like all UN resolutions, it lacked the power of law, which, in this case, meant it was meaningless.

The failure of *Cosmos 954* apparently triggered a reorientation in the entire Soviet ocean-surveillance program. It was put on hold for 15 months. In that interval, the ELINT ocean-surveillance satellites were redesigned and their operating profile was changed. *Cosmos 1094* was launched into orbit on 18 April 1979. Then, unlike past practice with the ELINT ocean-surveillance satellites, it was followed on 25 April by *Cosmos 1096*. Both went into a 284- by 272-mile (457- by 439-kilometer) orbit.[31] One satellite followed behind the other; when *Cosmos 1096* was placed into orbit, it was positioned 60 degrees behind *Cosmos 1094*. Each satellite carried a low-power thruster, which allowed the Russians to adjust the orbital spacing between the two satellites and to compensate for atmospheric drag. This was the same orbital pattern used by the pairs of nuclear satellites. Certain implications are obvious: first, the two satellites are working together, rather than solo as before; also, since the flight was made after *Cosmos 954*'s accident, it would seem to follow that the dual profile was a reaction to the crash.

The launch of the dual ELINT mission coincided with major Soviet naval maneuvers. The task force consisted of the aircraft carrier *Minsk,* two Kara-class guided-missile cruisers, the landing ship *Ivan Rogov,* and an oiler. One area of operation was near the Persian Gulf astride the shipping lanes used by Western tankers. This would be an important target in the event of a U.S.-Soviet conflict.[32]

About a year later, on 14 March 1980, another ELINT ocean-surveillance satellite—*Cosmos 1167*—went into orbit. When the first dual ELINT mission had gone up, there was major naval activity in the Indian Ocean. This time, the naval activity was an American force. The previous November, an Iranian mob had stormed the U.S. embassy and taken the diplomats hostage. A carrier task force had been assembled while the Carter administration tried to negotiate their release. *Cosmos 1167* would allow the Russians to monitor the task force's position off the Iranian coast.

At this point, more than two years had passed since the fall of *Cosmos 954*. In that interval, no further nuclear-powered satellites had been launched. Some thought the Russians might have abandoned them in favor of the ELINT type. The dual ELINT mission seemed to support this. On 29 April 1980, an F-1-m launched *Cosmos 1176* into a 165- by 162-mile (265- by 260-kilometer) orbit. When the NORAD computers determined its orbital characteristics, the U.S. government knew it was a nuclear-powered radar satellite. During the years of apparent inactivity, the Russians had actually been redesigning the nuclear satellite. One goal was to double the operational lifetime from 75 to 130–140 days. A more significant change was in the reactor; the core could now be separated from the reactor unit. This served two functions. Once the reactor was in its high final orbit, separating the core would delay its reentry for 50 to several hundred additional years beyond the 600 previously attained. This would allow further decay of the radioactive waste products. If the boost maneuver should fail, separating the core would insure it would burn up. *Cosmos 1176* operated for 135 days before the reactor was boosted into a final orbit and the core separated. It proved out the redesign.[33]

Despite this, the redesign was only a ''Band-Aid fix.'' The fatal flaw remained—the boost rocket *had* to work. If not, radioactive debris would reach the earth's surface.

On 4 November 1980, about 6 weeks after

Cosmos 1176 was boosted, an ELINT satellite, *Cosmos 1220,* was launched. Since the previous ELINT ocean surveillance satellite (*Cosmos 1167*) had been launched 8 months before, *Cosmos 1220* seemed to be a solo mission. A closer look implied it was the second part of a dual mission including *Cosmos 1167*. The ground track of *Cosmos 1220* was positioned exactly halfway between that of *1167*.[34]

The year's ocean surveillance activity ended on an ironic note; on 7 December 1980, some unfinished business was settled. The Soviets agreed to pay the Canadians $3 million for the damage caused by *Cosmos 954*—just in time too, since a repeat was in the works.

Activity Step-up

The year 1981 saw both a step-up in activity and the recurrence of old problems. Three nuclear satellites and three ELINT satellites were launched, alternating with each other.

- *Cosmos 1249*   5 March
  Nuclear satellite (operated 106 days).
- *Cosmos 1260*   20 March
  ELINT satellite (replaced *Cosmos 1167*).
- *Cosmos 1266*   21 April
  Nuclear satellite (operated 8 days).
- *Cosmos 1286*   4 August
  ELINT satellite (replaced *Cosmos 1220*).
- *Cosmos 1299*   24 August
  Nuclear satellite (operated 13 days).
- *Cosmos 1306*   14 September
  ELINT satellite (replaced *Cosmos 1260*).

Cosmos 1260 and *Cosmos 1286* were in an orbital arrangement like that of *Cosmos 1167* and *Cosmos 1220*. Their ground tracks were positioned midway between each other.[35] Although pairs continued to be used, 1 or more months would separate one launch from the other. The fast replacement of *Cosmos 1260* by *Cosmos 1306* indicated the ELINT ocean-surveillance satellites also had their problems. *Cosmos 1306* itself got off to a rocky start. Its initial orbit was 263 by 106 miles (424 by 171 kilometers), apparently because of a booster problem. Over the next 8 days, maneuvers raised the orbit to the operational altitude of 275 by 267 miles (443 by 430 kilometers).[36]

In 1982 there was a further increase in activity as the program hit its stride.

- *Cosmos 1337*   11 February
  ELINT satellite.
- *Cosmos 1355*   29 April
  ELINT satellite. (These two missions were not replacement launches, but an expansion of the network.) [37,38]
- *Cosmos 1365*   14 May
  Nuclear satellite (operated 136 days).
- *Cosmos 1372*   2 June
  Nuclear satellite (operated 71 days).
- *Cosmos 1402*   30 August
  Nuclear satellite.
- *Cosmos 1405*   4 September
  ELINT satellite.
- *Cosmos 1412*   2 October
  Nuclear satellite (operated 39 days).

As the year ended, of the four nuclear satellites launched, only *Cosmos 1402* was still operational. The Soviets could look back at a good year; although *Cosmos 1372* was only marginally successful and *Cosmos 1412* was a failure, it could be said that the Soviets had finally overcome the stigma of *Cosmos 954*. Events also gave the Soviets' ocean-surveillance satellites an opportunity to participate in a real war, if only as a spectator.

Cosmos 1355 (ELINT) and *Cosmos 1365* (nuclear) were launched during the Falklands war.

During the early morning hours of 28 December 1982, on *Cosmos 1402*'s 1,925th orbit, the Soviets radioed up the command for the satellite's reactor to separate. The maneuver failed.

Chicken Little II

U.S. photos of *Cosmos 1402* showed the reactor section still attached to the instrument unit, which prevented the booster rocket from firing. When they realized the problem, Soviet ground controllers immediately separated the core from the reactor unit.[39] The Reagan administration's reaction to the failure was different from that of the Carter administration. Whereas the latter had kept *Cosmos 954*'s failure a secret, the Reagan administration publicly announced, on 5 January 1983, that *Cosmos 1402* was coming down. Concern about possible radiation hazards centered not on the reactor core itself, but on the satellite's structure. During the *Cosmos 954* accident, the core material had virtually all burned up even though it had been protected by the reactor structure. The reactor structure, however, had been in close contact with the nuclear core for 4 months, which had caused it to become radioactive itself. Moreover, such structural materials were likely to survive reentry, as they had in the *Cosmos 954* accident.[40]

In preparation for the fall, the NEST team was put on alert and local emergency agencies were notified. In San Diego, California, for instance, calls were made to all local fire and police departments, the Red Cross, the county board of supervisors, and the coroner's office.[41] The public at large was told that if the satellite impacted in their area, they were to remain indoors and listen to the radio for information. Especially, they should not pick up any metallic particles or fragments they might find.[42] There was none of the panic the Carter administration had feared. Even in nuclear-sensitive Japan, there was no general fear. One Tokyo resident was quoted as saying, "It is a satellite that's falling but not the sky." In the U.S., a common feeling was a sincere desire to have it come crashing down on Russia. The odds of that were placed at 15 percent. Odds it would land on the U.S. were 2 percent; Canada, 3 percent; and the ocean, 70 percent. There were even lighter moments. One radio station offered $500 in personal injury insurance to listeners. A local precious metal dealer took out $1 million worth of damage insurance for Bakersfield, California. It covered the entire Bakersfield area and the 250,000 people who lived there. The premium was $1,000.[43]

The Soviet reaction was instructive. At first, they denied *Cosmos 1402* was in trouble. Then they said it had been separated in a preplanned maneuver and that it would burn up completely. Next, they said the satellite had malfunctioned, but any material that survived reentry would not exceed the normal background radiation.[44,45] Finally, they decided to blame the West, saying that the reports about the danger were a "stream of impudent lies and slander" meant to divert attention "from the unprecedented arms race that has been launched by the United States."[46]

By this time, Western tracking stations had determined *Cosmos 1402* would reenter during the afternoon of 23 January 1983 (U.S. time). Just where was unknown, given the uncertainties of atmospheric drag and the possibility the satellite would skip during reentry and come down farther downrange. The U.S.,

Canada, Italy, Japan, Australia, Spain, Portugal, West Germany, Belgium, and the Netherlands were all on alert. The West German government announced it planned to send the Russians a bill for their preparations.[47] It was assumed the Russians were also preparing. The satellite's final orbit took it across the eastern U.S. and Canada; over the North Atlantic and Scandinavia; then across western Russia and Iran; and finally over the Indian Ocean. At 5:21 P.M. EST, the satellite hit the upper atmosphere and began to burn and break up. U.S. personnel on the island of Diego Garcia saw it as a bright streak in the night sky. Whatever debris survived reentry impacted in the Indian Ocean about 1,100 miles (1,770 kilometers) southeast of Diego Garcia.[48]

Air Force WC-135 fallout-sampling aircraft and U.S. Navy ships began checking for any detectable radiation. The units on alert stood down.

Two weeks later, it was the core's turn to reenter. Its demise had none of the anxiety that accompanied the main part of the satellite's fall. Still, the NEST team was on standby. Finally, on 7 February, the core hit the atmosphere over the south Atlantic between Brazil and Africa. Any radioactive particles were lost in the cold waters of the Atlantic.[49]

Another Halt

Once more, nuclear satellite flights halted. Unlike the situation after *Cosmos 954*, however, the ELINT satellites continued to go up—*Cosmos 1461* on 7 May 1983, and *Cosmos 1507* on 29 October 1983, for instance. It is to be expected the Soviets' ocean-surveillance program will continue. These satellites are too important to the Soviet military. In the event of a war in Europe, the U.S. would attempt to resupply the NATO forces with a convoy. The Soviets would use their ocean-surveillance satellites to supply target information in order to destroy the convoy. If successful, the Soviets could do in a third world war what the Germans could not in the first two—win the battle of the Atlantic and, by extension, the whole of Europe.

The ocean-surveillance satellites are also unique; of all Soviet military space systems, their data alone is transmitted to earth in real time, directly to the combat units that will launch the attacks. It was the Soviet ocean-surveillance program that rekindled interest in antisatellite weapons. The U.S. had abandoned its only operational system, the Program 437 Thor ASAT, in 1975. To counter the ocean-surveillance satellite, the U.S. is developing an ASAT launched from an F-15 fighter. Given this, some in the Department of Defense have suggested trading the U.S. ASAT for the Soviet ocean-surveillance satellites. Unfortunately, the emphasis in the arms-control community and the U.S. Congress is on keeping space free from weapons, even if this means more dangerous things are given a sanctuary.[50]

The Soviet halt in nuclear satellite operations lasted 1.5 years—shorter than the previous interruption, but then the Soviets do have more practice. On 29 June 1984, *Cosmos 1579* was launched into the standard orbit. The reemerged nuclear satellite also had a new profile. *Cosmos 1579* works with an ELINT ocean-surveillance satellite—*Cosmos 1567* (launched 30 May). The orbital planes are separated by 150 degrees. The result is that the orbits intersect over the North Atlantic. The two satellites make their passes about 45 minutes apart. This arrangement allows the

Soviets to cover a limited but critical ocean area with two types of satellites over a brief time interval.[51] It is also significant that in spite of two near misses, the Soviets seem unwilling to find a safer alternative.

In the future, one can expect to see the Russians expanding their efforts into the field of antisubmarine warfare from space. They have apparently already made some efforts in this direction. Another possibility is the use of orbital radar for air defense. Such radar will require a power source. It is unlikely that *Cosmos 954, Cosmos 1402,* and any future failures will discourage them from using a reactor. In the past, when Soviet interests have come into conflict with those of the wider world, the Russians have seldom hesitated.

NOTES Chapter 15

1. P. H. Viger et al, *The Soviet War Machine,* (London: Salamander Books, 1976), Chapter 8.
2. Robert Berman & Bill Gunston, *Rockets & Missiles of World War III,* (New York: Exeter, 1983), 167.
3. Captain Mitsuo Fushida et al, *Illustrated Story of World War II,* (Pleasantville: Reader's Digest, 1969), 136–142,205.
4. R. Townsend Reese and Charles P. Vick, "Soviet Nuclear Powered Satellites," *Journal of the British Interplanetary Society,* (October 1983): 457–462.
5. Phillip S. Clark, "The Scarp Programme," *Spaceflight,* (May 1981): 149–152.
6. R. Townsend Reese and Charles P. Vick, "Soviet Nuclear Powered Satellites," *Journal of the British Interplanetary Society,* (October 1983): 457–462.
7. *Soviet Space Programs 1971–1975,* (Library of Congress, 1976), 430,431.
8. "Soviets Improve Ocean Satellites," *Aviation Week & Space Technology,* (9 September 1974): 26.
9. R. Townsend Reese and Charles P. Vick, "Soviet Nuclear Powered Satellites," *Journal of the British Interplanetary Society,* (October 1983): 457–462.
10. Daniel V. Gallery, *U-505,* (New York: Paperback Library, 1967), 51.
11. *Soviet Space Program 1971–1975,* 432,433.
12. Phillip S. Clark, "The Scarp Programme," *Spaceflight,* (May 1981): 149–152.
13. Gerald L. Borrowman, "Soviet Orbital Surveillance—The Legacy of Cosmos 954," *Journal of the British Interplanetary Society,* (February 1982): 67–71.
14. *Operation Morning Light—Canadian Northwest Territories/1978,* (Department of Energy, September 1978), 1–5,64,65.
15. "Cosmos 954 Incident," *Spaceflight,* (May 1978): 183,184.
16. Leo Heaps, *Operation Morning Light,* (New York: Paddington Press, Ltd, 1978), 32–42.
17. Heaps, *Operation Morning Light,* 54–56.
18. Department of Energy, *Operation Morning Light,* 67.
19. Heaps, *Operation Morning Light,* 116.
20. DOE, *Operation Morning Light,* 14.
21. Gerald L. Borrowman, "Operation Morning Light," *Spaceflight,* (July 1979): 302–307.
22. DOE, *Operation Morning Light,* 22,23.
23. DOE, *Operation Morning Light,* 33.
24. *Los Angeles Times,* 30 January 1978, 1.
25. DOE, *Operation Morning Light,* 49.
26. *Los Angeles Times,* 3 February 1978, 1.
27. *Soviet Space Programs 1971–1975,* 320.
28. *Los Angeles Times,* 5 February 1978, 1.
29. DOE, *Operation Morning Light,* 67–74.
30. Alton K. Marsh, "Agencies' Dispute Delays Power Effort," *Aviation Week & Space Technology,* (15 November 1982): 24,25.
31. "Soviet Satellite Orbiting Over South Atlantic," *Aviation Week & Space Technology,* (30 April 1982): 24.

32. Gerald L. Borrowman, "Soviet Orbital Surveillance—The Legacy of Cosmos 954," *Journal of the British Interplanetary Society,* (February 1982): 67–71.
33. R. Townsend Reese and Charles P. Vick, "Soviet Nuclear Powered Satellites," *Journal of the British Interplanetary Society,* (October 1983): 457–462.
34. Satellite Digest 144, *Spaceflight,* (March 1981): 90.
35. Satellite Digest 151, *Spaceflight,* (February 1982): 80,81.
36. Satellite Digest 152, *Spaceflight,* (March 1982): 136,137.
37. Satellite Digest 155, *Spaceflight,* (June 1982): 282.
38. Satellite Digest 157, *Spaceflight,* (September/October 1982): 372.
39. Craig Covault, "Soviet Nuclear Spacecraft Poses Reentry Danger," *Aviation Week & Space Technology,* (10 January 1983): 18,19.
40. Craig Covault, "U.S. Assesses Hazard of Cosmos Fuel," *Aviation Week & Space Technology,* (31 January 1983): 20,21.
41. *San Diego Union,* 22 January 1983, A-2.
42. *San Diego Union,* 21 January 1983, A-2.
43. *San Diego Union,* 17 January 1983, A-3.
44. *San Diego Union,* 7 January 1983, A-2.
45. *San Diego Union,* 16 January 1983, A-1.
46. *San Diego Union,* 22 January 1983, A-2.
47. *San Diego Union,* 23 January 1983, A-1.
48. *San Diego Union,* 24 January 1983, A-1.
49. *San Diego Union,* 8 February 1983, A-4.
50. *San Diego Union,* 17 June 1984, C-4.
51. Craig Covault, "Spaceplane Called a Weapons Platform," *Aviation Week & Space Technology,* (23 July 1984): 75.

CHAPTER 16

U.S. Ocean Surveillance Satellites

There are various reasons behind the U.S. Navy's slowness to fly ocean-surveillance satellites. First of all, the U.S. emerged from World War II in a dominant naval position. The Soviet navy, in contrast, was primarily a coastal force with only a few large surface ships. The U.S. Navy had a number of ways to keep watch on the Soviets. Surface ships and submarines could wait outside Soviet ports. When Soviet ships ventured forth, they could be discreetly shadowed and their position relayed to shore. Aircraft from carriers and large shore-based patrol bombers could also be used for surveillance. Shore stations, as in World War II, could back up these ships and aircraft by homing in on radio transmissions. A procedure developed after World War II deployed arrays of sonar detectors on the seafloor. Called SOSUS (Sound Surveillance System), these arrays can detect engine noise from Soviet submarines at ranges up to "several hundred kilometers" and locate their position to a 9.3-mile- (15-kilometer-) radius circle.[1] The SOSUS arrays are deployed in the Pacific and Atlantic as counters to Soviet ballistic missile submarines.

A more important reason that ocean-surveillance satellites weren't developed has to do with geography. U.S. ships and airplanes could use bases all around the world. During the 1950s and most of the 1960s, the Soviets had no such facilities; their naval forces had to operate from their home territories. This made things more difficult. Also, Soviet port facilities are poorly located because the severe climate makes year-round operations difficult or impossible, and the ports do not have free access to the open ocean. A Soviet ship sailing from the Baltic has to go through a very narrow gap between the Danish Island of Zeeland and the Swedish coast, then around the tip of Denmark. In the Far East, the main Soviet ports of Vladivostok, Sovyetskaya Gavan, and Petropavolsk all lack direct access to the Pacific. South Korea and Japan block the routes. The Soviet navy could be bottled up in port if the West remained in control of these "choke points." In peacetime, it would be relatively easy to keep track of Soviet naval activities by simply watching the choke points.

With these advantages, the U.S. Navy apparently did not feel the need to invest in a costly new ocean-surveillance system. Also, like many large organizations, the U.S. Navy may have simply continued to perform its traditional mission in traditional ways and overlooked new possibilities. Unlike the Air Force, the Navy had not been particularly space-minded, having limited its activities to navigation satellites for more than a decade.

Early Concepts

The first public mention of a space-based, ocean-surveillance system appeared in the mid-1960s—at a time when the Soviet naval buildup was only starting. Ocean-surveillance tests were to be performed aboard the Manned Orbiting Laboratory. Naval space enthusiasts pressed for an all-Navy MOL. The fourth MOL would have a Navy crew and be outfitted for ocean surveillance and antisubmarine warfare. The crew would use the equipment to spot, classify, and track the world's shipping.[2] For all-weather coverage, the MOL would use radar. When skies were clear, things were much simpler—the crew just had to look. Early astronauts had shown they were able to see the wakes of ships with the naked eye.[3]

The MOL flights were experimental; the Navy was also considering an operational follow-on in early 1968. Four proposals were under consideration by the Navy: one was an unmanned MOL canister, another was based on the Agena spacecraft, and the other two were proposed Air Force spacecraft. The Navy ocean-surveillance satellite would have to meet the same requirements as the Russian nuclear satellite that had just begun testing: the satellite would have to detect ships in all weather conditions and plot their course and speed with precision. Like the Soviets, the Navy selected a radar unit although apparently not with a nuclear power source. Design of the radar system was demanding, since it would have to operate for a year. The MOL experiment would have a crew to perform on-orbit maintenance, which would ease the problem of reliability. The unmanned ocean-surveillance satellite would have no such advantage.[4]

By mid-1969, the effort was given the designation Program 749. Two teams were assembled for the initial studies—Westinghouse/General Electric and Hughes/Lockheed. The Program 749 satellite would use either conventional forward-pointing radar or a comparatively new type called side-looking radar, to penetrate cloud cover.[5] The second type was developed in the early 1950s. As its name implies, the radar sends its beam to the side and down rather than forward. Because of the set's sophisticated design, the echoes form a high-resolution image much like an oblique aerial photo. By the late 1960s, side-looking radar was in common use aboard reconnaissance aircraft. Whatever the radar used, it was hoped that the satellite's on-board systems would be able to sort out the Soviet ships from the raw data, which would greatly speed up analysis. A second mission for the Program 749 satellite was the detection of low-flying, sea-launched missiles fired against U.S. ships.[6]

Ocean Surveillance from the U-2

Even as the Navy was studying the Program 749 satellite, it was also examining new possibilities for aircraft surveillance. From the available evidence, it appears the Navy got into the field of orbital surveillance in the same way the Air Force did, via the U-2. The roots extend back to the 1950s. On 27 September 1956, Francis Gary Powers, in a U-2 from Turkey, flew over the Mediterranean. He was looking for any group of two or more ships. His course took him as far west as the island of Malta before he returned to Turkey. The flight was looking for British, French, and Israeli ships that were being prepared for the invasion of Egypt. When the U-2's film was processed, it was possible to identify the ships.[7] This was possibly the first such high-

A U-2R on final approach to the aircraft carrier, USS *America*. A series of takeoff and landing trials were held on November 21–23, 1969, flown by Lockheed test pilot, Bill Parks. They were to test the U-2R's carrier compatibility and its usefulness for ocean surveillance. *Lockheed Photo*.

altitude, unconventional, ocean-surveillance mission.

The aircraft the Navy used in its studies during the late 1960s and early 1970s was not the early-model U-2 that Powers had used for his ocean-surveillance flight more than a decade before. Rather, it was an all-new version called the U-2 R. Its development began in August 1966 under CIA sponsorship. Kelly Johnson, with Ben Rich and Fred Cavanaugh, did the redesign. The aircraft that emerged was larger and could carry a heavier payload than the U-2. The U-2 R also had somewhat better flight characteristics and easier landing behavior. The first flight was made from North Base at Edwards Air Force Base on 28 August 1967, with Lockheed test pilot Bill Park at the controls. By late 1968, the U-2 R was being introduced into CIA service. Two aircraft each were stationed at Taiwan, North Base, and McCoy Air Force Base, Florida.

At about this time, the Navy had become interested in the U-2 R as part of its ocean-surveillance studies. The first step was to test the ability of the U-2 R to operate from an aircraft carrier. With a 102-foot (31-meter) wingspan, the U-2 R would seem out of place in the postage-stamp confines of a carrier deck. Yet, this was not the first time a carrier had played host to the long wings of a U-2. Six years earlier, in 1963, the CIA made arrangements for two U-2 As to shoot touch-and-go landings on the USS *Kitty Hawk*. The U-2 As were subsequently fitted with arrester hooks, special spoilers, and landing gear modifications. Redesignated U-2 G, the two air-craft, in February and March 1964, made take-offs and landings from the USS *Ranger*. Landings were made in a two-step procedure. When the U-2 G crossed over the edge of the deck, the pilot cut the throttle. As the aircraft passed over the arrester cables, the spoilers were extended and the U-2 G thumped down and caught the hook. Subsequently, operational flights were made by CIA aircraft from carriers.

In preparation for the U-2 R tests, pilot Bill Park underwent Navy carrier landing

A U-2R on the deck of the USS *America*. Landings and takeoffs proved easy. The U-2 has flown from aircraft carriers since 1963, including operational CIA flights. *Lockheed Photo.*

training and tested various combinations of approach speeds and flap and air brake settings to find the best one for a carrier landing. The actual carrier tests were made on 21–23 November 1969, aboard the USS *America*. Park made a series of landings and takeoffs. He found the landing procedure was easy, so much so that he decided the arrester hook wasn't really needed. Touchdown speed was a rather leisurely 72 knots (133 kilometers per hour). The deck crew also tested how best to move around the large aircraft. They even took it below to the hangar deck. The outer 70 inches (178 centimeters) of the wing could be folded and with careful positioning, the U-2 R could fit on the elevator. Takeoffs were accomplished as easily as the landings. The catapult was not needed. With a 20-knot (37-kilometer-per-hour) wind over the deck, the U-2 R needed only a 300-foot (91-meter) takeoff roll before it was airborne.[8] The Navy also looked at possible designs for a future ocean-surveillance aircraft. Boeing proposed a two-man aircraft that could fly at 100,000 feet (30,480 meters). This would put it above the range of Soviet shipboard radar.[9]

Borrowed Satellites

In the early 1970s, the Navy was also using existing Air Force satellites to perform ocean-surveillance experiments. The first of these were made aboard the USS *Constellation* in 1971. Naval officers were reluctant to be specific, noting "the details of the interactions between the Air Force and the Navy on ocean surveillance sensors systems are beyond the secret classification." The satellites and sensor equipment for these tests had been developed as part of other programs. The Air

Force handled launch and control of the satellites. Observers believed the satellites used in these tests were Titan III B–launched reconnaissance satellites. During the late 1960s, they had been used for high-resolution missions and had continued to fly after the introduction of the Big Bird. Apparently, they were also carrying the Navy ocean-surveillance experiments. According to published accounts, the satellites were equipped with an infrared imaging scanner. The hot "stack gases" pouring out of the ships' funnels would be an excellent infrared source. Resolution would not be a problem. Photos from the infrared scanner on the *Landsat 1* earth resources satellite showed small pleasure boats. *Landsat 1* orbited at 600 miles (965 kilometers), six times the altitude of the Titan III B satellite. Presumably, the infrared scanner was equipped with filters optimized for the detection of ships, their wakes, or smoke.[10] The signals were recorded on magnetic tape for transmission to earth, so the Navy's piggyback experiment would not cut into the satellite's supply of film earmarked for higher priority targets inside Russia. It is unknown just how extensive the Navy experiments were. Only a few Air Force satellites may have carried the ocean-surveillance packages, or they may have been a standard feature. Also not known is whether only Titan III B satellites carried them or if a variety of platforms were used as ELINT satellites.

In addition to borrowing Air Force satellites, the Navy also launched a test mission of its own. On 14 December 1971, an LTTAT Agena D launched a satellite into a 621- by 611-mile (999- by 983-kilometer) orbit, inclined at 70 degrees. Over the next few days, three subsatellites were released from the Agena into nearly identical orbits, their characteristics differing by only 1.2 miles (2 kilometers). It was a flight profile that would become familiar.[11]

New Directions

Despite pilot Bill Park's landing on a carrier and the space tests, the Navy was still unable to make up its collective mind about future ocean-surveillance programs. In early 1972, there were three options—aircraft only, satellites only, or a combination of both. Part of the reason for this indecision was difficulty in the Program 749 studies. It was discovered that a side-looking radar would not work. The rolling and pitching of the ship disrupted the echo and prevented the radar from picking up a reliable image.[12] A conventional radar would work, but it would need a large antenna and considerable electrical power (a problem the Russians also faced). This would require a larger spacecraft and boosters, which translates into higher operating costs. Talking about Program 749, Dr. Robert Frosch, assistant secretary for research and development, said, ''We have never been sure that the development would be successful enough that we would certainly want to deploy.'' [13] Another difficulty was the stringent coverage requirements for ocean surveillance. With strategic reconnaissance, it may not matter if a particular target is not photographed for months or even years—it wouldn't be going anywhere. Carrier task force commanders wanted to have near-continuous coverage over a 500-mile (805-kilometer) radius from their ship. This degree of coverage would be impossible unless the Navy ''blacken[ed] the skies'' with satellites. Aircraft could provide round-the-clock surveillance, but it would require a number of aircraft and overseas basing.

Before a decision about future ocean-surveillance satellites could be made, the Navy believed additional experiments were needed.[14] The airborne segment of this advanced series of tests was the Electronic Patrol-Experimental Program (EP-X)—a U-2 R outfitted with specialized ocean-surveillance equipment. In late 1972, the Navy borrowed a pair of U-2 Rs from the CIA. The first set of experiments fitted was a modified RCA weather radar for surface search, an ELINT receiver for picking up ships' transmissions, and an infrared scanner for spotting heat sources. To carry this equipment, one of the U-2 Rs was fitted with a shorter nose and pods under each wing. The aircraft was flown from North Base, with the tests being made off the southern California coast. One goal was to test the ability of the radar to pick out the target ship against the background of sea clutter. There were doubts it could.[15] The U-2 R was an excellent vehicle to serve as a prototype for an advanced ocean-surveillance aircraft. At 75,000 feet (22,860 meters) (twice the altitude of a normal patrol aircraft), it could cover targets within a radius of about 300 miles (483 kilometers). Also, it carried sufficient fuel to stay aloft for a pilot-numbing 15 hours. Data from the on-board experiments was transmitted from the aircraft to Navy ground stations. Testimony to Congress indicated the data would then be relayed to Navy command centers.[16] In an operational system, this would allow the highest levels of the Navy to keep watch on worldwide Soviet naval movements as they happen.

The EP-X test flights were underway by February 1973. In one early mission, the modified weather radar picked up a surfaced submarine 150 miles (241 kilometers) away.[17] The tests went so well that by late summer 1973, the Navy had decided to use both aircraft and satellites. One report indicated that the satellites would be used to locate Soviet ships; once found, aircraft could keep a close watch on them. The satellite portion had three phases. The first would use existing technology—apparently, a continuation of the use of Air Force satellites. Only limited coverage could be provided. The Titan III B satellites were launched, in the early 1970s, three times a year at most. They stayed aloft for 6 or 7 weeks. Naturally, in between flights, there would be gaps. More importantly, the infrared scanner cannot penetrate cloud cover, which is a problem in perpetually stormy areas such as the North Atlantic. Balancing these shortcomings was the fact the Titan III B satellites were only a first and interim step. The next phase would be an advanced ELINT satellite. The final stage would be a satellite with a radar, which would enable it to work in all weather conditions.[18]

White Cloud

The initial phase lasted about 3 years. Air Force U-2 Rs were deployed in England, Cyprus, and Thailand as well as at bases in the U.S. They were available to undertake ocean-surveillance flights if requested by the Navy.

The Titan III B flights were also continuing—two in 1973, three in 1974, and two in 1975. Behind the scenes, the second phase was taking shape. The satellite, code named "White Cloud," was developed by the Naval Research Laboratory. Like the Russian ELINT satellites, White Cloud satellites use radio direction finding to locate their targets. The system involves three satellites flying in parallel orbits. Also, they use a more sophisticated technique called "interferometry," first tested in 1971. The transmission from a ship arrives at the individual satellites at slightly different times because of their separation. The satellites record the incoming signals along with the time they are received. The time source is a clock—a very accurate one—since, traveling at the speed of light, a radio transmission takes only 0.00047 second to travel the distance separating the individual satellites. The signals and their arrival time are transmitted to earth. Individuals who have listened to White Cloud telemetry have noted the large amount of data being transmitted. When the arrival time for two satellites is combined, a line on the earth's surface is generated. The ship is somewhere along the line. When the third satellite's data is added, a second line is formed. Where the two lines cross is the ship's location.

In assessing the accuracy of the White Cloud's position fixes, it is worth noting a similar procedure was once proposed for a navigation satellite's system.[19] Present navigation satellites allow a ship to locate its position to an accuracy of 200 feet (61 meters). To put this in perspective, the amount of error is less than the ship's own length. Ocean surveillance would probably not need this degree of accuracy. Aircraft pilots and antishipping missiles can find their targets, once they are directed to the correct area. The White Cloud satellites are also believed to carry infrared sensors, which may give them the capability to track nuclear submarines. For years it has been suggested that an infrared scanner could detect the warm-water discharged from a nuclear submarine's reactor.

The first White Cloud was orbited on 30 April 1976, from Vandenberg Air Force Base. The launch vehicle was an Atlas F. The dispenser unit went into a 701- by 679-mile (1,128- by 1,092-kilometer) orbit, inclined 63.4 degrees. Then, one by one, the three White Cloud satellites separated from the dis-

An Atlas F stands ready for launch on its Vandenberg pad. This booster is used to launch the White Cloud ocean surveillance satellites. The Atlas F boosters are converted ICBMs built in 1961. After the Atlas ICBMs were withdrawn from service, the missiles' electrical and mechanical systems were standardized and adapted for space use. This improved reliability, turnaround time, and flexibility. Use of the old ICBMs saves $20 million per launch over a newly constructed booster. By mid-1986, 84 converted Atlases had been launched with eight failures. Twelve more remain for launches through 1992. *U.S. Air Force Photo.*

penser unit and went into their separate orbits. Each satellite is 8 feet (2.4 meters) long, 3 feet (0.9 meter) wide, and 1 foot (0.3 meter) thick. The upper surface is covered with solar cells; the opposite side has an array of antennas. The subsatellites are arranged in a dogleg pattern. Although they fly in parallel orbits, the distance from one satellite to the others will change gradually over time.[20]

Presumably, the first few months after launch would be taken up with tests. The 5-year gap between the 1971 tests and the first White Cloud cluster would imply that although the test mission showed the idea had promise, more development was needed. Another indication was the difference in orbital characteristics. White Cloud satellites went into higher orbits and had a lower inclination. The latter may have been to prevent the satellites from drifting apart because of gravitational effects from the earth.

Eighteen months later, on 8 December 1977, the second White Cloud cluster was orbited. Its orbital plane was positioned approximately 120 degrees west of the 1976 cluster. Both of these clusters had been built by the Naval Research Laboratory. Subsequent launches, however, were to be built by Martin Marietta. The ELINT receivers and antennas were subcontracted to E-Systems. The first Martin Marietta–built White Cloud was originally expected to be launched in early 1979. As things turned out, it was not launched until 3 March 1980, positioned 120 degrees east of the first 1976 cluster. This completed the operational three-cluster White Cloud network.[21]

From their 700-mile- (1,126-kilometer-) high orbit, a cluster could detect transmissions from ships as far as 2,000 miles (3,218 kilometers) away. This meant that at the equator, a ship could be picked up on two consecutive passes. As a satellite passed over the same spot on earth twice each day, a single cluster could provide four position reports each day. With three clusters, Navy commanders would have twelve position reports spread over 24 hours. This is the situation at the equator, where subsequent ground tracks are farthest apart. At higher latitude, with the ground tracks closer together, the ship's location can be noted on four consecutive passes—thus, twenty-four ship locations per day from the complete network. The satellites transmit their position data separately to earth. A large volume of data requires a wide band–width telemetry signal.[22] Presumably, the signals are processed and the data sent on to Navy command posts and finally to ships at sea.

By the end of 1980, the first White Cloud cluster had been operating for 4.5 years. A replacement launch was therefore planned; it was made on 9 December 1980. Lift-off went well; but 7 minutes into the flight, the Atlas F booster veered off its normal trajectory and had to be destroyed by the Vandenberg range safety officer. To make matters worse for White Cloud, this was the second Atlas F failure in 6 months. Accordingly, the Air Force halted further launches until the problems were isolated and corrected. Their efforts were apparently insufficient, since still another Atlas F failed in December 1982. It was not until early 1983, 2 years after the White Cloud failure, that the next Atlas F stood on the launch pad. Clearly, this long a delay must have adversely affected coverage (by 1983, the first two clusters would be 7 and 6 years old). This possibility was further underlined by the launch sequence—two White Clouds in 4 months. The first went up on 9 February 1983. The 5:47 A.M. PST launch was seen throughout southern California. One witness, an El Cajon street sweeper, said, "It was

white and it looked like a large spotlight. . . . it was moving at the speed of a jet liner. It looked like a jet, only the light was in the back and it appeared to have smoke-like rings coming out the back."[23] Such sightings of predawn and early-evening Vandenberg launches are common occurrences and regularly generate frantic calls to the police.

The year's second launch came on 10 June—this time, at about 4:55 P.M.[24,25] Presumably, complete coverage was restored.

Clipper Bow Runs Aground

The third phase of the Navy's satellite program was a system using radar to track Soviet ships in spite of weather or their maintaining radio silence. The program became known as "Clipper Bow" (a suitable nautical term describing the steeply raked bow on modern warships). The program was a continuation of the earlier Program 749 studies. Early Clipper Bow studies were done by Lockheed, Martin Marietta, McDonnell Douglas, and TRW. The plan was for an austere research and development effort aimed at showing the program's feasibility. The first Clipper Bow satellite would be launched in 1983, with the second following in 1984.[26] The Navy would then evaluate the results and decide if the move to operational status was justified. The early studies were favorable, and in the summer of 1978, the Navy announced it was beginning full-scale development of Clipper Bow. The areas to be covered by the initial development funding included evaluating the satellite's actual usefulness, detailing the system specifications, testing a ground radar power module, and studying methods to counter Soviet jamming efforts.

The go-ahead decision raised some questions among members of the Senate Armed Services Committee. The senators were worried about possible duplication. In the years since Program 749 began, orbiting radars had branched out into fields other than ocean surveillance. The Air Force was studying their possible use for tracking aircraft and cruise missiles. The CIA was interested in a reconnaissance radar. Many areas of Russia were almost constantly cloud-covered. There was concern the Soviets could build military installations without their being photographed by reconnaissance satellites. The Navy tried to reassure them Clipper Bow would only detect surface ships, which was much easier than trying to detect the smaller radar return from an airplane.[27] Soon, however, the Navy was having its own doubts. By mid-1979, it was decided not to go ahead with full-scale development. Studies were to continue with left-over money. In mid-1980, even this was cancelled. The reasons given for Clipper Bow's demise were its high projected cost and the low usefulness of the system. It is also worth remembering that by this time, the three-cluster White Cloud network was operational; White Cloud may be able to perform many of the functions that previously were believed to require a radar unit.

The Navy effort was redirected into a joint Navy-Army-Marine Corps program called the Integrated Tactical Surveillance System (ITSS). As before, radar was the main sensor. However, the ITSS was to perform not only ocean surveillance but also to follow the movement of Warsaw Pact armored units. The Air Force also considered joining the program; it would use the ITSS to track aircraft.[28,29] Thus, it would be an ocean-surveillance satellite able to warn of airborne threats. The circle is now complete—returning back to the original Program 749 concept.

NOTES Chapter 16

1. Col. William V. Kennedy, *Intelligence Warfare,* (New York: Crescent, 1983), Chapter 9.
2. "Industry Observer," *Aviation Week & Space Technology,* (18 October 1965): 13.
3. Ken Gatland, *Manned Spaceflight Second Revision,* (New York: Macmillan, 1976), 164,170.
4. "Orbiting Radar Sought by Navy," *Aviation Week & Space Technology,* (19 February 1968): 16.
5. "Industry Observer," *Aviation Week & Space Technology,* (7 July 1969): 13.
6. "Ocean Surveillance Satellite Concept to be Studied," *Aviation Week & Space Technology,* (18 September 1972): 12.
7. Francis Gary Powers and Curt Gentry, *Operation Overflight,* (New York: Holt Rhinehart Winston, 1970), 308,309.
8. Jay Miller, *Lockheed U-2,* (Austin: Aerofax Inc., 1983), 33–35,90,91.
9. Philip J. Klass, "Aircraft Ocean Surveillance Role Studied," *Aviation Week & Space Technology,* (8 May 1972): 26.
10. Philip J. Klass, "Soviets Push Ocean Surveillance", *Aviation Week & Space Technology,* (10 September 1972): 12.
11. Anthony Kenden, "Recent Developments in U.S. Reconnaissance Satellite Programmes," *Journal of the British Interplanetary Society,* (January 1982): 41,42.
12. Anthony Kenden, "U.S. Reconnaissance Satellite Programmes," *Spaceflight,* (July 1978): 257,258.
13. Philip J. Klass, "Aircraft Ocean Surveillance Role Studied," *Aviation Week & Space Technology,* (8 May 1972): 26.
14. "Navy Plans Satellite/Aircraft Ocean Surveillance," *Aviation Week & Space Technology,* (20 August 1973): 66.
15. Barry Miller, "Lockheed to Flight Test U-2 for Navy Surveillance Role," *Aviation Week & Space Technology,* (29 January 1973): 17,18.
16. Philip J. Klass, "Aircraft Ocean Surveillance Role Studied," *Aviation Week & Space Technology,* (8 May 1972): 26.
17. "U-2 Flight Tests," *Aviation Week & Space Technology,* (19 February 1973): 15.
18. "Navy Plans Satellite/Aircraft Ocean Surveillance," *Aviation Week & Space Technology,* (20 August 1973): 66.
19. TRW Space Log, Summer 1967, 10–12.
20. "Expanded Ocean Surveillance Effort Set," *Aviation Week & Space Technology,* (10 July 1978): 22,23.
21. Anthony Kenden, "Recent Developments in U.S. Reconnaissance Satellite Programmes," *Journal of the British Interplanetary Society,* (January 1982): 41,42.
22. "Expanded Ocean Surveillance Effort Set," *Aviation Week & Space Technology,* (10 July 1978): 22,23.
23. *Daily Californian,* 9 February 1983, A-1.
24. Satellite Digest 165, *Spaceflight,* (July/August 1983): 334.
25. Satellite Digest 167, *Spaceflight,* (November 1983): 424.
26. Anthony Kenden, "Recent Developments in U.S. Reconnaissance Satellite Programmes," *Journal of the British Interplanetary Society,* (January 1982): 41,42.

27. "Navy Will Develop All-Weather Ocean Monitor Satellite," *Aviation Week & Space Technology,* (28 August 1978): 50.

28. "Space Reconnaissance Dwindles," *Aviation Week & Space Technology,* (6 October 1980): 19.

29. Anthony Kenden, "Recent Developments in U.S. Reconnaissance Satellite Programmes," *Journal of the British Interplanetary Society,* (January 1982): 41,42.

CHAPTER 17

U.S. Early Warning Satellites

Every nation, to a certain extent, is trapped by its own history. For the past 40 years, much of U.S. military thought and planning has been dominated by the events of one Sunday morning—the destruction of the U.S. Pacific fleet at Pearl Harbor. From the failure to anticipate that the Japanese might launch a preemptive strike to the ignoring of the radar detection of the attacking aircraft just before the bombs began to fall, Pearl Harbor is a true horror story. A repetition is every military man's private nightmare. With nuclear weapons, it would not be just the fleet but the country that would be destroyed. One measure of U.S. anxiety was the fact that in the 1950s, the first question the CIA asked any defector was if he knew of any plans for an attack. Most were amazed by the question.[1]

As the Soviets built up their bomber forces, the U.S. was improving its air defenses to meet this threat. A network of radar sites, called the Distant Early Warning (DEW) line, was constructed across Canada. It would provide 2 hours' warning of a jet bomber attack. Additional radars were built in the U.S. Interceptors fitted with air-to-air missiles were deployed; they could destroy Soviet bombers at night or in bad weather. SAM sites were built around major U.S. cities. Tying the radars, interceptors, and SAMs together was a network of ground control centers. Using

computers, the centers would track the enemy bombers, then direct their interception.

With the development of the SS-6 ICBM, the Soviets could bypass the U.S. air defense network. They could now launch a near-instantaneous surprise attack on the U.S. The specter of Pearl Harbor returned. The U.S. was very vulnerable to such an attack. There were no radars able to detect incoming Soviet missiles, and U.S. bombers were not on alert for a quick takeoff. It was estimated in the late 1950s that to destroy U.S. nuclear forces, the Soviets would need 300 missiles, half of them ICBMs.

It was not certain how long the U.S. would have to prepare until the Soviets had built such a force. The U.S. undertook several steps to reduce its vulnerability. One was to more widely disperse its aircraft. Formerly, up to ninety B-47 bombers and forty KC-97 tankers, nearly 10 percent of the total U.S. force, would be crowded onto a single base— a tempting target to anybody considering a preemptive strike. Another action was to place a percentage of the bomber force on 15-minute ground alert. The aircraft were parked at the end of the runway with their crews in trailers nearby. Bombers were fueled and armed with nuclear weapons. They could take off and be far enough away to escape the blast if their base was destroyed.[2] To give them this precious 15 minutes of warning of a Soviet attack, the U.S. began work on the Ballistic Missile Early Warning System (BMEWS)—huge radar stations at Thule, Greenland; Clear, Alaska; and Fylingdales Moor, England. They covered the incoming flight paths of Soviet missiles. Missiles, however, would be detected only about 15 minutes *after* launch. The reason for this delay was both technical and geographic. The BMEWS radar beams went straight out into space; they couldn't ''see'' a missile until it rose above the station's horizon. If a missile could be detected at launch, the warning time would be doubled to 30 minutes, which would allow more bombers to take off. The Soviets, knowing there was no way to catch the bombers on the ground, would be deterred from launching an attack in the first place. Unfortunately, there was no way for radar to provide this extra time. It was not then possible to build a radar with a beam that could bend around the horizon. A satellite, however, could provide the added minutes of warning time. A rocket engine's exhaust forms a large plume as it climbs into space. Because the exhaust gases are very hot, they give off infrared energy. A satellite with an infrared detector could pick up this plume and radio a warning to the ground.

Midas

The idea of an early warning satellite emerged soon after work on reconnaissance satellites began. It was included in the blanket project WS 117L. After authority for WS 117L was transferred to the Advanced Research Project Agency, it was split into the separate Discoverer, SAMOS, and Midas Early Warning Satellite Programs. Midas

stood for Missile Defense Alarm System. In November 1959, control was transferred to the Air Force. At this time, the first Midas was being prepared. A ten-launch development program was planned—two in 1960 and four each in 1961 and 1962. The Midas system would be operational in 1963. The operational system was to be eight satellites, equally spaced, in two orbital rings. The Agena engines would be used to maintain separation. With this network, the probability of detecting an ICBM launch in southern Russia was in the upper 80 percent range. In the middle third of Russia, the odds were in the high 90 percent range, and in northern latitudes, the odds were 100 percent. The infrared detectors were in the nose of the Agena, which during launch was covered by a shroud. Once in orbit, the Midas would be oriented so the scanner faced the earth. It would not only detect the launch but also indicate the number of missiles and their approximate location and direction. Only infrared sources that rose above most of the atmosphere would be detected. The early test scanners were built by either Baird-Atomic or Aerojet-General.[3]

Midas 1 was launched on 26 February 1960, from Cape Canaveral. At first the launch went well, but when the Agena A separated from the Atlas D, there was an explosion and the debris fell into the Atlantic. The next month, the initial data on the infrared signatures of rockets was gathered. Two U-2s were outfitted with infrared radiometers and spectrometers to measure the amount of infrared emissions at selected wavelengths across a wide band. The U-2s would fly at high altitude, about 100 miles away from the pad. From this point, the equipment could follow the missile from launch through the boost phase. Measurements included the amount of infrared energy emitted by the rocket and how

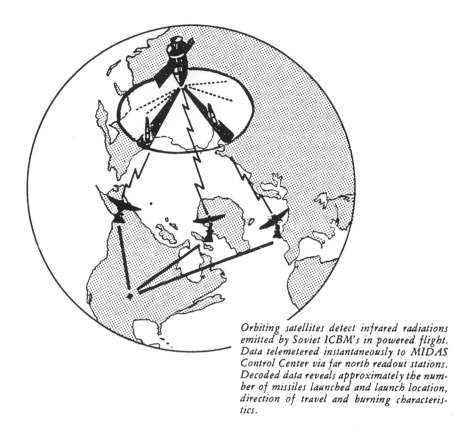

Orbiting satellites detect infrared radiations emitted by Soviet ICBM's in powered flight. Data telemetered instantaneously to MIDAS Control Center via far north readout stations. Decoded data reveals approximately the number of missiles launched and launch location, direction of travel and burning characteristics.

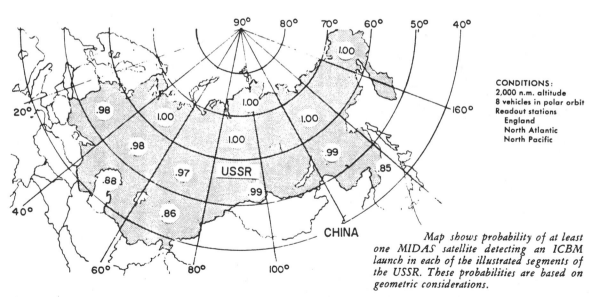

CONDITIONS:
2,000 n.m. altitude
8 vehicles in polar orbit
Readout stations
    England
    North Atlantic
    North Pacific

Map shows probability of at least one MIDAS satellite detecting an ICBM launch in each of the illustrated segments of the USSR. These probabilities are based on geometric considerations.

Top: A Midas detects Soviet ICBM launches, and a warning is radioed to ground stations in the far north. This is relayed to the Midas control center. Its data indicates the number of missiles launched, the site, direction of travel, and burning characteristics.

Bottom: The probability of at least one Midas spotting an ICBM launch in specific areas of the USSR. Plesetsk had 100 percent probability, Tyuratam 86 percent, and sites in central Russia and along the Trans-Siberian Railroad 97 to 99 percent. *DoD Drawing.*

much was absorbed by the atmosphere. The data was recorded on long paper charts.[4]

On 24 May 1960, the Air Force tried again. This time the Atlas Agena A worked perfectly. Final engine shutdown occurred within 1.1 seconds of the scheduled time. *Midas 2* was in a 322- by 292-mile (519- by 470-kilometer) orbit. It had battery power for 28 days. The infrared payload was built by Aerojet-General and weighed 3,246 pounds (1,472 kilograms). The mission's purpose was to collect data on the infrared background. The scanner had filters to adjust its sensitivity. Unfortunately, after only sixteen orbits, the telemetry system failed, preventing more advanced tests. Large sodium flares, providing a source of known intensity, were to have been ignited at Edwards and Vandenberg as *Midas 2* passed overhead. Also, a Titan ICBM was scheduled to be launched from the Cape.[5]

In July, two additional Midas-type missions were added to the schedule. These two flights, RM-1 and RM-2, were modified Discoverers without reentry capsules. They would not carry the Midas detection scanners but, rather, a background infrared radiometer to measure differing conditions, such as between the Arctic and Tropics, and other background data.[6] The first of these infrared measurement flights, *Discoverer 19,* was launched on 20 December 1960, into a 400- by 130-mile (644- by 209-kilometer) orbit. The flight

Launch of a Midas early warning satellite from Cape Canaveral, Florida. Infrared detection equipment was fitted to the nose of the Agena, and the Midas orbited nose toward the earth to pick up the exhaust plume from a rocket's engines. Detecting ICBM launches proved more difficult than expected and success was not achieved until the program's end. *U.S. Air Force Photo.*

lasted 4 days.[7] The second was *Discoverer 21* on 18 February 1961. This flight was the first to restart the Agena engine in orbit. Although not a full-scale, dual burn of the engine, it was a milestone. The operational Midas would need to make periodic engine firings to preserve the satellites' spacings. Reconnaissance satellites would also need to raise their orbits to prevent decay from atmospheric drag.

More than a year after *Midas 2*'s short flight, the next phase of the program began. Before it did, however, word was made public that not all was well with Midas. In the spring of 1961, Defense Secretary Robert S. McNamara, in his testimony before Congress, said about Midas—"There are a number of highly technical, highly complex problems associated with this system. . . . The problems have not been solved and we are not prepared to state when, if ever, it will be operational." Simply put, Midas was subject to chronic false alarms and had poor reliability.[8] The development plan for Midas had changed considerably over the earlier, optimistic, ten-launch effort. A total of twenty-four launches would be needed, and Midas would not be operational until early 1964. The launches planned for 1961, called Series II, were to demonstrate the 2,000-nautical-mile- (3,704-kilometer-) high polar orbit of the operational system.[9] They would be launched from Vandenberg Air Force Base and use the Agena B upper stage. This had twice the propellent of the Agena A used in the first two Midas flights. At 1,641 pounds (744 kilograms), the infrared package was lighter than the *Midas 2*'s unit. Another change was in the power supply. For

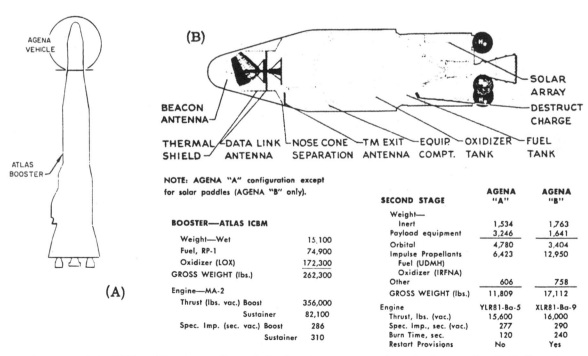

Cutaway of the Midas. The Agena B carried a heavier scanner unit and solar panels. *DoD Drawing.*

a longer life (up to a year in the operational vehicle), the Series II Midas used two solar panels attached to the engine. They could rotate to follow the sun.

Midas 3 was launched on 12 July 1961, into a 2,196- by 2,087-mile (3,534- by 3,358-kilometer) orbit, inclined 91.2 degrees. *Midas 4* followed on 21 October into a similar orbit. The infrared scanners for both were built by Baird-Atomic. Both were to be tested against ICBMs launched from the Cape and Vandenberg, as well as against flares. Much was riding on them. The Air Force was pressing for a change from development to production status for Midas. Obviously, successful missions would support this effort. It was reported that *Midas 4* detected a Titan ICBM 90 seconds after its launch at the Cape on 25 October, but, in fact, the launch was not picked up.[10] Indeed, according to one report, neither Midas lasted long enough to view an ICBM launch.[11]

What data they did return doomed Midas. The scanner was still subject to false alarms. The problems were traced to a basic lack of infrared data. There had been no satellite or

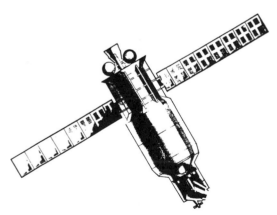

Artist's concept of the Midas satellite. The Midas orbited nose down with the infrared scanner pointing towards the earth. Two solar panels supplied power. *DoD Drawing*.

instruments available during Midas's development to find this data. The U-2 measurement program actually followed Midas's development rather than coming before. At first, the task had seemed simple. The infrared scanner would view the cool earth. A hot ICBM exhaust should be a major increase in heat and be readily detectable. In practice, it was much more complex. The exhaust itself was primarily hot carbon dioxide and water vapor. These, however, are also constituents of the upper atmosphere. The only difference is that the upper atmosphere is cooler and at lower pressure. Seen against the background of the lower atmosphere, also containing carbon dioxide and water vapor at still lower temperatures, it becomes a matter of shading rather than distinctly separate things. To further complicate matters, natural phenomena can mimic ICBM plumes. A particular problem was high-altitude clouds reflecting sunlight.[12] A further problem was that different boosters had different spectra (solid-fuel boosters' exhausts are cooler and harder to detect). The infrared characteristics also changed with altitude.

Midas also faced management problems. The Air Force and Lockheed were under tremendous pressure to get Midas operational to protect Strategic Air Command (SAC) bombers. (It will be remembered these were the years of the missile gap.) They were preoccupied with lesser details rather than with the basic physics of infrared detection. They were also overly optimistic about the reliability of the very complex (more than 60,000 components) Midas payload.[13]

In late 1961, Midas was cut back to a test effort with a designation of Program 461. Plans for an operational system were abandoned. The Air Force request for $190 million was cut back to $35 million. To justify this decision, Dr. Harold Brown, director of Defense Research and Engineering, who recommended the cuts, said, ''The way the program was going, it would never produce a reliable, dependable system. . . . Within a year or two reorientation of the program may result in obtaining the basic information which will then enable us to go to some other system.'' He continued on to say that this new system ''would be a very different system from the one originally proposed.'' Dr. Brown estimated that about half the $423 million spent on Midas was wasted on ''premature system-oriented hardware.'' The other half provided ''data which are quite necessary for any system . . . we may develop in the future.''

Another reason for the Midas cutback was the more diversified U.S. nuclear force that had developed during the missile gap. With missiles in hardened silos, Polaris submarines, and bombers on airborne alert, the added 15 minutes of warning time was less important, particularly in light of the handful of ICBMs the Soviets had.[14]

The first Midas to be launched after the change went up on 9 April 1962. Since the Kennedy administration's secrecy order had gone into effect in the meantime, the Midas was not named. But its 2,102- by 1,749-mile (3,382- by 2,814-kilometer) orbit clearly labeled it as a Midas. An Atlas Agena B failed on 17 December 1962; presumably, it was a Midas.

Midas's fortunes were at a low ebb. The competency of both the Air Force and Lockheed was being questioned. Then came the Midas launch of 19 May 1963. The satellite operated for 6 weeks and detected nine Titan II, Atlas E, Minuteman, and Polaris launches. The spotting of the solid-fuel Minuteman and Polaris launches was important, since the infrared system had been designed to detect only the hotter, longer burning, liquid-fuel ICBMs.

The Minuteman, Titan, and Atlas launches were detected in real time. The Polaris required additional analysis. Midas had finally achieved success.

A second Midas satellite was a launch failure on 12 June. The program bounced back with the 18 July 1963 launch. The Midas test satellite operated for only ninety-six orbits before starting to tumble. In that time, however, it detected an Atlas E launch on 26 July. It also detected Soviet missile launches. The two successful Midas flights of 1963 showed satellite early warning was feasible.[15] In January 1964, President Lyndon Johnson announced their success.[16,17]

On this note, Midas launches stopped for 3 years. This reflected a reorientation of the program. On 7 November 1963, further Midas test flights were dropped in order to develop a system to detect submarine-launched missiles as well as IRBMs, and to pinpoint their launch sites.

The Midas follow-up program had two phases. The first phase had three parts—January–May 1966, initial tests; January 1967–April 1968, improved lifetime and the ability to perform both early warning and intelligence; and January–July 1968, limited operational status. In the first two parts, the satellites would be placed into the Midas's 2,000-nautical-mile- (3,704-kilometer-) high orbit. The third part would see the satellites put into 6,000-nautical-mile- (11,112-kilometer-) high orbits. Eleven Atlas Agena D launches would be made.

The second phase had two parts—March–November 1969, initial tests; and January–August 1970, the operational system in a 6,500-nautical-mile- (12,038-kilometer-) high orbit. The second phase would use Titan III boosters and seven launches. As with Midas, a network of satellites would be used.[18]

The first part of the Midas follow-on program began in the summer of 1966. The resumption had a shaky start. On 9 June, the Midas was left stranded in a 2,247- by 108-mile (3,616- by 174-kilometer) transfer orbit. The Midas is first placed into an elliptical orbit; it then coasts halfway around the world until the orbit's high point is reached. The Agena engine is then fired again to circularize the orbit. On the 9 June flight, the second burn failed. The other two flights, on 19 August and 5 October, went into the familiar 2,000-nautical-mile- (3,704-kilometer-) high orbit. The launch vehicle was the Atlas Agena D. The new flights may have tested a new concept in early warning sensors. By the mid-1960s, there seemed no way to completely eliminate false alarms. A solution was to combine the infrared detector with a television camera. If the scanner should detect an infrared source, the television picture would show ground controllers if it was from high-altitude clouds or a real launch.

Program 949

By the end of 1966, the Midas follow-on program had been cancelled and the U.S. early warning program once more reoriented. Flights halted for 2 years. The problem was that Midas required a network of from eight to twelve or fifteen satellites to keep the Soviet landmass under constant surveillance.[19] This was necessary because each satellite would be over the Soviet Union for a limited time each day, and no gaps could be allowed to develop. Replacement launches would be necessary as each satellite failed, which meant a high operating cost of $100 million per year.[20]

A similar problem was faced by the devel-opers of U.S. civilian communications satel-lites. Two dozen low-orbit satellites would be needed, which was time consuming and costly. In 1963, however, it was realized all of this could be avoided by putting a single satellite into geosynchronous orbit. This orbit had a period of 24 hours; the satellite's orbital motion exactly matched the rotation of the earth. The satellite, therefore, appeared to stay in one place in the sky.[21] A single early warning satellite, in geosynchronous orbit, could keep the entire Russian landmass under constant surveillance. Two other satellites could cover the ocean areas used by Soviet missile submarines. The advantages over Midas are obvious; however, the geosynchronous orbit approach is not without difficulty. The major penalty is in booster payload. An Atlas Agena D could put 8,800 pounds (3,992 kilograms) into low orbit but only 2,920 pounds (1,325 kilograms) into geosynchronous orbit.[22]

The geosynchronous early warning satellite was given the designation Program 949. The infrared scanner payload, built by Aerojet-General, was described as weighing about 1,000 pounds (454 kilograms) and having a telescope diameter of 3 feet (0.9 meter). The infrared detector itself was cooled by liquid hydrogen, reducing its temperature to 5 degrees Kelvin. Cooling was important since otherwise the weak infrared energy from the missile would be swamped by the detector's own heat. The package was mounted on the Agena D's nose at a 5 degree angle. As the Agena spun, the detector would scan in a circular pattern. The package also carried a television camera for launch confirmation.

The first Program 949 launch was reportedly scheduled for spring 1968. It was apparently delayed. Finally, on 6 August 1968, an Atlas Agena D launched from the Cape put the payload into a 24,769- by 19,686-mile

(39,860- by 31,680-kilometer) near-geosynchronous orbit. It had an important difference from the geosynchronous orbit used by communications satellites. Program 949's orbit was inclined 9.9 degrees, which had an interesting consequence for a satellite in geosynchronous orbit. As seen from the earth, the satellite would trace out a figure eight. At the northern end, the satellite would have a better view of the Soviet Union—important, given its northerly location. This would last 12 hours. For the remaining 12 hours of the day, the satellite would be south of the equator and accordingly have a poorer view.

Given the poor track record of early warning satellites and the fact that this satellite was the first of its kind, tests probably occupied most of its operating life. It is probable the satellite was parked over the Pacific so it could observe ICBM launches from Vandenberg. It could measure the infrared signature of both liquid- and solid-fuel missiles under all conditions. The U.S. was very sensitive about the satellites. No orbital characteristics were released, just the statement: "Current Elements Not Maintained." This implied it had failed, was only debris, and not worth tracking. (The correct elements were released by the Royal Aircraft Establishment in England.) [23]

A second Program 949 early warning satellite was launched on 13 April 1969, into a 24,402- by 20,301-mile (39,270- by 32,670-kilometer) orbit, again inclined 9.9 degrees. As with the first mission, NORAD listed only, "Current Elements Not Maintained." This concealment of orbital characteristics went beyond anything previously done. Although reconnaissance satellites were not named, their orbital characteristics were published—if belatedly. A possible explanation would be to prevent the Soviets from learning the area covered and when the satellites were in a poor viewing position.

In late 1969, heavily censored congressional testimony was released. In it, Dr. John S. Foster, Jr., director of Defense Research and Engineering, said that $157 million was being requested in fiscal year 1970 for "continued development and initial deployment" of the early warning system. He also gave an indication of Program 949's purpose: "We are accelerating the [deleted] program which will provide early warning by satellite . . . to insure adequate warning for our bomber force against the possibly increasing threat of Soviet sea launched missiles." [24] Program 949 satellites were not watching for ICBMs from Russia. BMEWS took care of this. Rather, they were to be stationed over the Pacific and Atlantic patrol grounds of Soviet missile submarines. Program 949 thus was to fill a gap in the U.S. early warning network. BMEWS covered the northern approach, but to the west and east, coverage was spotty or nonexistent. Soviet missile submarines submerged off the U.S. coast were not covered by radar. Their missiles were targeted against U.S. bomber bases. The detection of an attack from these submarines at launch was important, since their missiles had a flight time of only 5 to 10 minutes. Even with maximum warning time, this was barely long enough for the B-52s to take off.

Two Program 949 launches were made in 1970. According to one report, these were larger, more advanced satellites. They used an Atlas Agena D with an additional third stage, which was probably a kick stage to transfer it from the elliptical initial orbit into the final inclined geosynchronous orbit. During the first launch, on 19 June, the third-stage engine apparently failed. The satellite was left stranded in the 20,932- by 111-mile

A Program 647/Defense Support Program early warning satellite undergoing checkout at Cape Kennedy. The satellite is the result of a decade of studies, test flights, and setbacks. The task of detecting ICBM launches proved more difficult than originally thought due both to atmospheric physics and technical shortcomings. Clearly visible is the large infrared telescope. The small box at its base holds the nuclear detection sensors. The two cone-shaped objects sticking out from the side of the spacecraft are star sensors used to stabilize the spacecraft. *U.S. Air Force Photo.*

(33,685- by 178-kilometer) transfer orbit. The second launch on 1 September 1970 was successful, putting the satellite into the standard inclined geosynchronous orbit.[25] With this launch, the U.S. had an interim operational satellite warning system against submarine-launched missiles. Launches continued in the years to follow. An Atlas Agena D failed in December 1971. A year later, on 20 December 1972, another payload went into the inclined geosynchronous orbit used by Program 949 satellites. This flight brought the program to a close. In retrospect, Program 949 was an interim step. Its purpose was as much to develop reliable infrared scanners and to gain experience as it was to watch for submarine-launched missiles and fill gaps in radar coverage. Even as Program 949 was reaching operational status, the definitive U.S. early warning satellite was making its initial flights.

Program 647

On 6 November 1970, a Titan III C climbed into the Florida sky, trailing smoke and flames from its two solid-fuel boosters. Once in orbit, the transtage and its payload began the long, slow climb towards geosynchronous altitude. Reaching it, the transtage engines reignited. Before they could complete the burn, the bad luck that had plagued previous early warning satellites raised its head. The engines malfunctioned and left the payload in a 22,300- by 16,187-mile (35,886- by 26,050-kilometer) orbit.[26] This inauspicious note marked the debut of the Program 647 early warning satellite. TRW, the prime contractor, had begun work on it in late 1966 and early 1967. They were to build four satellites (one for ground tests and three flight models), which were phase 1 satellites. The phase

2 satellites to follow were the operational versions.[27] The Program 647 satellite is a large object, 21.5 feet (6.5 meters) long and 9 feet (2.7 meters) across its base, weighing 2,500 pounds (1,134 kilograms). The four solar panels span 23 feet (7 meters). The spacecraft is in two parts—the cylindrical equipment section, which contains the electronics and power system, and a 24-inch (61-centimeter) in diameter momentum wheel used to stabilize the satellite. The outside of the 9-foot- (2.78-meter-) long cylinder is covered with solar cells. Attached to it is a 12-foot- (3.63-meter-) long Schmidt telescope. The infrared energy enters the telescope and passes through a corrector lens and is reflected off a 3-foot (0.91-meter) in diameter mirror. It then strikes the spherical infrared detector positioned midway between them. This detector is fitted with 2,000 lead sulfide cells, each one covered by a narrow band filter that allows only the peak wavelength of a rocket's exhaust to pass through. From geosynchronous orbit, each cell covers an area on the earth less than 2 nautical miles (4 kilometers) across. When infrared light strikes a cell, its electrical resistance changes. The difference in current is then measured.

The Schmidt telescope is mounted on the satellite at a 7.5 degree angle, and the satellite is spun at 5 to 7 revolutions per minute. The telescope scans a particular area on the earth once every 8 to 12 seconds as it traces out a circular path. This is one of the ways the designers hoped to reduce false alarms. The rotating telescope would "see" a stationary infrared source such as a forest fire or the infamous high clouds remain in the same spot over several scans. A rocket, however, would be in a different position on each scan. Another technique is to adjust the threshold of the cells so they would not be triggered below a certain level. Also, the on-board computer

is programmed to reject false signals. The data from the infrared cells is passed on to the electronics located behind the main mirror, where they are digitized. The information transmitted to earth includes that from the individual sensors, which pick up the source, the time/frame number, and, most importantly, the voltage reading from the cell. This voltage is proportional to the amount of infrared energy striking the cell: a weak source will produce a low voltage, a strong source an accordingly higher voltage. The signals are received by two ground stations—one at Alice Springs, Australia, in the Outback and the other at Buckley Airfield in Colorado. From there, they are relayed to NORAD headquarters under Cheyenne Mountain, Colorado.[28]

Because of the transtage failure, the first Program 647 satellite was left in a 20-hour orbit rather than in a geosynchronous one. As seen from a spot on earth, the satellite would drift slowly across the sky, taking 3 days to go from west to east. Another 3 days would pass before it would again appear over the western horizon. Because the satellite was not in a fixed position, the amount of scanner testing was severely limited. Indeed, the only data may have been from housekeeping systems.[29]

The second launching was made on 5 May 1971. It was successful, and the satellite was parked over the Indian Ocean to monitor the Soviet landmass. The third phase 1 satellite, launched on 1 March 1972, also went into geosynchronous orbit and was stationed over Panama to watch for submarine-launched missiles.[30]

The satellites succeeded beyond the Air Force's hopes. The phase 1 satellites were originally designed only for testing purposes, but in 1972 the network was declared operational and was turned over to the Air Defense

Command. Part of the reason for their success was the use of well-established, proven technology. The detector cells are a case in point. Lead sulfide is one of the oldest materials used for infrared detection. Also, it does not require cryogenic cooling. Lead sulfide has good sensitivity at temperatures as high as 193 degrees Kelvin.[*] Only passive cooling is needed; heat can simply be radiated away into space, which avoided the need for complicated plumbing and tanks for supercool liquids. The telescope proved to have unsuspected added capability. The Air Force soon found it could not only detect missiles but also provide tracking data. The telescope's successive scans traced out the missile's flight path and from this, the launch azimuth could be determined. Only certain azimuths can be used to attack the U.S. Thus, satellite launches can, in most cases, be quickly separated from an ICBM attack. From the satellite's tracking data, the target of a missile can be determined within 6 minutes of launch. In certain cases, satellite and ICBM launch azimuths do overlap. The difference is in the rocket's altitude above the earth. The 647 satellite cannot indicate this. As the missile nears final stage burnout, the amount of infrared energy it emits drops off and the satellite loses track. By this time, however, the missile will be above the horizon of one of the BMEWS stations. Its exact flight path can then be determined.[31,32]

While standing watch for an attack, the Program 647 satellites also perform an intelligence function. By the end of 1974, the satellites had observed more than 1,000 Russian, Chinese, U.S., and French rocket launchings. From the satellites' data, estimates could be made of the ICBMs' capabilities. Observations of U.S. launches served as control samples to assess the estimates' accuracy. At this time, the first flight tests of the Soviet SS-18, SS-19, and SS-N-8 missiles were being made. The Program 647 satellites corrected erroneous estimates of certain Soviet missiles' capabilities.

Phase 2 Launches

After 10 years of disappointment and failure, the hopes originally expressed for Midas were finally being realized. The Air Force placed orders with TRW for ten phase 2 early warning satellites. The main difference in these production satellites was the addition of nuclear radiation detectors to take over monitoring of the nuclear test ban treaty from the Vela satellites. (Nuclear detection will be dealt with in a later chapter.) The first phase 2 satellite was delivered to the Air Force in February 1973. The next seven were scheduled to follow at 3-month intervals. On 12 June 1973, the first phase 2 satellite went into a near-geosynchronous orbit. With its debut, a remarkable change took place. Normally, a satellite was referred to only by its project number (i.e., 647) and all details, even its appearance, were classified. However, in 1973 and 1974, many details of the Program 647 early warning satellite were released, including not only its shape but also how it worked. The satellite was also renamed the Defense Support Program (DSP). A possible reason for such uncharacteristic openness was to let the Russians know the U.S. could detect an attack.

---

[*] The freezing point of water is 273 degrees Kelvin.

Two years would pass before another Defense Support Program satellite was launched. The two launchings of 1975 had rather distinctive individual orbital characteristics. The first, launched on 18 June 1975, went into a 25,353- by 18,766-mile (40,800- by 30,200-kilometer) orbit, inclined 9 degrees, similar to the Program 949 inclined geosynchronous orbit. This was the greatest inclination and most elliptical orbit of any DSP satellite up to that date. The previous three had inclinations less than 1 degree. The second launch, on 14 December, went into a 22,237- by 22,166-mile (35,785- by 35,671-kilometer) orbit, typical of the circular orbits used previously. (The inclination was an atypical 3 degrees, however.) The pattern would repeat with subsequent launches. Satellites with very low inclinations went into circular 21,749-mile (35,000-kilometer) orbits. Those with large inclinations (the 10 June 1978 launch had a 12 degree inclination and the 1 October 1979 satellite had a 7.5 degree inclination) went into 24,856- by 18,642-mile (40,000- by 30,000-kilometer) elliptical orbits. The inclined and elliptical orbits would allow the Defense Support Program satellites to cover a wider area as they traced out a figure eight ground track.[33,34]

The phase 2 satellites underwent regular improvements and upgrading. The first dealt with a problem that appeared in the phase 1 satellites—a gradual loss of sensitivity in the infrared sensors. It was speculated that the last Program 949 launch was to test an improved array.

Another improvement was an expansion of the network. Originally, the DSPs were positioned over the Indian Ocean (to watch Russia) and Panama (to detect submarine-launched missiles from the Pacific and Atlantic). With the increase in range of Soviet submarine missiles, a larger ocean area had to be covered. The Panama station was replaced by two new stations: one was over the east central Pacific (about 135 degrees west latitude); the other was over the rain forest of Brazil (about 70 degrees west). These satellites used the circular, low-inclination, geosynchronous orbit. Throughout the 1970s and early 1980s, DSPs were launched at an average rate of one per year. Individual satellite's lifetimes were measured in years, which was much better than the Air Force expected. Accordingly, replacement launches were not required as often as the delivery schedule envisioned. A backlog of DSP satellites accumulated. By the mid-1970s, more than half a dozen satellites were in storage. Plans for ordering additional satellites were delayed. The final satellite of the original series was funded in fiscal year 1975, and another 7 years would pass before the next was requested. The Air Force at first sought two DSPs for fiscal year 1982. This was subsequently cut to a single satellite, the fourteenth DSP built, due to their reliability.[35]

Vulnerability

If the Russians should actually make a nuclear attack on the U.S., a first step would be to blind or disable U.S. early warning satellites; while the ground controllers worked on what they thought was a malfunction, the Soviets could launch their missiles undetected.

The DSPs were considered highly vulnerable. Their position in geosynchronous orbit was a fixed point in space. Accordingly, several changes were made starting in 1974–1975 to improve their survivability. One change was to add an impact detector, which radioed a warning to the ground if the satellite was

hit by a foreign object. The Soviet ASAT killed by hitting its target with shrapnel. The device also acted as an airline flight recorder. From its information, the Air Force was able to determine that one malfunctioning DSP was disabled by a ruptured propellent line.[36,37]

Another area of vulnerability was the two ground stations. They were vulnerable to attack, sabotage, and in the case of the Alice Springs station, political upheavals. (Some segments of the Australian Labour Party have objected to the presence of U.S. bases.) Their loss would sever the link between the satellite and command authority. To correct this, six mobile ground stations were developed. These vans, carrying the antennas and data processing and relay equipment, could be moved by air to any place they were needed and safe. Similar equipment also could be placed aboard E-3A and E-4B aircraft.[38]

Another way DSPs could be disabled is attack by Soviet ground lasers, which has a long history. As early as 1974, Martin Marietta studied a passive laser warning device— a silicon detector inside the telescope that would either sound an alarm or close a protective shutter if the satellite was illuminated by a laser. Optical systems are particularly vulnerable to laser damage.[39]

A year later, an event occurred which seemed to be a Soviet laser attack on an early warning satellite. On 18 October 1975, a DSP satellite over the Indian Ocean detected a strong source of infrared energy, which effectively blinded it. This occurred five times during October and November, in one case lasting 4 hours. Nothing like this had occurred before.

Some in the Defense Department concluded the Soviets had been using a laser to blind the satellite.[40] Although the laser beam would spread over the 22,000-mile (35,398-kilometer) distance to the satellite, the telescope would collect and concentrate the light. Subsequent analysis indicated it was actually gas pipeline fires that had caused the incidents, which was reportedly confirmed by reconnaissance satellite photos of the pipeline breaks.[41,42] In the wake of the incident, concern about laser attacks dropped off. The Martin Marietta detector never went past the breadboard stage. In the early 1980s, with increased awareness of the dimension of the Soviet beam weapons program, this changed. Some areas on the DSPs were covered with heat-resistant ablative materials. New optical materials were used that were not damaged when exposed to laser light, and electronic filters were added to prevent damage to the infrared sensor elements.[43]

Current and Future Developments

In the early 1980s, the DSP satellite system continued to evolve, including better communications and an improved data processing system to give more precise tracking information. Another change was in the sensor

---

Night launch of a Titan III C in 1973. Similar boosters were used to orbit Program 647/Defense Support Program early warning satellites. The satellite and attached transtage are initially placed in a highly elliptical transfer orbit. When the satellite reaches the orbit's high point, the transtage engines restart to circularize the orbit and geosynchronous altitude; the satellite then separates from the transtage. Once the satellite is operational, its small engines can be used to adjust the orbit. The satellite can then drift to a new station and be stablized there. *U.S. Air Force Photo.*

systems, in order to cope with the Soviet SS-N-6 submarine–launched missile. Its exhaust plume was difficult to detect by the old system. Production of DSP satellites resumed in June 1981. The fourteenth satellite was readied for launch in early 1982, and work began on the fifteenth, sixteenth, and seventeenth DSPs.[44] Also in 1982, control of the DSP satellites was transferred to the new Air Force Space Command. The network, in the same year, was again expanded from three to four satellites, a process that began with the launch of the fourteenth DSP on 6 March 1982. It was positioned over Brazil to cover the Atlantic Ocean. The old Atlantic satellite, the twelfth DSP (launched 16 March 1981), fired its small, on-board engines and maneuvered out of geosynchronous orbit, which caused it to drift across the sky. It was finally positioned over the Pacific Ocean. The old Pacific satellite, the tenth DSP (launched 10 June 1979), was also similarly maneuvered, finally taking up a position over a spot at 86.5 degrees west. This is about 435 miles (700 kilometers) off the coast of Ecuador—a position not previously used. Various reasons were suggested. Earlier, the Air Force had proposed the network be expanded to five satellites for better coverage; this four-position network using an older satellite to fill the new slot may be a low-cost alternative. The new position may be for laser testing of countermeasures aboard the satellite; the position would be visible from the White Sands test site. It may be to provide a target for deep space satellite tracking systems. And finally, it may simply be in semi-dormant storage.

The shifting around tends to imply the watch over the Atlantic Ocean has a higher priority than the Pacific. The new satellite would be more capable than the older ones.[45]

Subsequent launches would be from the space shuttle. The DSPs have been modified with laser protection, higher sensitivity, and a greater electrical supply from an extended equipment section. The satellites will be boosted into geosynchronous orbit by an IUS. The first shuttle-launched DSP was scheduled for delivery to the Cape for initial checkout in late 1985 or early 1986. Its flight was tentatively scheduled for late 1986 or early 1987, depending upon the health of the satellites already in orbit.[46] The loss of *Challenger* puts this into question. Launches will either be delayed or transferred to Titan 34 Ds.

The future directions of the U.S. early warning satellite program are uncertain at the moment. One possibility is to simply keep improving the DSPs. This new version would be able to function throughout a nuclear exchange and to transmit data directly to the user. A major new mission would be tracking aircraft from orbit. The present DSP can pick up aircraft using afterburners. The modified DSPs would have three times the number of sensor elements. With this, both bombers and transport aircraft could be detected just from their hot exhaust.

The alternative is a completely new system—the Advanced Missile Warning System. It would have the same goals as the modified DSP but would use an array of several thousand charged coupled devices. These would not spin like the DSP telescope but would stare at the earth, taking in a panoramic view. Several thousand detector elements would be on a single chip along with the readout and data processing circuits. Multiple integrated circuits could be strung together to form arrays mounted at the focal point of the telescope. This mosaic array has several advantages. Coverage and resolution are improved; with more elements, finer details could be detected. As the complete field of view is covered at all times, short-duration events can be spotted. One big advantage is cost. The infrared detectors in the DSP must be hand-assembled, -wired, and -installed, increasing the cost to $4 million per array. The integrated circuits, however, can be mass produced with a low unit cost.

The mosaic array is to be tested aboard the Teal Ruby satellite. The infrared package is understood to have 150,000 detectors in twelve subarrays, each subarray covering a different part of the infrared spectrum. Its purpose is to test the feasibility of detecting both bombers and smaller targets, such as cruise missiles, from space.[47,48]

NOTES Chapter 17

1. William Hood, *Mole,* (New York: W. W. Norton, 1982): 152,153.
2. David A. Anderton, *Strategic Air Command,* (New York: Scribners, 1975), 53–58.
3. Carrollton Press DoD 36C 1980.
4. John Ball, Jr., *Edwards: Flight Test Center of the USAF,* (New York: Duell, Sloan and Pearce, 1962), 119–125.
5. "Midas Infrared Measurements Cut Short by Telemetry Failure," *Aviation Week & Space Technology,* (6 June 1960): 31.
6. Carrollton Press DoD 36D 1980.
7. "Discoverer XIX Gathering Midas Data," *Aviation Week & Space Technology,* (2 January 1961): 46.
8. Philip J. Klass, *Secret Sentries in Space,* (New York: Random House, 1971), 175.
9. Carrollton Press White House 194A 1977.
10. James Trainor, "MIDAS Faces Re-orientation," *Missiles and Rockets,* (16 July 1962): 12.
11. "Schriever Says Over-Optimism Hurt Midas," *Aviation Week & Space Technology,* (6 August 1962): 28.
12. Philip J. Klass, "Lack of Infrared Data Hampers Midas," *Aviation Week & Space Technology,* (24 September 1962): 55–57.
13. "Schriever Says Over-Optimism Hurt Midas," *Aviation Week & Space Technology,* (6 August 1962): 28.
14. "Half of Midas Spending Viewed as Waste," *Aviation Week & Space Technology,* (17 June 1962): 38.
15. The Air Force in Space Fiscal Year 1964, Microfilm roll #30730 U.S. Air Force Historical Research Center, Maxwell AFB, Alabama.
16. Heather M. David, "MIDAS Concept Showing Promise," *Missiles and Rockets,* (3 February 1964): 19.
17. Donald E. Fink, "New Missile Warning Satellite Succeeds," *Aviation Week & Space Technology,* (3 February 1964): 33.
18. The Air Force in Space Fiscal Year 1964 Microfilm roll #30730.
19. Reginald Turnill, *The Observers Book of Unmanned Spaceflight,* (New York: Warne, 1974), 74.
20. Donald E. Fink, "New Missile Warning Satellite Succeeds," *Aviation Week & Space Technology,* (3 February 1964): 33.
21. Sidney Metzer, "Geosynchronous Vs. Low Orbit, The First Big Technical Decision," *Communications Satellite Corporation Magazine* (#12, 1983): 24,25.
22. T. M. Wilding-White, *Jane's Pocket Book of Space Exploration,* (New York: Macmillan, 1976), 217.
23. Klass, *Secret Sentries in Space,* 180,181.
24. "DOD Accelerates Plan to Deploy Early Warning Satellite System," *Aviation Week & Space Technology,* (12 January 1970): 12.
25. Anthony Kenden, "Recent Developments in U.S. Reconnaissance Satellite Programmes," *Journal of the British Interplanetary Society,* (January 1982): 36–38.
26. "Satellites Provide Early Warning of ICBMs," *Aviation Week & Space Technology,* (25 October 1971): 14.

27. "Additional Warning Satellites Expected," *Aviation Week & Space Technology,* (14 May 1973): 17.
28. Barry Miller, "U.S. Moves to Upgrade Missile Warning," *Aviation Week & Space Technology,* (2 December 1974): 16–18.
29. *Astronautics and Aeronautics 1970,* (NASA SP-4015), 376.
30. Anthony Kenden, "Recent Developments in U.S. Reconnaissance Satellite Programmes," *Journal of the British Interplanetary Society,* (January 1982): 36–38.
31. Clarence A. Robinson, Jr., "Space-Based System Stressed," *Aviation Week & Space Technology,* (3 March 1980): 25,26,33,35.
32. Herbert J. Coleman, "NORAD Broaden Operational Horizons," *Aviation Week & Space Technology,* (16 June 1980): 233–237.
33. Anthony Kenden, "Recent Developments in U.S. Reconnaissance Satellite Programmes," *Journal of the British Interplanetary Society,* (January 1982): 36–38.
34. Anthony Kenden, "Military Maneuvers in Synchronous Orbit," *Journal of the British Interplanetary Society,* (February 1983): 88–91.
35. "Washington Roundup," *Aviation Week & Space Technology,* (3 November 1980): 27.
36. Barry Miller, "USAF Pushes Satellite Survivability," *Aviation Week & Space Technology,* (28 March 1977): 52–54.
37. Barry Miller, "U.S. Moves to Upgrade Missile Warning," *Aviation Week & Space Technology,* (2 December 1974): 16–18.
38. Edgar Ulsamer, "In Focus . . . ," *Air Force Magazine,* (July 1983): 19.
39. Barry Miller, "U.S. Moves to Upgrade Missile Warning," *Aviation Week & Space Technology,* (2 December 1974): 16–18.
40. Philip J. Klass, "Anti-Satellite Laser Use Suspected," *Aviation Week & Space Technology,* (8 December 1975): 12,13.
41. "DOD Continues Satellite Blinding Investigation," *Aviation Week & Space Technology,* (5 January 1976): 18.
42. *Astronautics and Aeronautics 1975,* (NASA SP-4020), 244.
43. "Improvements for Early Warning Satellites Set," *Aviation Week & Space Technology,* (16 February 1981): 18,19.
44. Craig Covault, "USAF Initiates Broad Program to Improve Surveillance of Soviets," *Aviation Week & Space Technology,* (21 January 1985): 15.
45. Anthony Kenden, "Military Maneuvers in Synchronous Orbit," *Journal of the British Interplanetary Society,* (February 1983): 88–91.
46. Craig Covault, "USAF Initiates Broad Program to Improve Surveillance of Soviets," *Aviation Week & Space Technology,* (21 January 1985): 15.
47. Edgar Ulsamer, "In Focus . . . ," *Air Force Magazine,* (July 1983): 19.
48. Anthony Kenden, "Recent Developments in U.S. Reconnaissance Satellite Programmes," *Journal of the British Interplanetary Society,* (January 1982): 36–38.

CHAPTER 18

Soviet Early Warning Satellites

Like the U.S., the Soviet Union was also the victim of a surprise attack in World War II. Circumstances, however, were different. The success of the German invasion was due entirely to the blindness of one man—Josef Stalin. He ignored repeated warnings, even one from a German deserter who gave the date and exact time of the attack. From their experiences in what they call the "great patriotic war," the Soviets have come to place great emphasis on surprise and initiative. If war seemed inevitable, the best course would be to launch a sudden, violent attack to disrupt the enemy's preparations. If the enemy should strike first, then the Soviet military must be ready to launch a counteroffensive to regain the initiative. In either case, the enemy is to be destroyed on his own territory. This doctrine covers both conventional and nuclear forces.

There has never been any doubt about who the enemy is. From the days of the revolution, it has been a basic cornerstone of Soviet policy that the capitalist West was seeking every opportunity to destroy communism and reestablish capitalism in Russia. Since the early 1960s, the Soviet people have been told the U.S. was plotting to launch a surprise nuclear attack against the USSR. It is only the might of the Soviet military that has prevented it.[1]

Given this professed fear of a Western attack, early warning satellites would seem likely to be a high priority. Indeed, a 1962 Stanford Research Institute study said, "Of all the conceivable military satellites, one of the most likely to exist in the post-1965 era will probably be the missile warning satellite." (Reconnaissance satellites were rated only a medium probability.)[2] The Soviets built their own equivalent of the BMEWS (called Hen House) in the early 60s. But it, too, had limits on coverage. The Soviets would have known about Midas from U.S. press reports in the late 1950s. The Soviet early warning satellite effort was probably underway by the mid-1960s, roughly the same time as Midas was being cut back. Presumably, the Soviets also made measurements of rocket plumes. The early history of the program, however, is obscure. It is not even certain which was the first Soviet early warning satellite. The reason is its orbit. Normally, different types of satellites go into distinctive orbits. Soviet early warning satellites, however, use orbits nearly identical to those of their Molniya communications satellites. It requires careful analysis to separate failed Molniya satellites and the early warning vehicles. The giveaway is the shape of the ground track, the points at which the orbit crosses the equator, and whether the satellite fits into the regular orbital spacing of the Molniya satellites.

The early warning satellites' orbital characteristics are approximately 24,458 by 391 miles (39,360 by 630 kilometers). Molniyas go into orbits of approximately 25,353 by 311 miles (40,800 by 500 kilometers). Individual satellites can vary greatly. Both orbits have periods of 12 hours. For a long time, it was thought that *Cosmos 174* and *Cosmos 260,* launched in 1967 and 1968, respectively, were Soviet early warning satellites. A more careful analysis indicated they were, in fact, failed *Molniya 1* satellites. The first was replaced by the sixteenth *Molniya 1* one month after it was launched. The other was apparently to be a replacement for the eighth *Molniya 1* satellite.[3]

Another candidate for the first Soviet early warning satellite was *Cosmos 159,* launched on 17 May 1967. It went into a 37,657- by 236-mile (60,600- by 380-kilometer) orbit. Some have suggested this mission was actually a test of the Soyuz propulsion module.[4] It is an ambiguous situation at best.

Between December 1968 and September 1972, there were no launches for which an early warning role was even suspected. The first Soviet mission that can be clearly identified as an early warning satellite is *Cosmos 520.* It was launched on 19 September 1972, into a 24,433- by 405-mile (39,319- by 652-kilometer) orbit from Plesetsk. The booster was an A-2e. The next such mission, *Cosmos 606,* followed in November 1973. Then came *Cosmos 665* in June 1974 and *Cosmos 706* in January 1975. These four missions were clearly development flights. Their infrared sensors were presumably tested against both U.S. and Soviet ICBMs. Their vulnerability to false alarms and the systems' overall reliability could also be assessed by these flights.

Mode of Operation

After *Cosmos 706* was launched, there was a halt of such flights for 21 months. Certainly, the Soviets would want to look over the data. This long a delay, however, implied that there may have been Midas-like problems. Tests resumed with *Cosmos 862* (22 October 1976). One peculiarity of this flight was that its ground track was positioned farther east than was normal, which would have limited its usefulness for early warning over the U.S. One possible reason was so it could have a better view of Soviet test launches. The satellite was later blown up.

The move to operational status came in 1977. *Cosmos 903* went up on 11 April. *Cosmos 917* followed in June, and *Cosmos 931* a month later in July. The satellites were positioned 80 degrees apart.

The satellite's distinctive orbit served two functions. First, it maximized the amount of time the satellite was over the northern hemisphere, because the satellite's speed changes as it orbits the earth. Its highest velocity occurs closest to the earth—over the southern hemisphere. As the satellite climbs away from the earth, gravity pulls it back and its speed diminishes. The lowest velocity comes when the satellite is farthest from the earth, which means that of each 12-hour orbit, the satellite is north of the equator about 85 percent of the time.[5] Second, its high looping orbit allows the satellite to monitor the central U.S., where the missile silos are located, while also having direct communications with Russian ground stations. The satellite makes two orbits each day. One of them passes off the eastern U.S. coast; the other takes place over eastern Russia and the western Pacific. In the first orbit, the satellite can keep U.S. missile fields under watch for just over 9 hours. During the second orbit, it can also keep watch on the U.S. for another 6 hours. This orbit also occurs almost directly over China, so the Soviets can monitor Chinese missiles.[6]

Because of the similarity in orbits, the early warning satellites are often presumed to use a modified version of the Molniya. If correct, the satellite is a cylinder topped by the cone-shaped maneuvering engine housing, with a diameter of 5 feet (1.5 meters) and a height of 11 to 13 feet (3.4 to 4.1 meters). The engine is used to correct its orbit so the ground track repeats each day. Around its base are six large solar panels spanning about 30 feet (9 meters). The infrared scanner would replace communications relay equipment. The Molniya is stabilized in all three axes rather than spin-stabilized like the DSP. Assuming the Soviet early warning satellites were actually modified Molniyas, the scanner would have to be movable, sweeping back and forth over the earth below.[7]

For the first 3 years of operation, the Soviets maintained three early warning satellites on station. Launches were not to expand the network but to replace existing satellites that had malfunctioned. *Cosmos 1024* replaced *Cosmos 931*, for instance. These replacements indicated an average lifetime of 13 months.[8] There were two such replacement launches each in 1978 and 1979.

The Network Expands

A major step-up in early warning activity came in 1980, with six launches being made. The real importance was not immediately apparent, however. The first launch of the year, *Cosmos 1164* on 12 February, was a failure.

A Soviet Molniya communication satellite. Some analysts have suggested a modified system is used as the Soviet early warning satellite. Both use similar orbits and the same booster. If true, the infrared scanner replaces the communications equipment in the base of the satellite. *Soviet Photo via James E. Oberg.*

It was to replace *Cosmos 1109;* but after being placed into a 360- by 139-mile (580- by 223-kilometer) initial orbit, the "e" stage that was to place it into the final elliptical orbit failed to fire, leaving it stranded. The gap in coverage was corrected by *Cosmos 1172,* launched 2 months to the day after the failure. *Cosmos 1188* went up in June as a replacement for *Cosmos 1024,* launched in 1978. *Cosmos 1188* had a short operating lifetime, being replaced by *Cosmos 1217* in October. The reason was another peculiarity of high, elliptical orbits. They are adversely affected by both irregularities in the earth's gravitational field and by the attraction of the sun and moon. These caused *Cosmos 1188*'s ground track to wander, and it was apparently unable to make corrections. *Cosmos 1191,* in July, was also a replacement launch. The most important launch was *Cosmos 1223* on 27 November. Unlike the others, it was not a replacement but, in fact, went into a new orbital slot. Its launching heralded a major expansion of the network.

The next launch, *Cosmos 1247* in February 1981, was also stationed in a new slot. Both of these were 80 degrees from their nearest neighbor. Four more launches followed in 1981. This high launch rate continued in 1982 with another five satellites. The Soviets were methodically expanding the network. At first, the satellites were 80 degrees apart. Then the Soviets began placing them between the earlier pairs. By the end of 1982, the final nine-satellite operational network was in place. Moreover, the entire constellation was shifted to the east to further refine the coverage.[9]

An interesting sidelight on this activity was the fact that on several occasions, the satellite boosters were observed from Chile and Argentina as "fuzzy" halos. The fourth stage makes its transfer burn over the South Pacific. As the booster and satellite begin to climb away from the earth, they pass over the tip of South America. At this point, roughly 1 hour after launch, the booster begins to vent propellent, forming a cloud that, back-lit by the sun, could be seen from the ground, which is in darkness. The first combination of proper launch timing, clear weather, and sun angle occurred with *Cosmos 1164* on 12 February 1980. It happened again on 14 June 1980 (*Cosmos 1188*) and 31 October 1981 (*Cosmos 1317*). The first was seen and photographed at the Cerro Tololo Observatory in Chile. The other two, which occurred at 7 P.M. and 9 P.M., respectively, on Saturday evenings, frightened tens of thousands of people in Argentina.[10,11]

Another aspect is the relationship between Soviet early warning satellites and the U.S. planetary program. U.S. planetary spacecraft transmit on a radio frequency of 2.29 to 2.30 GHz—the same frequency range used by some of the Soviet early warning satellites. Their signals are 100 to 1,000 times stronger than the signals from the more distant U.S. probes. When a conflict is noted, NASA informally asks the Soviets to shut down the satellites for a brief period. The Soviets have complied when asked. This is not a theoretical concern. During the *Pioneer 11* encounter with Jupiter on 3 September 1979, about 45 minutes of data on the moon Titan was lost because of interference from *Cosmos 1124*. It had been shut down during the requested times on 1 and 2 September, but the controllers had not realized there would also be a conflict on 3 September.[12,13]

Despite the expansion of the network, the Soviet early warning satellites continued to show mediocre reliability. Between 1976 and 1983, twenty-eight launches were made. Of these, fourteen (exactly half) either failed to reach the proper orbit or were exploded. The early warning flights of 1983 were symbolic of the problem.

Cosmos 1456 (25 April) was a replacement for *Cosmos 1191,* which failed in the spring of 1981. *Cosmos 1456* operated until 12 August 1983, when several fragments were detected and the satellite began to drift. The next early warning satellite, *Cosmos 1481* (8 July), failed within a day of launch. Ironically, the previous two satellites launched into this orbital slot also failed. *Cosmos 1261* broke up within a month of launch, and *Cosmos 1285* was left stranded, unable to make the transfer burn into the elliptical orbit. Only the third early warning launch of 1983, *Cosmos 1518* (28 December), was successful.[14]

The program's misfortunes continued in

both 1984 and 1985. The Soviets launched seven early warning satellites in each year. This was nearly enough to replenish the complete network each year.[15]

The Two Systems Compared

In developing their early warning satellites, the U.S. and Russia used completely different approaches. It is worthwhile to step back a moment and see how well the two systems fulfill their goals. The U.S. Program 467/DSP satellites, with their multiyear life span and diverse capabilities, are an obvious success. More importantly, they have overcome the problems that wrecked Midas. The Soviet results are mixed. With a nine-satellite network, more are required; also, each satellite has a much shorter life span. Both of these factors increase operating costs. In a way, the Soviet system, except for the elliptical orbit, resembles Midas. Both required multiple satellites to keep watch. Clearly, a geosynchronous orbit system would be more effective.

There are several reasons why the Soviets did not use a geosynchronous orbit. Most obviously, such an orbit involves a major payload penalty for a given rocket. For the Soviets, a geosynchronous orbit also had geographic problems, due to Russia's northerly location. (In some areas, an antenna would have to be pointed nearly horizontally to track the satellite.) Accordingly, the Soviets developed the highly inclined elliptical orbit.

Although the satellite would not appear to stay fixed in the sky, each satellite would provide several hours of coverage per day. There may be another, more subtle, reason: the U.S. originally used geosynchronous orbits for communications satellites; early warning and ELINT missions came later. When the Soviets began flying early warning satellites, they simply adapted the Molniya orbit. It was not until 1974 that the Soviets put a satellite into a geosynchronous orbit (*Cosmos 637*).

Soviet early warning satellites, according to some reports, share another characteristic of Midas—infrared sensor problems.[16] One must remember, however, the years of often bitter experience the U.S. acquired with infrared sensors before the Soviets began flying their satellites. A lag would be expected.

The future direction of the Soviet program is unknown. Certainly, there will be improvements to the satellites. That is to be expected. The real question is whether the Soviets will go to a geosynchronous orbit. They may have already made a move in that direction. On 8 October 1975, *Cosmos 775* was placed into geosynchronous orbit. At the time, it was suspected that it was an early warning satellite.[17] Since then, however, no successors have been flown. The only Soviet satellites placed in geosynchronous orbit are communications satellites. One is left with two possibilities—that *Cosmos 775* was an early warning satellite, but its performance was too poor for the series to continue; or that it was a test bed for the communications satellites to follow.

NOTES Chapter 18

1. Graham D. Vernon, *Soviet Perceptions of War and Peace*, (National Defense University Press, 1981), 58.
2. Max Fishman, *A Study of the Anti-Satellite Capabilities of the Present Nike Zeus System*, (Stanford Research Institute, November 1962).
3. G. E. Perry, "Soviet Early Warning Satellites," *Journal of the British Interplanetary Society*, (February 1982): 72–74.
4. Nicholas L. Johnson, *Handbook of Soviet Manned Space Flight*, (San Diego: American Astronautical Society, 1980), 125,130.
5. The Soviet Space Threat Appendix 1, Aerospace Defense Command, 15 December 1980.
6. G. E. Perry, "Soviet Early Warning Satellites," *Journal of the British Interplanetary Society*, (February 1982): 72–74.
7. T. M. Wilding-White, *Jane's Pocket Book of Space Exploration*, (New York: Macmillan, 1976), 172,173.
8. Nicholas L. Johnson, "Estimated Operational Lifetime of Several Soviet Satellite Classes," *Journal of the British Interplanetary Society*, (July 1981): 282,283.
9. Satellite Digest 147, *Spaceflight*, (August/September 1981): 227,228,236.
10. Correspondence, *Spaceflight*, (September/October 1982): 381.
11. Visual Perceptions of Soviet Space Activities by Naive Eyewitnesses (Paper), James E. Oberg, June 1982.
12. G. E. Perry, "Soviet Early Warning Satellites," *Journal of the British Interplanetary Society*, (February 1982): 72–74.
13. "Soviet Aid Asked on Voyager Interference," *Aviation Week & Space Technology*, (8 November 1980): 32.
14. Nicholas L. Johnson, "The Soviet Year in Space: 1983 Part 3," *Space World*, (November 1984): 26,27.
15. Craig Covault, "Soviets Set to Make Big Gains in Outer Space Exploration," *Aviation Week & Space Technology*, (10 March 1986): 134.
16. Philip J. Klass, "U.S. Scrutinizing New Soviet Radar," *Aviation Week & Space Technology*, (22 August 1983): 19,20.
17. *Soviet Space Programs 1971–1975*, (Library of Congress, 1976).

CHAPTER 19

Nuclear Detection

Amid all the controversy over violations of the SALT I and SALT II treaties, questions arose as to the future of arms control, build down, zero options, and freezes. One treaty has been so successful as to be overlooked: the limited test ban treaty forbids nuclear explosions in the atmosphere, underwater, and in space. There are a number of reasons for its success—the directness of the treaty, the sophistication of underground testing, and the availability of data from earlier explosions. A major reason, however, is a network of various sensors to police the treaty. One part of this is a series of nuclear detection satellites called Vela.

Vela Hotel

In the late 1940s, the only means of detecting nuclear explosions was by picking up the radioactive fallout. From this debris, it was found that details about the design of the weapon and its yield could be found. Several B-29s were modified in the late 1940s to carry sampling equipment. The unit called AFOATS 1 became operational in early 1949. On 3 September, one of its aircraft flying over the Sea of Japan collected fallout from the first Soviet atomic bomb.[1] As the pace of Soviet nuclear testing picked up in the 1950s,

new sensors were developed. Both CIA and Air Force U-2s were equipped with air scoops and filters for picking up fallout. Seismograph networks were set up to detect the earthquake generated by the explosion. Precise air pressure sensors were designed to pick up the atmospheric shock wave caused by the blast. As discussions for a comprehensive test ban treaty—one that would prohibit all nuclear tests—began in 1958, it became apparent that there was a glaring blind spot. Tests underground, in the atmosphere, and underwater could be detected, but there was no way to pick up a nuclear explosion in space. A technical working group that was part of the Geneva Conference on the Discontinuance of Nuclear Weapons Tests proposed a system that would use five or six large satellites at an altitude of more than 18,000 miles (28,962 kilometers). Additional equipment would be placed in the 170 manned ground stations that were planned to monitor the ban. Even with this system, there were costs and technical problems. If the weapon was encased in radiation shielding, it could very well escape detection. An effective monitoring system was not then feasible.[2] It was not a trivial concern. One of the major questions of the late 1950s and early 1960s was antimissile defenses. The ABM warheads would be exploded in space. It was necessary to know what effects the detonation would have on the target warhead. Another unknown was the effect of the fireball on radar tracking systems. The ionized gas could black out radar and radio signals. The explosion in the earth's magnetic field also generated a tremendous electromagnetic pulse, whose current could burn out electronic components. If the Soviets could find the answer to these questions through clandestine tests and develop an ABM defense, the effectiveness of the U.S. deterrent would be diminished.

Work on solving this detection problem began in the fall of 1959. It was part of an intensive examination of the whole question of detecting nuclear explosions. Run by the Atomic Energy Commission (AEC) and Defense Department, the program was called Vela (Spanish for watchman); it had three parts. Vela Uniform looked into detecting underground and surface tests using seismic means. Vela Sierra was concerned with surface detection of nuclear explosions at high altitude. The final part, Vela Hotel, involved the development of satellites that would spot any detonation in space.[3] It was a significant technical challenge. The satellites would have to pick up the burst of X rays, neutrons, and gamma rays emitted by the fireball. The problem was that space was filled with radiation. There were cosmic rays from galaxies; solar flares, great eruptions on the sun that generate vast quantities of X rays; the solar wind, a constant stream of particles emitted from the sun; and, of course, the Van Allen belts, the complex system of trapped radiation that encircles the earth. The Vela satellites would have to pick out the radiation from the bomb amid this natural background radiation. Moreover, it would have to do this for different test locations. These ranged from 20 miles (32 kilometers) or more above the earth to

hundreds of millions of miles away, as well as behind the moon or close to the sun. There was, however, a large amount of basic data from which nuclear-detection instruments could be designed. Part of this came from satellites measuring the natural radiation of space. Other data was from actual, high-altitude nuclear tests.

In 1958, the U.S. made five high-altitude nuclear tests. Their altitudes ranged from 141,000 feet (42,977 meters) to 300 miles (483 kilometers); yields varied from 1 megaton to 1 kiloton. The tests were monitored from orbit by the *Explorer IV* satellite.[4] After 1958, both the U.S. and Soviets began a test moratorium while negotiating a permanent test ban treaty.

The sensors the Vela satellite would carry were planned by the Lawrence Radiation Laboratory and were built by the Los Alamos National Laboratory and the Sandia Corporation. Work progressed rapidly during 1960 and 1961.

On 22 June 1961, the Advanced Research Project Agency (ARPA) approved a five-launch program. Each of the launches would carry two Vela satellites. The first pair was to go into orbit in April 1963, with additional launches to follow at 3-month intervals. Two months after Vela was given the go-ahead, the Soviets broke the 3-year moratorium on nuclear testing. On 5 September 1961, they exploded a nuclear warhead at high altitude above the Sary Shagan ABM test site to test radar blackout.[5] In the Soviets' 1962 test series, three more high-altitude tests were made. The explosions took place at an altitude of 250 miles (402 kilometers).

At the same time, the U.S. was also conducting high-altitude tests from Johnston Island in the Pacific. The most significant test was the STARFISH shot of 9 July 1962. A

What Vela was all about—a photo of a high-altitude nuclear explosion. This was the Kingfish test made on November 1, 1962. The warhead was launched by a Thor IRBM. The yield was less than one megaton and it exploded tens of kilometers above Johnston Island in the Pacific. Observers on the island first saw a yellow-white circle with intense purple streamers that seemed to twist. A large pale green patch then appeared. Although it faded to dull gray after an hour, it remained visible for three hours. The test was for data on the effects of nuclear explosions on the ionosphere, radio transmission, and radio and radar performance. This was critical to the design of antiballistic missile systems. *Defense Nuclear Agency Photo.*

Thor missile carried a 1.4-megaton warhead to 250 miles (402 kilometers), where it exploded.[6] Although these tests were primarily for data on ABM effects, they would also have provided information on detection systems. Both U.S. and Soviet satellites detected the radiation belts caused by the tests. While the Soviet tests were beginning, the first Vela hardware tests were made. Six groups of experiments were fitted to the *Discoverer 29* and *31* satellites. The tests were moderately successful, and about 35 hours of data were obtained from each flight. In December 1961, TRW won the development contract. The work went smoothly and, in very early 1963, testing on the first pair was finished. At the same time, in the wake of the tests and the Cuban missile crisis, the U.S. and Russia resumed test-ban negotiations. The comprehensive ban on nuclear testing proved too difficult. There were political problems with on-site inspection, and there seemed no way to reliably separate underground nuclear blasts and earthquakes. The ban on underground testing was dropped and the limited test ban treaty was signed on 5 August 1963, in Moscow.

At this same time, the policing means were also being readied for launch. The Vela satellite was a curious-looking object. It was a regular icosahedron (a sphere with twenty triangular faces) 56 inches (1.42 meters) in diameter, covered with 14,000 solar cells. Twelve square detectors—X-ray and two diagnostic detectors—were placed at the vertices of the triangular faces.[7] So located, each had a field of view greater than a complete hemisphere. Inside the Vela were gamma ray and neutron detectors. The eighteen detectors could spot an unshielded explosion as small as 10 kilotons and as far away as 100 million miles from earth.[8] The detector package weighed about 80 pounds (36 kilograms). Once in its final 68,000-mile- (109,412-kilometer-) high circular orbit, the Vela weighed about 300 pounds (136 kilograms).

On 10 October 1963, the limited test ban treaty entered into force. Six days later, an Atlas Agena D roared aloft from the Cape. At 231 seconds after launch, the fiberglass shroud separated and exposed the two gleaming Vela satellites. The Atlas shut down and the Agena ignited. When the burn was completed, the Agena was in a 64,388- by 129-mile (103,600- by 208-kilometer) orbit and the two satellites separated. Eighteen hours after launch, a radio signal from Vandenberg Air Force Base fired the first satellite's engine. It went into a 68,905- by 63,301-mile (110,868- by 101,851-kilometer) orbit, with a period of 105 hours. On 19 October, the second satellite was placed into a similar orbit, with a period of 109 hours. The two satellites were 180 degrees apart, so they looked into space in opposite directions simultaneously.[9]

The Velas used the highest orbit of any U.S. military satellite. They were three times higher than geosynchronous orbit and above the Van Allen radiation belt. They will still be in orbit one million years from now.

The second Vela pair was launched on 17 July 1964. After the satellites separated, the Agena's engine continued to burn briefly. This added velocity caused it to bump one of the Velas. Damage was minor—a few of the detectors were knocked out and the spin axis was slightly shifted (the Velas were spin-stabilized at 2 revolutions per second). The third pair was launched on 20 July 1965, in a night launch. They carried an improved instrument package, which included an optical flash detector.

These satellites were a major success. Intended to operate for only 6 months, they were still going strong 5 years later. The fourth and fifth pairs were not launched, since they were not needed.[10]

Advanced Vela

In October 1964, the People's Republic of China exploded its first atomic bomb. This resulted in a change to the emphasis of the Vela program. With suitable instruments, a Vela could detect nuclear tests in the atmo-

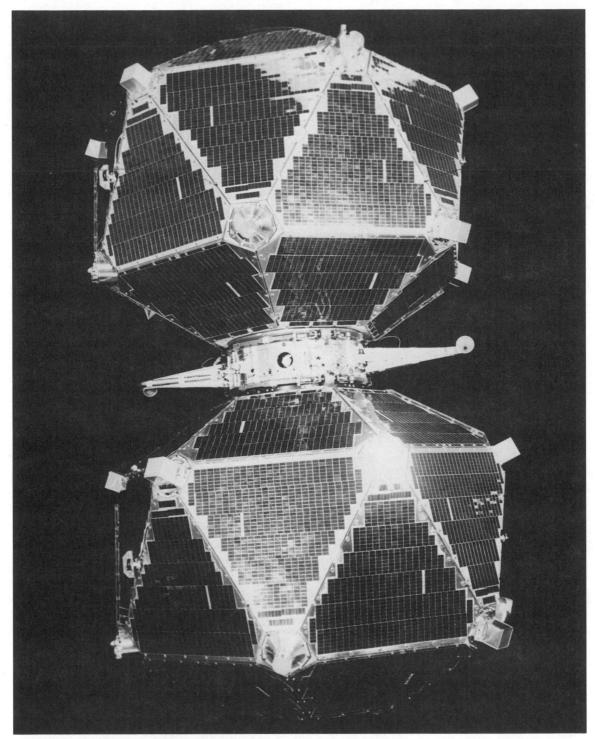

A pair of advanced Vela nuclear detection satellites awaiting launch. The small boxes at each corner are detection instruments. From its orbit, a quarter of the distance to the moon, the Vela can detect both atmospheric nuclear explosions and tests in deep space. The satellites also measure solar flares and radiation from other stars and galaxies. *U.S. Air Force Photo.*

sphere as well as in space. The yield of the weapon could be measured along with other characteristics. In March 1965, TRW received a contract to develop an advanced Vela for such dual monitoring. The new satellites retained some of the characteristics of the early Vela. They were also launched in pairs, used the same high orbit, and retained the same general appearance. (Although the advanced Velas were twenty-six-sided spheres, this is not apparent at first glance.)

The main difference was inside. The advanced Vela's final orbital weight was approximately 500 pounds (227 kilograms). Of this, 138 pounds (63 kilograms) was the detector package—a 40 percent increase over the earlier Vela payload.[11] One addition was a pair of optical flash detectors called Bhangmeters, which determined the explosive yield of a nuclear blast by the brightness of its flash. The two Bhangmeters had different sensitivity levels. From the Velas' orbital altitude, the Bhangmeters could take in half of the earth's surface.[12] The advanced Velas also carried eight X-ray detectors, four gamma-ray detectors, an electron-proton spectrometer, one neutron detector, two heavy-particle detectors, one X-ray analyzer, a solid-state spectrometer, two Geiger counters, an extreme

Launch of an early Titan III C from Cape Kennedy. The booster was based on the Titan II ICBM with the addition of two 120 inch (305 centimeter) diameter solid rocket boosters and a transtage which carried the payload and could maneuver in orbit. The strap-on provided 2.4 million pounds (1,088,640 kilograms) of thrust. Among other payloads, Titan III Cs carried the advanced Vela nuclear detection satellites into their very high orbits. *U.S. Air Force Photo.*

ultraviolet radiation detector, and electromagnetic pulse detectors.

About half of the satellites' increase in weight was from a new attitude control system. The early Velas had been spin-stabilized. The advanced Velas, however, would have to keep the Bhangmeter flash detectors constantly facing the earth. To do this, the advanced Velas were stabilized in all three axes by a reaction wheel, rate gyro, and nitrogen gas jets.[13] To provide the power needed by the added instruments, each advanced Vela carried 22,500 solar cells, producing 120 watts. (This is a little more power than required by a household light bulb.)[14]

The first pair of advanced Velas was orbited on 28 April 1967. The launch vehicle was a Titan III C. Also riding along were three small scientific satellites. The next Vela launch took place 2 years later, on 23 May 1969. A third pair, numbers 11 and 12, followed on 8 April 1970. These carried 162 pounds (74 kilograms) of instruments.

The advanced Velas were as successful as the earlier flights. Designed for a lifetime of 18 months, several were still operating 10 years later. At the time of *Vela 11* and *12*'s launch, it was announced that they would be the last of the series. What was left unsaid was that the U.S. early warning satellites would take over their test-ban monitoring role. The Program 647/DSP satellite carried proton counters and X-ray detectors mounted on two of their solar panels. Optical flash detectors were located on the telescope housing. They had cone-shaped sun shields and were pointed directly at the earth.[15]

In all the systems covered so far, there are both U.S. and Soviet equivalents. Nuclear detection is the one apparent exception. No Soviet satellite goes into Vela's far-ranging orbit. There are no unusual satellites that might be considered nuclear-detection missions by the process of elimination. If the Soviets have an equivalent for Vela, it would have to be in the form of detectors riding along on other satellites. It would be unlikely for the Soviets to simply trust the U.S. to abide by the treaty. Another motivation for a Soviet Vela would be to seek data on Chinese nuclear tests—something important to the Russians, due to their ideological conflict with the Chinese.

Scientific Results

Although the Velas were military satellites built for military purposes, their most valuable results may be scientific. During the Apollo missions to the moon, the Velas kept watch on radiation from the sun. An outburst could have endangered the crews.[16]

The Velas were also the start of orbital X-ray astronomy. Until then, the only data on X-ray stars had come from brief sounding-rocket flights. *Vela 10* provided a decade of data on the X-ray star Cygnus X-1. This was the first black-hole candidate—a collapsed star with gravity so strong that even light is dragged back to its surface. Cygnus X-1 is a double star system. One of the stars is a blue supergiant with a mass fifteen to thirty times that of the sun. Orbiting it every 5.6 days is a black hole only about 5 miles (8 kilometers) in diameter but with a mass of ten to fifteen times that of the sun. Its gravitational force is so strong that it pulls hydrogen gas from the surface of the supergiant star. As the gas spirals toward the black hole, it forms a disk. When the gas finally vanishes into the black hole, it has been heated to tens of millions of degrees, causing it to emit X rays.[17]

The Vela data has confirmed that the gas

flowing into the black hole is very turbulent, varying from second to second in intensity. Cygnus X-1 can go from a low level of activity to a high one in a day or so. Such periods of high activity can last for weeks or even years. The *Vela 10* data from May 1969 until its X-ray detectors failed on 19 June 1979, indicated an overall 294-day period in its X-ray emissions regardless of the overall level of activity.[18]

Two other mysteries the Velas examined were X-ray and gamma-ray Bursters. They were first noted from Vela data in 1967 by Ian B. Strong at Los Alamos. These were very powerful, extremely brief waves of particles. The time the waves reached a Vela satellite could be determined to 0.05 second. Timing measurements from several satellites showed the bursts were coming from outside the solar system.[19,20] The bursts were equivalent to an object 100,000 times brighter than the sun, appearing one second and disappearing the next. The two phenomena had different characteristics. The X-ray Bursters were periodic, occurring every few hours or days. The greater the burst, the longer the interval between them. By the early 1980s, about thirty X-ray Bursters had been found. The source was a double star system—a normal star and a neutron star. A neutron star is the remnant of a supernova explosion: a star that has run out of nuclear fuel explodes, blasting most of its material out into space. The star's core is crushed until it is only 10 miles (16 kilometers) in diameter. While not as powerful as a black hole, the gravitational field of the neutron star can also pull hydrogen from the primary star. The gas piles up on the surface of the neutron star until temperatures and pressures are high enough for the hydrogen atoms to fuse and cause a thermonuclear explosion. In effect, X-ray Bursters are natural hydrogen

bombs. The irony is that Velas were launched to watch for H-bomb explosions in deep space.[21]

The gamma-ray Bursters are more mysterious. They never seem to repeat and are more numerous than the X-ray Bursters. The most powerful occurred on 5 March 1979; it was 100 times more powerful than any previous one. Nine satellites—the three operational Velas, *ISEE-3*, the Pioneer Venus orbiter, *Helios 2,* and the Soviet *Prognoz 7* and *Veneras 11* and *12* spacecraft—all observed it. The gamma-ray bursts were traced from their measurements to a neutron star in the Large Magellanic Cloud, a satellite galaxy of the Milky Way, about 186,000 light years distant.[22] The cause of these gamma-ray Bursters is still not completely understood. They may be a more powerful thermonuclear explosion (gamma rays are more energetic particles than X rays). Alternatively, they could be a star quake on a neutron star. The leading theory at present is the impact of an asteroid on the surface of a neutron star. As the asteroid approaches, the powerful gravitational and magnetic fields of the neutron star distort it, stretching it into a long ribbon. The asteroid, with a mass of perhaps a billion tons, hits the surface at a large fraction of the speed of light. The debris of the impact is directed and channeled by the neutron star's magnetic field into a searchlight beam of gamma rays.[23]

Closer to home, the Velas have also provided information on lightning bolts. Their optical sensors pick up the flashes. From their measurements, it was found that about 1 percent of the bolts had energy levels 100 times the average. These "superbolts" were found primarily in winter storms east of Japan.[24]

For 10 years, the Velas patiently kept watch, little noticed by the world below, until one night when a Vela detected a sudden flash of light.

A Clandestine Test?

The flash occurred in the early morning hours of 22 September 1979. It was located somewhere within a 4,500-square-mile (11,655-square-kilometer) area of South Africa, the South Atlantic, and Antarctica. From the Bhangmeter readings, the yield was estimated to be 2 to 4 kilotons. The reason the flash was believed to be a nuclear test was its profile. A lightning bolt generates only a single flash; a nuclear explosion, on the other hand, causes two distinctive pulses separated by one-third of a second. The first is caused by the triggering device; the second flash is the main detonation.[25] The 22 September event showed this profile.

A controversy that would last nearly a year was caused by the flash. The first question everybody asked was, "Who did it?" Suspicions initially settled on South Africa and Israel. Both were known to have nuclear research programs, but, more importantly, both were viewed by many as international pariahs. South Africa, for its part, denied any responsibility, calling the charge "complete nonsense."[26] The South African navy chief suggested it was an explosion aboard a Soviet nuclear submarine. An Echo II class submarine had been spotted near the Cape of Good Hope in September. A White House spokesman said the possibility had been looked into but was rejected.[27] From the first, however, there were questions about the incident. U.S. aircraft sent into the area failed to find any radioactive fallout. No other U.S. sensors picked up acoustic signals. The only other evidence came from outside sources; the initial evidence came and went quickly. In mid-November, a health physics laboratory in New Zealand reported detection of fresh fallout. The National Radiation Laboratory in New Zealand did not confirm the results, and a U.S. laboratory was unable to find any radioactivity. The instruments used in the first test had apparently been contaminated. The U.S. also checked several countries in the southern hemisphere but could not find any traces. This lack of fallout, though not final, was a major blow, since it was unlikely such radiation could escape detection.

There were two other bits of evidence. During the early morning hours of 22 September, astronomers at the Arecibo radio telescope in Puerto Rico were monitoring the ionosphere. The scientists observed a ripple in the ionosphere coming from a southeasterly direction—the first time they had seen such a disturbance moving from this direction. They calculated the speed of the ripple at 1,969 to 2,297 feet (600 to 700 meters) per second, which coincided with the time and place of the Vela flash. It was suggested that the electrons released by the blast caused the ionosphere to "bob up and down," generating the ripple.[28] The other evidence was hydroacoustic signals recorded at two navy sites. These signals correspond to direct rays from a source near Prince Edward Island and to reflections of the sound from the Antarctic ice shelf.[29] One Naval Research Laboratory (NRL) scientist said that these hydroacoustic signals were the strongest he had seen and were comparable to those generated by French nuclear tests in the Pacific.[30]

To settle the question, the Carter administration organized a nine-member committee of scientists under Dr. Jack Ruina of MIT. At first, the committee was confident that there had been a secret nuclear test. One member later said, ''We were betting four-to-one that there was an explosion.'' [31] However, as they looked deeper into the evidence, the less certain they became. The panel found from weather satellite photos that there had been a tropical storm only a few hundred miles south of the Arecibo telescope. A storm could also generate an ionospheric ripple. Additionally, there were doubts about the scientists' estimates of the disturbance's speed and direction.[32] There were also questions as to whether an explosion as small as the Vela data indicated could generate such a large disturbance.[33] As for the hydroacoustic signals, the committee decided that the Naval Research Laboratory analysis was ambiguous and not complete enough to be proof that a test occurred. Another blow to the hydroacoustic data was a difficult, complex examination of the positions and viewing angles of the other operational Velas and early warning satellites. This was combined with local weather conditions in the areas covered. It was found that if an explosion had taken place near Prince Edward Island, as the NRL analysis suggested, it would have been seen by another satellite. This would have removed any doubt about whether a test took place. The only area where a detonation could have been detected by only the one Vela was near the Palmer Peninsula of Antarctica.

In the end, the committee was left with only the Vela data. The first possibility was a satellite malfunction, but this was soon eliminated. Both before and after the flash, the satellite had responded properly to tests. When the committee compared the characteristics of the flash to forty-two previous atmospheric nuclear tests observed by the Velas, they noted a curious difference. The Velas carry two Bhangmeter flash detectors, one more sensitive than the other, to allow the satellite to detect a wide range of nuclear yields. The ratio of their readings remains constant. If, for a particular test, the more sensitive detector should measure the first flash as 20 units and the less sensitive one read only 10, then this ratio of 2:1 would also apply to the second flash. This is a standard characteristic of all the nuclear tests measured by the Vela. The 22 September data did not have this distinctive characteristic. During the second flash, the less sensitive Bhangmeter detected a greater volume of light than the more sensitive one, which had never been seen in a nuclear test. This difference in light ratios was, however, a characteristic of phenomena called Zoo Events. These were unexplained light flashes picked up by the Vela sensors that did not correlate with either nuclear explosions or such sources as ground-based laser tests or the sun coming out of an eclipse.

Several thousand Zoo Events had been detected over the years. Most of them involved only a single flash of light. A few (sixty or seventy in 10 years), however, did show a twin flash like that of a nuclear explosion, the difference being, as in the 22 September event, the light ratios did not remain constant between the two flashes. The White House panel concluded that the 22 September flash was actually a Zoo Event that mimicked a nuclear blast unusually well. The committee was still faced with the question of what was causing the Zoo Events. The fact that the flash had appeared brighter in one detector than the other implied that it came from an object close to the satellite. Some DOD analysts had long suggested that Zoo Events were caused by sunlight reflecting from meteoroids passing the Vela's detectors. The panel found, however, that the meteoroids would be traveling too fast. This line of reasoning opened another possibility. The committee suggested that a small, dust-sized meteoroid had hit the Vela. The impact threw off very small bits of aluminum, paint, or glass solar cell covers at relatively slow velocities. As one passed near the Bhangmeters, the spinning debris reflected the sunlight, causing the first flash, then the dip, and finally the second peak. Because the particle was closer to one of the Bhangmeters, it saw the flash as being brighter than the other one did.[34] The ionospheric ripple and the hydroacoustic signals were simply coincidences. There was no cause and effect relationship with the flash.

The White House explanation was widely attacked by the "believers," those who felt the evidence proved that there had been a clandestine test or that the Carter administration was trying to avoid facing the fact that its nuclear nonproliferation policy would be vaporized if there had been a test. The Defense Intelligence Agency and the Naval Research Laboratory both issued secret reports endorsing the believers' opinions.[35]

In the end, the question was not definitively solved; rather, it quietly drifted away. In retrospect, two factors seem to stand out. However strained the White House explanation may be, the fact remains that the 22 September flash does show characteristics like those of earlier Zoo Events and unlike those of real nuclear blasts. More importantly, there is the question of pulling it off. The analysis of the coverage of the Vela and early warning satellite detectors was so complex that it is doubtful any country could have known beforehand the precise time and place, so that *only* one Vela would spot it. Additionally, there were the factors that could not be controlled, such as fallout. When the evidence for an event always remains short of that needed for proof, especially when such proof should exist and when the proof is actively

sought, then it becomes evidence the event is not real.

The Future of Nuclear Detection

The most important aftereffect of the mysterious flash may be to highlight shortcomings in the existing detection network. Ideally, false alarms (which are inevitable given the existence of mimics such as Zoo Events) should be quickly identified as such and eliminated. For the 22 September event, this could not be done and a year of controversy was the result. The program to improve the U.S. detection network, called Forest Green, actually predates the flash. One part is to increase the number of detection satellites by fitting Navstar navigation satellites with Bhangmeters and EMP sensors.[36]

Studies began in early 1975, and a contract was awarded to Ford Aerospace in October. As with the Vela systems, Sandia Laboratories were to develop the sensors and data-processing equipment. Rockwell International, the Navstar prime contractor, took care of the communications system and fitting the package on the satellite. The sensors themselves weigh 60 pounds (27 kilograms). The first to carry the package was the sixth Navstar, launched on 26 April 1980, by an Atlas F from Vandenberg Air Force Base.[37,38] By the late 1980s, a total of eighteen Navstars will make up the operational network. The satellites are placed into a 12,500-mile (20,112-kilometer) orbit, lower than either the Velas or early warning satellites. The important difference is not altitude but inclination. The Velas' orbits were inclined only 33 degrees; Navstars' are inclined 63 degrees, which means better coverage of the polar areas, both north and south. Another increase in coverage is a result of the Navstar's operating mode. Measurements from several satellites are required for a navigational fix, which means the sensors on multiple satellites will be viewing each spot on earth. If one satellite sees a flash, the others should also detect it. We would then know if it was the electronic equivalent of spots before one's eyes or if it was the sinister signal that another nation has the bomb.

NOTES Chapter 19

1. John Prados, *The Soviet Estimate*, (New York: The Dial Press, 1983), 19,20.
2. Glenn T. Seaborg, *Kennedy, Khrushchev and the Test Ban*, (Berkeley: University of California Press, 1981), 19–23.
3. *The Aerospace Corporation—Its Work: 1960–1980*, (Los Angeles: Times Mirror Press, 1980), 69,70.
4. Walter Sullivan, *Assault on the Unknown*, (New York: McGraw-Hill, 1961), Chapter 8.
5. Prados, *The Soviet Estimate*, 153.
6. Seaborg, *Kennedy, Khrushchev and the Test Ban*, Chapter 12.
7. Reginal Turnill, *The Observer's Book of Unmanned Spaceflight*, (New York: Warne, 1974), 77,78.
8. *Astronautics and Aeronautics, 1963*, (NASA SP-4004), 299.
9. *Astronautics and Aeronautics, 1963*, (NASA SP-4004), 299.
10. *The Aerospace Corporation*, 70,71.
11. Philip J. Klass, *Secret Sentries in Space*, (New York: Random House, 1971).
12. Philip J. Klass, "Clandestine Nuclear Test Doubted," *Aviation Week & Space Technology*, (11 August 1980): 67–69, 71,72.
13. George S. Hunter, "USAF to Orbit 2 Advanced Velas in April," *Aviation Week & Space Technology*, (13 February 1967): 71,73,77.
14. Turnill, *Observer's Book of Unmanned Spaceflight*, 78.
15. Barry Miller, "U.S. Moves to Upgrade Missile Warning," *Aviation Week & Space Technology*, (2 December 1974): 17.
16. T. M. Wilding-White, *Jane's Pocket Book of Space Exploration*, (New York: Macmillan, 1976), 151.
17. Herbert Friedman, *The Amazing Universe*, (Washington, D.C.: National Geographic Society, 1975), 126–129.
18. "Cygnus X-1: The Longest-Period Celestial X-Ray Source," *Sky & Telescope*, (March 1983): 221,222.
19. Friedman, *The Amazing Universe*, 130.
20. "Gamma-Ray Bursts from Deep Space," *Sky & Telescope*, (September 1973): 146.
21. "Nature's Own H-Bombs," *Time*, (15 August 1983): 40.
22. "What Happened on March 5, 1979?", *Sky & Telescope*, (April 1980): 294,295.
23. Ronald A. Schorn, "The Gamma-Ray Burster Puzzle," *Sky & Telescope*, (June 1982): 560–562.
24. Joel W. Powell, "Lightning Research from Space," *Spaceflight*, (June 1983): 280,281.
25. *New York Times*, 1 November 1979, D-22.
26. *New York Times*, 27 October 1979, A-1, A-5.
27. *New York Times*, 28 October 1979, A-14.
28. Eliot Marshall, "Scientists Fail to Solve Vela Mystery," *Science*, (1 February 1980): 504–506.
29. Philip J. Klass, "Clandestine Nuclear Test Doubted," *Aviation Week & Space Technology*, (11 August 1980): 67–69, 71,72.
30. Eliot Marshall, "Navy Lab Concludes the Vela Saw a Bomb," *Science*, (29 August 1980): 996,997.
31. "Debate Continues on the Bomb That Wasn't," *Science*, (1 August 1980): 572,573.
32. Philip J. Klass, "Clandestine Nuclear Test Doubted," *Aviation Week & Space Technology*, (11 August 1980): 67–69,71,72.

33. Eliot Marshall, "Scientists Fail to Solve Vela Mystery," *Science,* (1 February 1980): 504–506.
34. Philip J. Klass, "Clandestine Nuclear Test Doubted," *Aviation Week & Space Technology,* (11 August 1980): 67–69,71,72.
35. Eliot Marshall, "Navy Lab Concludes the Vela Saw a Bomb," *Science,* (29 August 1980): 996,997.
36. "Industry Observer," *Aviation Week & Space Technology,* (13 December 1982): 13.
37. "South Atlantic Flash," *Spaceflight,* (January 1981): 28.
38. "GPS to Test Nuclear Detonation Sensor," *Aviation Week & Space Technology,* (27 August 1979): 51.

CHAPTER 20

Conclusions

The story of military space systems is not simply one of technology. The satellites are a way of acquiring knowledge—the best yet developed. Information that agents had to struggle and often die for can now be found almost effortlessly. Yet one must not be swept away by these technological marvels. Satellites are not omnipotent. They cannot watch everything all the time. Moreover, the information must be interpreted by fallible humans. Despite their training, men can fool themselves, be deceived, or simply refuse to accept the truth. And even with satellites, sometimes what is needed is the special knowledge only an agent can acquire. In the end, governments and the public must act on the information the satellites provide. The effectiveness of technology depends on social factors.

For the U.S., these social factors are mixed. The years of the reconnaissance satellites coincide with a traumatic period in U.S. history—Vietnam. The reasons for U.S. involvement are complex and will be argued for decades to come. In part, however, it was the experiences of the years leading up to World War II that played a major role—specifically, the attempts by Britain's Prime Minister Chamberlain to appease Hitler. One interpretation of that experience is that the West, especially the U.S., must never again compromise with aggression—that it must resist even

if that means force. Containment, the cold war, Korea, and ultimately, Vietnam were the result. The passions released by Vietnam tore the U.S. apart and left it weakened, divided, and guilt-ridden. The last helicopter out of Saigon marked a turning point in U.S. history. The lessons of World War II, which had been a guide to U.S. policymakers for 30 years, were replaced by the Vietnam experience.

The U.S. defeat in Vietnam generated varying opinions as to how the U.S. should act in the future. Some held that, far from being omnipotent, the U.S. was actually powerless to influence world events—that it was, in any event, immoral to use force to coerce another sovereign nation and it was not worth the price to resist aggression. In the end, the U.S. was better served by being passive—submitting to the inevitable. In this view, there is no issue, no cause, no crisis that really matters; there is nothing worth dying for. The lesson that still others draw from Vietnam is that American society is utterly corrupt—an evil empire worthy of only contempt. Accordingly, any action, any policy of the U.S. government is inherently wrong and must be opposed.

The most subtle effect of the Vietnam experience is one of national fatigue. Soldiers are not the only ones to be worn down after seeing too much combat and experiencing too much stress. Some of those who saw the war every night on the network news grew tired and disillusioned, tired of the responsibilities and demands placed on the U.S. They just didn't care; self-indulgence became the guiding star. It was, perhaps, inevitable that the pain of Vietnam would be followed by the "me generation."

All this challenges the U.S. leadership. It poses the question of how to defend the U.S. when a certain percentage of its citizens deem it unworthy of preservation. And face this question the U.S. must, for an entire generation had its first political experiences during these years. The attitudes and beliefs shaped by Vietnam will not readily fade away. The World War II experience lasted 30 years—Vietnam has been with us 10 years and the exact duration of its power is unknown. Of course, the world turns, times change, the trauma of Vietnam will, in time, be replaced by some new historical era—hopefully, not by, say, Pearl Harbor II.

The U.S. changes; Russia doesn't. While the U.S. has a legal revolution every 4 years, in Russia, Stalin's system still rules. But it is more than that. Russia's heritage of tyranny, orthodoxy, reaction, militarism, expansionism, and hostility to freedom in any form goes back centuries. What Marxism has done is to add a crusading mission. Russia's destiny, as seen by its new czars, is to lead the march of history towards the shining future of communism. This makes Russia different from other nations—not just a state, but a crusade. It also means that Russia can never accept a stable relationship based on equality. That is what Kennedy sought from Khrushchev during the 1961 Vienna Summit—a mutual recognition of the other's vital interests, leading to a real and lasting coexistence. Kennedy saw the roots of the U.S.-USSR rivalry in nationalism. Khrushchev turned down flat this stabilized world order. To guarantee the safety of the U.S., its government, and its fundamental beliefs meant to Khrushchev betraying Russia's role as leader of world revolution. More than that, it meant betraying the Communist ideology of the inevitable forces of history and the laws of class struggle. To do so would also remove the legitimacy of the Soviet regime. For Khrushchev and his successors, the issue is not nations but the clash of two implacably hostile ideologies for control. The status quo for Khrushchev was not the Western idea of stability but the further uninterrupted triumphs of revolutionary communism.

The U.S., historically, has had difficulty in coping with Russia. At times, the West behaves as an ordinary person does when confronted by a sociopath. The ordinary person constantly reassures the sociopath he means no harm, passively submitting even when resistance would be successful. His submission, however, only encourages the sociopath to greater demands and violence. So the victims of Hitler, Stalin, Pol Pot, and uncounted numbers of street thugs go quietly to their deaths. This position argues that U.S. weakness will reassure the Russians. Thus, any attempts by the U.S. to develop new weapons is condemned as destabilizing and an extension of the arms race. Any Soviet moves are apologized for as self-defense or denied as a "mythical threat." So, once more, the voices of appeasement and unilateral disarmament ring loud. If the West gives up its weapons, they say, it would not be a threat to the Soviets so would not be attacked. Better Red than dead, they say—never mind the numerous historical examples of being both Red *and* dead.

Other voices argue that safety is found in paper. A treaty, any treaty, can bring a utopia of love and brotherhood, a world free of nuclear weapons, can end the arms race and bring "peace in our time." Treaties become an end in themselves. The past abuse of solemn treaty agreements gets lost in the shuffle. Reconnaissance satellites have acted as the policemen of such treaties. This monitoring is a technical matter—a certain resolution is needed for a given treaty provision. The satellites' ability to watch for violations, however, has obscured a more pressing question—what to do if a violation is detected. If the Soviets are cheating on one or more provisions, should the U.S. abrogate the treaty, thereby losing all the benefits the treaty provides? If not, then the U.S. accepts a certain amount of cheating as permissible. The treaty then becomes a unilateral restraint on the U.S. Russia is free to violate any part it wishes, while the U.S. must continue to observe the treaty in full.

Yet, at the same time, a hard line seems no more successful. Its effectiveness is limited by the American people's need for "good relations" as well as by the Soviets' military strength and their own belief in communism's inevitable triumph. They see fatal flaws in American society that will force the U.S. to again weaken. Thus, if the West is soft, the Soviets exploit it; if hard, the Soviets know it will pass and dismiss it as part of the crisis of capitalism.

The broad sweep of history shows there are many ways nations can die. Carthage ended in violence—its population killed, city destroyed, rubble plowed under, and the ground covered with salt. Nuclear weapons are no more effective, just faster. Rome fell just as completely, but its end came not in sudden violence but with a slow, insidious decline—a gentle slope with few landmarks to measure the ebbing of resolve and power. It was apparent only when looking back at where Rome had been. Carthage, Rome, and all the great empires scattered through history each had its own sources of intelligence, and ways and means they thought were beyond improvement. Yet each ultimately fell. For us, satellites provide information, but that is all they do. They cannot guarantee our safety. The photos and tapes must be acted on wisely, without illusions, wishful thinking, or blindness. Human weaknesses can turn the satellites' technological achievements into little more than interesting snapshots and tapes of funny noises.

The question, therefore, is whether democracy, with all its crosscurrents, can survive in a world of militant, aggressive dictatorships—if America's experiment with self-government, its belief in the worth of the individual and in the rights of man, is just a brief interruption in the totalitarian pattern of history. On the will of the American people rests the answer.

Appendix A

Orbital Mechanics Made Easy

The mechanics of spaceflight are outside everyday experiences, so this section is included as a nontechnical guide for those not familiar with some of the concepts mentioned in the text.

The first factor to consider is the rocket booster. It has a given fuel capacity, weight, efficiency, and performance. This translates into how much payload it can place into a given orbit. The payload drops off as the height of the orbit increases because it requires more energy to lift the payload higher. A rocket may be able to place several thousand pounds into low orbit but only a few hundred in a geosynchronous orbit. The payload can be increased by stretching the stages so they can carry more fuel, adding new upper stages, using higher thrust, using more efficient engines, or adding strap-on solid-fuel boosters.

The rotation of the earth affects booster performance and payload. Since the earth rotates to the east once in 24 hours, a rocket launched due east can take advantage of this rotation. Most military satellites, however, cannot be launched to the east. Photo reconnaissance satellites must be launched into high inclination orbits (inclination is the angle between the orbit and the equator). A satellite in a 30° inclination orbit will go no farther north and south than 30° latitude; this means it would not pass over Russia at all and only over the southern part of the U.S. A 51° orbit would cover all of the U.S. but not the northern part of Russia. For this reason, U.S. reconnaissance satellites are launched into orbits of 80° to 100° inclination; Soviet satellites range from 65° to 81°. Because these orbits pass over the poles, they are called "polar orbits." The problem is that high inclination launches from Tyuratam must be directed to the northeast, which loses much of the earth's rotational velocity. The situation for Vandenberg AFB is worse. Because of the shape of the coastline, launches must be made to the southwest—against the earth's rotation. A polar launch from the Kennedy Space Center has to be directed to the east, then turned south—this to avoid flying over populated areas of the U.S. and Cuba. The maneuver results in a major loss in payload. Only from Plesetsk can a satellite go into a polar orbit with a near-easterly launch (of course because of its northerly location, the earth's rotation is less).

Another consequence of inclination is the "sun-synchronous orbit." If a satellite is placed into an orbit with an inclination of 96°, the satellite will be over a specific latitude at the same local time each day. This means the sun angle, in the photos, will stay constant over time. This is an advantage for long-dura-

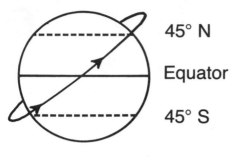

Orbit Inclination of 45°

tion reconnaissance satellites such as the Big Bird and KH-11.

It should also be mentioned that once the inclination of an orbit is established, it is extremely difficult to change. The satellite must in effect be "re-launched" into a new orbit. The maneuver requires large amounts of fuel, which cuts into payload weight (simply raising

or lowering the orbit involves speeding up or slowing down the satellite).

Military satellites are put into three types of orbits depending on their mission—low circular (where the high and low points do not differ by a significant amount), elliptical (where they do), or a high orbit. An orbit of whatever type is a balancing act between the satellite's forward velocity and the pull of the earth's gravity. The satellite tries to go in a straight line; gravity pulls it towards the earth. The net result is a curved path around the earth.

Photo reconnaissance satellites are placed into low circular orbits. This both maximizes payload and resolution (the closer you are to an object, the better it can be seen). What targets can be photographed depends on the ground track, which is the orbit projected onto the earth's surface. If the period of the orbit

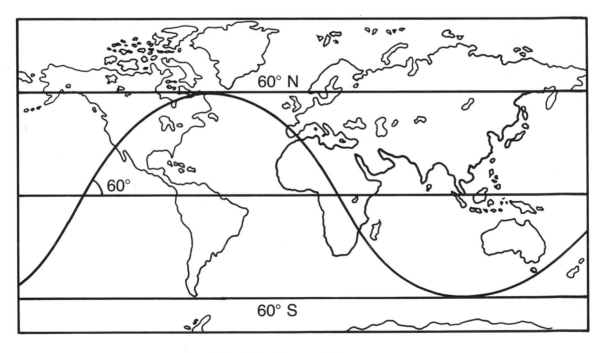

Satellite Ground Trace

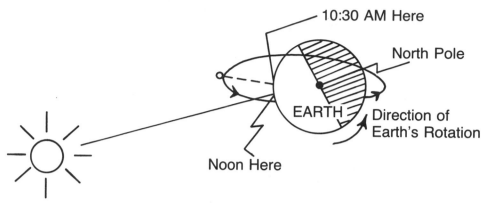

A Sun-Synchronous Orbit

is an even multiple of 24 hours (e.g., 90 minutes), the ground track will repeat each day (i.e., pass over the same points on the earth). If the period is not an even multiple (91 minutes), the ground track will seem to drift across the earth's surface. The first is preferred if the satellite is to photograph a specific target on the earth; the other is to survey multiple targets.

Some missions require the satellite be placed into elliptical orbits. The prime example of a satellite that uses this type of orbit is the Soviet early warning satellite. The orbit allows the satellite to monitor U.S. ICBM silos for several hours during a single orbit. This is a somewhat more complex process. The payload is placed into an initial low orbit.

After it has coasted half way around the earth from the launch site, a "kick stage" is fired. This speeds up the satellite and sends it climbing away from the earth. Its speed slows as the earth's gravity tries to pull it back; the lowest velocity comes at the satellite's farthest point from earth. As the satellite heads back towards the earth, gravity speeds it up (it recovers the velocity lost on the "uphill" part of the orbit). The satellite's highest speed comes at its closest approach to the earth.

In other cases, such as the U.S. Midas, Vela, and geosynchronous satellites, a high circular orbit is needed. This allows the satellite to observe much or all of a hemisphere at a time. It can therefore constantly watch

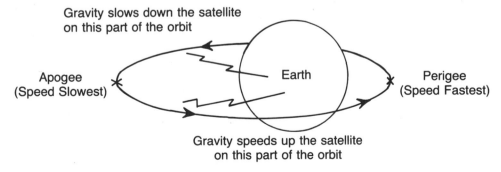

Effects of Elliptical Orbits

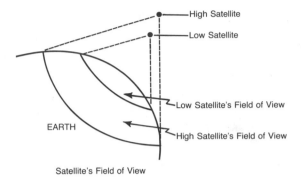

Satellite's Field of View

Various factors act on a satellite as it coasts around the earth. For low orbit reconnaissance satellites, the most important factor is air drag. The satellites are orbiting in the upper-most fringes of the atmosphere. The satellite collides with air molecules, slowing it down and lowering the orbit. This increases the air drag, accelerating the process. Left uncorrected, the satellite would descend into the denser atmosphere and burn up. The satellite must either be placed in an orbit high enough so air drag will not be significant over the course of the flight, or fitted with a small rocket engine for periodical orbital boosts.

Satellites in high orbits are not affected by air drag. Rather it is the gravitational fields of the earth, sun and moon, which build up over a long period of time and cause the satellite to drift out of position. The corrective measure is to fire small attitude-control jets.

activities that are scattered over a wide area. To accomplish this, the satellite is placed into an elliptical transfer orbit. When it reaches the maximum altitude, an engine firing is made. This increases the speed of the satellite and raises the low point of the orbit to match that of the high point. A geosynchronous orbit is just a special kind of high circular orbit. It has a period of 24 hours, thus *as seen from the earth,* the satellite seems to remain fixed in the same point in the sky.

A variation of this is the inclined geosynchronous orbit. U.S. early warning satellites sometimes use this type of orbit. The satellite does not remain fixed but traces a figure "8" in the sky. The size of the "8" depends on the inclination. This allows a better view of areas north and south of the equator.

With some reconnaissance satellites, it is necessary to physically return film to earth. For this, re-entry capsules are used. The re-entry sequence begins with the firing of a retro-rocket. The rocket is pointed against the capsule's direction of motion to slow it down. This causes the capsule to go into a lower orbit, which takes it into the earth's atmosphere (if the retro was pointed with the direction of motion, it would speed up the capsule, sending it into a higher orbit). As the capsule enters the denser parts of the earth's atmosphere, shock waves form, compressing the air and raising its temperature to several thousand degrees. The energy used to accelerate the capsule into orbit is now being given up in the form of heat. The capsule's plastic heat shield now begins to melt, carrying off some of the heat. Some of the plastic turns into a thick glassy substance which absorbs energy as it flows back along the capsule. During the final stage of re-entry, the plastic begins

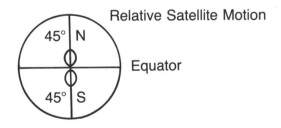

Ground Trace for a Synchronous Satellite

to char forming an insulation layer. This procedure is called "ablation." Once the heating phase is complete, the parachute is deployed and recovery is made.

The final fate of a satellite varies. In many cases, no action is taken. In others, however, there is concern that if the satellite is allowed to make an uncontrolled re-entry, some debris may survive and reveal information about the satellite. The Soviet technique, to avoid this, is to blow up the satellite; the resulting fragments are too small to survive. The U.S. method is to de-orbit the satellite so any debris lands in the Pacific and sinks out of reach.

A special case is the Soviet nuclear ocean surveillance satellite. To prevent radioactive debris from reaching the earth, the reactor is separated and a rocket sends it into a high orbit. After the failure of Cosmos 954, which scattered radioactive debris over Canada in 1978, a modification was made. The core could now be separated from the reactor. When done in high final orbit, this prolongs its orbital life further. If the boost maneuver fails, separating the core lowers the chance that highly radioactive material will survive re-entry and contaminate the earth's surface.

Appendix B

Military Satellite Launches, 1959–1985

U.S. Discoverer and Program 162 Missions

Name	Date	Booster	Orbit (km)	Inclination (degrees)	Remarks
1959					
D-1	Feb 28	Thor Agena A	968 – 163	89.7	First launch.
D-2	Apr 13	Thor Agena A	346 – 239	89.9	Capsule recovery failed.
D-3	Jun 3	Thor Agena A	–	–	Failed to orbit.
D-4	Jun 25	Thor Agena A	–	–	Failed to orbit.
D-5	Aug 13	Thor Agena A	739 – 217	80.0	Capsule boosted into higher orbit.
D-6	Aug 19	Thor Agena A	848 – 212	84.0	Capsule recovery failed.
D-7	Nov 7	Thor Agena A	847 – 159	81.6	Satellite tumbled.
D-8	Nov 20	Thor Agena A	1,679 – 187	80.6	Capsule recovery failed.
1960					
D-9	Feb 4	Thor Agena A	–	–	Failed to orbit.
D-10	Feb 19	Thor Agena A	–	–	Failed to orbit.
D-11	Apr 15	Thor Agena A	589 – 170	80.1	Capsule recovery failed.
D-12	Jun 29	Thor Agena A	–	–	Failed to orbit.
D-13	Aug 10	Thor Agena A	683 – 258	82.8	First successful capsule recovery; after 1.11 days.
D-14	Aug 18	Thor Agena A	805 – 186	79.6	Photographed Plesetsk. Capsule recovered after 1.12 days.
D-15	Sep 13	Thor Agena A	761 – 199	80.9	Capsule lost at sea.
D-16	Oct 16	Thor Agena B	–	–	Failed to orbit.
D-17	Nov 12	Thor Agena B	984 – 190	81.7	Capsule recovered after 2.08 days. Photos poor.
D-18	Dec 7	Thor Agena B	661 – 243	81.5	Capsule recovered after 3.12 days. Photos poor.
D-19	Dec 20	Thor Agena B	631 – 209	83.4	Did not carry capsule.
1961					
D-20	Feb 17	Thor Agena B	786 – 288	80.9	Capsule recovery not attempted.
D-21	Feb 18	Thor Agena B	1,069 – 240	80.7	Did not carry capsule.
D-22	Mar 30	Thor Agena B	–	–	Failed to orbit.
D-23	Apr 8	Thor Agena B	651 – 295	82.3	Capsule went into higher orbit.
D-24	Jun 8	Thor Agena B	–	–	Failed to orbit.
D-25	Jun 16	Thor Agena B	409 – 222	82.11	Capsule recovered after 2.08 days.
D-26	Jul 7	Thor Agena B	808 – 228	82.9	Capsule recovered after 2.11 days.
D-27	Jul 27	Thor Agena B	–	–	Failed to orbit.
D-28	Aug 3	Thor Agena B	–	–	Failed to orbit.
D-29	Aug 30	Thor Agena B	542 – 152	82.1	Took photos of Plesetsk ICBM sites. Capsule recovered after 2.1 days.
D-30	Sep 12	Thor Agena B	546 – 235	82.6	Capsule recovered after 2.12 days.
D-31	Sep 17	Thor Agena B	396 – 235	82.7	Capsule recovery failed.

Name	Date	Booster	Orbit (km)	Inclination (degrees)	Remarks
D-32	Oct 13	Thor Agena B	395 – 234	81.6	Capsule recovered after 1.14 days.
D-33	Oct 23	Thor Agena B	–	–	Failed to orbit.
D-34	Nov 5	Thor Agena B	1,011 – 227	82.5	Capsule recovery not attempted.
D-35	Nov 15	Thor Agena B	278 – 238	81.6	Capsule recovered after 1.12 days.
D-36	Dec 12	Thor Agena B	484 – 241	81.2	Capsule recovered after 4.08 days.

1962

Name	Date	Booster	Orbit (km)	Inclination (degrees)	Remarks
D-37	Jan 13	Thor Agena B	–	–	Failed to orbit.
1962-Delta	Feb 21	Thor Agena B	374 – 167	81.9	First ''unannounced'' Discoverer (Program 162).
D-38	Feb 27	Thor Agena B	341 – 208	82.2	Capsule recovered after 4.06 days.
1962-Lambda	Apr 18	Thor Agena B	441 – 200	73.4	
1962-Rho	Apr 29	Thor Agena B	475 – 180	73.1	
1962-Sigma	May 15	Thor Agena B	634 – 305	82.3	Capsule recovered after 4.11 days.
1962-Phi	May 30	Thor Agena B	319 – 199	74.1	Capsule recovered after 3.1 days.
1962-Chi	Jun 2	Thor Agena B	385 – 211	74.2	
1962 A Beta	Jun 23	Thor Agena B	293 – 213	75.0	
1962 A Gamma	Jun 28	Thor Agena D	689 – 211	76.0	Capsule recovered after 4.09 days.
1962 A Eta	Jul 21	Thor Agena B	381 – 208	70.3	
1962 A Theta	Jul 28	Thor Agena B	386 – 225	71.0	Capsule recovered after 4.08 days.
1962 A Kappa	Aug 2	Thor Agena D	418 – 204	82.2	Capsule recovered after 4.09 days.
1962 A Sigma	Aug 29	Thor Agena D	400 – 187	65.2	Capsule recovered after 4.07 days.
1962 A Tau	Sep 1	Thor Agena B	669 – 300	82.8	
1962 A Phi	Sep 17	Thor Agena B	668 – 204	81.8	Capsule recovered after 1.10 days.
1962 B Beta	Sep 29	Thor Agena D	376 – 203	65.4	
1962 B Epsilon	Oct 9	Thor Agena B	427 – 213	81.9	Capsule recovered after 4.11 days.
1962 B Omicron	Nov 5	Thor Agena B	409 – 208	74.9	Capsule recovered after 4.08 days.
1962 B Rho	Nov 24	Thor Agena B	337 – 204	65.1	
1962 B Sigma	Dec 4	Thor Agena D	273 – 194	65.1	
1962 B Phi	Dec 14	Thor Agena D	392 – 199	70.9	Capsule recovered after 4.08 days.

1963

Name	Date	Booster	Orbit (km)	Inclination (degrees)	Remarks
1963–2	Jan 7	Thor Agena D	399 – 205	82.2	
None	Feb 28	TAT Agena	–	–	Failed to orbit.
None	Mar 18	TAT Agena	–	–	Failed to orbit.
1963–7	Apr 1	Thor Agena D	408 – 201	75.4	Capsule recovered after 3.08 days.
None	Apr 26	Thor Agena D	–	–	Failed to orbit.
1963–16	May 18	TAT Agena	497 – 153	74.5	
1963–19	Jun 13	TAT Agena	419 – 192	81.8	
1963–25	Jun 27	TAT Agena	396 – 196	81.6	
1963–29	Jul 19	Thor Agena D	387 – 194	82.8	
1963–32	Jul 31	TAT Agena	411 – 157	74.9	
1963–34	Aug 25	TAT Agena	320 – 161	75.0	
1963–35	Aug 29	Thor Agena D	324 – 292	81.8	Carried ELINT sub-satellite.

Name	Date	Booster	Orbit (km)	Inclination (degrees)	Remarks
1963–37	Sep 23	TAT Agena	441 – 161	74.9	
1963–42	Oct 23	TAT Agena	345 – 279	89.9	Carried ELINT sub-satellite.
None	Nov 9	Thor Agena D	–	–	Failed to orbit.
1963–48	Nov 27	Thor Agena D	386 – 175	69.9	
1963–55	Dec 21	TAT Agena	355 – 176	64.9	Carried ELINT sub-satellite.

1964

Name	Date	Booster	Orbit (km)	Inclination (degrees)	Remarks
1964–8	Feb 15	TAT Agena	444 – 179	74.9	
None	Mar 24	TAT Agena	–	–	Failed to orbit.
1964–22	Apr 27	TAT Agena	446 – 178	79.9	78th and last Discoverer/ Program 162 flight.

Notes: All launches from Vandenberg AFB. The capsule recovery times are those known; the others are still classified.

U.S. SAMOS Launches

Name	Date	Booster	Orbit (km)	Inclination (degrees)	Remarks
SAMOS 1	Oct 11, 1960	Atlas Agena A	–	–	Failed to orbit.
SAMOS 2	Jan 31, 1961	Atlas Agena A	557 – 474	97.4	Photos poor.
SAMOS 3	Sep 9, 1961	Atlas Agena B	–	–	Failed to orbit.
None	Nov 22, 1961	Atlas Agena B	–	–	Failed to orbit.
1962 A Lambda	Dec 22, 1961	Atlas Agena B	702 – 244	89.6	Last SAMOS satellite.

Notes: All launches from Vandenberg AFB. Satellites used radio transmission of photos.

U.S. Second-Generation High-Resolution Satellites

Name	Date	Booster	Orbit (km)	Inclination (degrees)	Life-Time (days)	Remarks
1962						
1962 Zeta	Mar 7	Atlas Agena B	676 – 251	90.9	457.1	First launch.
1962 Omicron	Apr 26	Atlas Agena B	?	74.1	2	Orbit not available.
1962 Psi	Jun 17	Atlas Agena B	?	?	1	Orbit not available.
1962 A Zeta	Jul 18	Atlas Agena B	236 – 184	96.1	9	
1962 A Kappa	Aug 5	Atlas Agena B	205 – 205	96.3	1	
1962 B Pi	Nov 11	Atlas Agena B	206 – 206	96.0	1	
1963						
1963–28	Jul 12	Atlas Agena D	164 – 164	95.3	5.2	
1963–36	Sep 6	Atlas Agena D	263 – 168	94.3	7.05	
1963–41	Oct 25	Atlas Agena D	332 – 144	99.0	4.0	
1963–51	Dec 18	Atlas Agena D	266 – 122	97.8	1.28	
1964						
1964–9	Feb 25	Atlas Agena D	190 – 173	95.6	4	
1964–12	Mar 11	Atlas Agena D	203 – 163	95.7	4.3	
1964–20	Apr 23	Atlas Agena D	336 – 150	103.5	5.2	
1964–24	May 19	Atlas Agena D	380 – 141	101.1	2.9	
1964–36	Jul 6	Atlas Agena D	346 – 121	92.8	2	Carried ELINT sub-satellite.
1964–45	Aug 14	Atlas Agena D	307 – 149	95.5	8.8	
1964–58	Sep 23	Atlas Agena D	303 – 145	92.9	4.78	
None	Oct 8	Atlas Agena D	–	–	–	Failed to orbit.
1964–68	Oct 23	Atlas Agena D	271 – 139	95.5	5.06	Carried ELINT sub-satellite.
1964–79	Dec 4	Atlas Agena D	357 – 158	97.0	1.2	
1965						
1965–19	Mar 12	Atlas Agena D	247 – 155	107.6	4.98	
1965–31	Apr 8	Atlas Agena D	259 – 180	95.6	5.14	Carried ELINT sub-satellite.
1965–41	May 27	Atlas Agena D	267 – 149	95.7	5.11	
1965–50	Jun 25	Atlas Agena D	283 – 151	107.6	4.9	Carried ELINT sub-satellite.
None	Jul 12	Atlas Agena D	–	–	–	Failed to orbit.
1965–62	Aug 3	Atlas Agena D	307 – 149	107.4	4.11	Carried ELINT sub-satellite.
1965–76	Sep 30	Atlas Agena D	264 – 158	95.6	4.70	
1965–90	Nov 8	Atlas Agena D	277 – 145	93.8	2.92	
1966						
1966–2	Jan 19	Atlas Agena D	246 – 154	93.8	3.88	
1966–12	Feb 15	Atlas Agena D	293 – 148	96.5	7.44	
1966–22	Mar 18	Atlas Agena D	284 – 152	101.0	4.92	
1966–32	Apr 19	Atlas Agena D	398 – 145	116.9	6	
1966–39	May 14	Atlas Agena D	358 – 133	110.5	6	Carried ELINT sub-satellite.
1966–48	Jun 3	Atlas Agena D	288 – 143	87.0	6.17	
1966–74	Aug 16	Atlas Agena D	358 – 146	93.2	7.5	Carried ELINT sub-satellite.
1966–83	Sep 16	Atlas Agena D	333 – 148	93.9	6	Carried ELINT sub-satellite.
1966–90	Oct 12	Atlas Agena D	258 – 181	90.8	8.46	
1966–98	Nov 2	Atlas Agena D	305 – 159	90.9	7.2	
1966–109	Dec 5	Atlas Agena D	388 – 137	104.6	8.2	
1967						
1967–7	Feb 2	Atlas Agena D	357 – 136	102.9	9	
1967–50	May 22	Atlas Agena D	293 – 135	91.4	8.18	
1967–55	Jun 4	Atlas Agena D	456 – 149	104.8	8.17	Last launch.

Notes: All launches from Vandenberg AFB. After photo run completed, the Agena engine was re-started, de-orbiting the entire satellite.

U.S. Second-Generation Area Surveillance Satellites

Name	Date	Booster	Orbit (km)	Inclination (degrees)	Remarks
1964					
1964–27	Jun 4	TAT Agena D	429 – 149	79.9	Replacement for Program 162 satellites.
1964–32	Jun 19	TAT Agena D	462 – 176	85.0	
1964–37	Jul 10	TAT Agena D	461 – 180	84.9	
1964–43	Aug 5	TAT Agena D	436 – 182	79.9	
1964–56	Sep 14	TAT Agena D	466 – 172	84.9	
1964–61	Oct 5	TAT Agena D	440 – 182	79.9	
1964–67	Oct 17	TAT Agena D	416 – 189	74.9	
1964–71	Nov 2	TAT Agena D	448 – 180	79.9	
1964–75	Nov 18	TAT Agena D	330 – 180	70.0	
1964–85	Dec 19	TAT Agena D	410 – 183	74.9	
1964–87	Dec 21	TAT Agena D	264 – 238	70.0	
1965					
1965–2	Jan 15	TAT Agena D	420 – 180	74.9	
1965–13	Feb 25	TAT Agena D	377 – 177	75.0	
1965–26	Mar 25	TAT Agena D	265 – 186	96.0	
1965–33	Apr 29	TAT Agena D	473 – 178	85.0	
1965–37	May 18	TAT Agena D	331 – 198	75.0	
1965–45	Jun 9	TAT Agena D	362 – 176	75.0	
1965–57	Jul 19	TAT Agena D	464 – 182	85.0	
1965–67	Aug 17	TAT Agena D	407 – 180	70.0	
None	Sep 2	TAT Agena D	–	–	Failed to orbit.
1965–74	Sep 22	TAT Agena D	364 – 191	80.0	
1965–79	Oct 5	TAT Agena D	323 – 203	75.0	
1965–86	Oct 28	TAT Agena D	430 – 176	74.9	
1965–102	Dec 9	TAT Agena D	437 – 183	80.0	
1965–110	Dec 24	TAT Agena D	446 – 178	80.0	
1966					
1966–7	Feb 2	TAT Agena D	425 – 185	75.0	
1966–18	Mar 9	TAT Agena D	432 – 178	75.0	
1966–29	Apr 7	TAT Agena D	312 – 193	75.0	
None	May 3	TAT Agena D	–	–	Failed to orbit.
1966–42	May 24	TAT Agena D	271 – 179	66.0	
1966–55	Jun 21	TAT Agena D	367 – 194	80.1	
1966–85	Sep 20	TAT Agena D	442 – 188	85.1	
1966–102	Nov 8	TAT Agena D	318 – 172	100.0	
1967					
1967–2	Jan 14	TAT Agena D	380 – 180	80.0	
1967–15	Feb 22	TAT Agena D	380 – 180	80.0	
1967–29	Mar 30	TAT Agena D	326 – 167	85.0	Last launch.

Notes: All launches from Vandenberg AFB. Satellites used a capsule to return film to earth. After recovery, the empty satellite continued to orbit for about 3 weeks.

U.S. Third-Generation High-Resolution Satellites

Name	Date	Booster	Orbit (km)	Inclination (degrees)	Life-Time (days)	Remarks
1966						
1966–69	Jul 29	Titan III B	250 – 158	94.1	7	
1966–86	Sep 28	Titan III B	296 – 151	93.9	9.06	
1966–113	Dec 14	Titan III B	368 – 138	109.5	9	
1967						
1967–16	Feb 24	Titan III B	414 – 135	106.9	10.15	
None	Apr 26	Titan III B	–	–	–	Failed to orbit.
1967–64	Jun 20	Titan III B	325 – 127	111.4	10.22	
1967–79	Aug 16	Titan III B	449 – 142	111.8	13	
1967–90	Sep 19	Titan III B	401 – 122	106.1	10.23	
1967–103	Oct 25	Titan III B	429 – 136	111.5	9	
1967–121	Dec 5	Titan III B	430 – 137	109.5	11.18	
1968						
1968–5	Jan 18	Titan III B	404 – 138	111.5	17.13	
1968–18	Mar 13	Titan III B	407 – 128	99.8	11	
1968–31	Apr 17	Titan III B	427 – 134	111.5	12	
1968–47	Jun 5	Titan III B	456 – 123	110.5	12.2	
1968–64	Aug 6	Titan III B	395 – 142	110.0	9	
1968–74	Sep 10	Titan III B	404 – 125	106.0	15	
1968–99	Nov 6	Titan III B	390 – 130	106.0	14	
1968–108	Dec 4	Titan III B	736 – 136	106.2	8	
1969						
1969–7	Jan 22	Titan III B	1,090 – 142	106.1	12	
1969–18	Mar 4	Titan III B	461 – 134	92.0	14	
1969–39	Apr 15	Titan III B	410 – 135	108.7	15	
1969–50	Jun 3	Titan III B	414 – 137	110.0	11.2	
1969–74	Aug 22	Titan III B	366 – 133	108.0	16	
1969–95	Oct 24	Titan III B	740 – 136	108.0	15	
1970						
1970–2	Jan 14	Titan III B	383 – 134	109.9	18	
1970–31	Apr 15	Titan III B	388 – 130	110.9	21	
1970–48	Jun 25	Titan III B	389 – 129	108.8	11	
1970–61	Aug 18	Titan III B	365 – 151	110.9	16	
1970–90	Oct 23	Titan III B	396 – 135	111.0	19	
1971						
1971–5	Jan 21	Titan III B	418 – 139	110.8	19	
1971–33	Apr 22	Titan III B	401 – 132	110.9	21	
1971–70	Aug 12	Titan III B	424 – 137	111.0	22	
1971–92	Oct 23	Titan III B	416 – 134	110.9	25	
1972						
None	Feb 16	Titan III B	–	–	–	Failed to orbit.
1972–16	Mar 17	Titan III B	409 – 131	110.9	25	
None	May 20	Titan III B	–	–	–	Failed to orbit.
1972–68	Sep 1	Titan III B	380 – 140	110.5	29	
1972–103	Dec 21	Titan III B	378 – 139	110.4	33	

Name	Date	Booster	Orbit (km)	Inclination (degrees)	Life-Time (days)	Remarks
1973						
1973–28	May 16	Titan III B	352 – 136	110.4	28	
None	Jun 26	Titan III B	–	–	–	Failed to orbit.
1973–68	Sep 27	Titan III B	385 – 131	110.4	32	
1974						
1974–7	Feb 13	Titan III B	393 – 134	110.4	32	
1974–42	Jun 6	Titan III B	394 – 136	110.4	47	
1974–65	Aug 14	Titan III B	402 – 135	110.5	46	
1975						
1975–32	Apr 18	Titan III B	401 – 134	110.5	48	
1975–98	Oct 9	Titan III B	356 – 125	96.4	52	New lower inclination and orbit.
1976						
1976–27	Mar 22	Titan III B	347 – 125	96.4	57	
1976–94	Sep 15	Titan III B	330 – 135	96.3	51	
1977						
1977–19	Mar 13	Titan III B	348 – 124	96.4	74	
1977–94	Sep 23	Titan III B	352 – 125	96.5	76	

No launches of this type during *1978*.

Name	Date	Booster	Orbit (km)	Inclination (degrees)	Life-Time (days)	Remarks
1979						
1979–44	May 28	Titan III B	285 – 131	96.4	90	

No launches of this type during *1980* or *1981*.

Name	Date	Booster	Orbit (km)	Inclination (degrees)	Life-Time (days)	Remarks
1982						
1982–6	Jan 21	Titan III B	531 – 141	97.3	122	Made a number of maneuvers; may have been a test of KH-12 systems.
1983						
1983–32	Apr 15	Titan III B	298 – 135	96.5	128	
1984						
1984–39	Apr 17	Titan III B	311 – 127	96.4	118	
USA-4	Aug 28	Titan III B	Data Withheld			
1985						
USA-9	Feb 8	Titan III B	Data Withheld			

U.S. Third-Generation Area Surveillance Satellites

Name	Date	Booster	Orbit (km)	Inclination (degrees)	Life-Time (days)	Remarks
1966						
1966–72	Aug 9	LTTAT	287 – 194	100.1	32.2	First launch. Satellite may have had technical problems.
1967						
1967–43	May 9	LTTAT	777 – 200	85.1	64.62	Carried ELINT sub-satellite.
1967–62	Jun 16	LTTAT	367 – 181	80.0	33.16	Carried ELINT sub-satellite.
1967–76	Aug 7	LTTAT	346 – 174	79.94	24.85	
1967–87	Sep 15	LTTAT	389 – 150	80.0	18.69	
1967–109	Nov 2	LTTAT	410 – 183	81.5	29.83	Carried ELINT sub-satellite.
1967–122	Dec 9	LTTAT	237 – 158	81.6	15	
1968						
1968–8	Jan 24	LTTAT	430 – 176	81.4	33.54	Carried ELINT sub-satellite.
1968–20	Mar 14	LTTAT	391 – 178	83.0	26.22	Carried ELINT sub-satellite.
1968–39	May 1	LTTAT	243 – 164	83.0	14	
1968–52	Jun 20	LTTAT	326 – 193	84.9	25	Carried ELINT sub-satellite.
1968–65	Aug 7	LTTAT	257 – 152	82.1	19.45	
1968–78	Sep 18	LTTAT	393 – 167	83.0	19.25	Carried ELINT sub-satellite.
1968–98	Nov 3	LTTAT	288 – 150	82.1	19.99	
1968–112	Dec 12	LTTAT	248 – 169	81.0	15.65	Carried ELINT sub-satellite.
1969						
1969–10	Feb 5	LTTAT	239 – 178	81.5	18.8	Carried ELINT sub-satellite.
1969–26	Mar 19	LTTAT	241 – 179	83.0	4.35	Carried ELINT sub-satellite.
1969–41	May 2	LTTAT	326 – 179	64.9	21.35	Unusually low inclination. Carried ELINT sub-satellite.
1969–63	Jul 24	LTTAT	220 – 178	74.9	30.44	
1969–79	Sep 22	LTTAT	253 – 178	85.0	19.74	Carried ELINT sub-satellite.
1969–105	Dec 4	LTTAT	251 – 159	81.4	36.26	
1970						
1970–16	Mar 4	LTTAT	257 – 167	88.0	21.98	Carried ELINT sub-satellite.
1970–40	May 20	LTTAT	247 – 162	83.0	27.53	Carried ELINT sub-satellite.
1970–54	Jul 23	LTTAT	398 – 158	60.0	26.99	Believed to have photographed Middle East.
1970–98	Nov 18	LTTAT	232 – 185	82.9	22.78	Carried ELINT sub-satellite.
1971						
None	Feb 17	LTTAT	–	–	–	Failed to orbit.
1971–22	Mar 24	LTTAT	246 – 157	81.5	18.81	
1971–76	Sep 10	LTTAT	244 – 156	74.9	25.02	Carried ELINT sub-satellite.
1972						
1972–32	Apr 19	LTTAT	277 – 155	81.4	22.77	
1972–39	May 25	LTTAT	305 – 158	96.3	10.2	Last LTTAT launch.

Notes: All launches from Vandenberg AFB. Satellites transmitted photos to earth by radio rather than via capsule.

U.S. Big Bird Reconnaissance Satellites

Name	Date	Booster	Orbit (km)	Inclination (degrees)	Life-Time (days)	Remarks
1971–56	Jun 15, 1971	Titan III D	300 – 184	96.4	52	
1972–2	Jan 20, 1972	Titan III D	331 – 157	97.0	40	Carried ELINT sub-satellite.
1972–52	Jul 7, 1972	Titan III D	251 – 174	96.8	68	Carried ELINT sub-satellite.
1972–79	Oct 10, 1972	Titan III D	281 – 160	96.4	90	Carried ELINT sub-satellite.
1973–14	Mar 9, 1973	Titan III D	270 – 152	95.7	71	
1973–46	Jul 13, 1973	Titan III D	269 – 156	96.2	91	
1973–88	Nov 10, 1973	Titan III D	275 – 159	96.9	123	Carried 2 ELINT sub-satellites.
1974–20	Apr 10, 1974	Titan III D	285 – 153	94.5	109	Carried ELINT sub-satellite.
1974–85	Oct 29, 1974	Titan III D	271 – 162	96.6	141	Carried ELINT sub-satellite.
1975–51	Jun 8, 1975	Titan III D	269 – 154	96.3	150	
1975–114	Dec 4, 1975	Titan III D	234 – 157	96.2	119	
1976–65	Jul 8, 1976	Titan III D	242 – 159	97.0	158	Carried ELINT sub-satellite.
1977–56	Jun 27, 1977	Titan III D	239 – 155	97.0	179	
1978–29	Mar 16, 1978	Titan III D	240 – 160	96.4	179	Carried ELINT sub-satellite.
1979–25	Mar 16, 1979	Titan III D	258 – 170	96.3	190	Carried ELINT sub-satellite.
1980–52	Jun 18, 1980	Titan III D	265 – 169	96.4	261	Carried ELINT sub-satellite.
1981–19	Feb 28, 1981	Titan III D	332 – 134	96.3	112	
1982–41	May 11, 1982	Titan III D	262 – 177	96.4	208	Carried ELINT sub-satellite.
1983–60	Jun 20, 1983	Titan 34 D	230 – 157	96.4	275	First West Coast Titan 34 D launch. Carried ELINT sub-satellite.
USA-2	Jun 25, 1984	Titan 34 D	260 – 170	96.4	113	Carried ELINT sub-satellite.
None	Apr 18, 1986	Titan 34 D	–	–	–	Launch failure; believed to be a Big Bird.

U.S. KH-11 Digital Imagery Satellites

Name	Date	Booster	Orbit (km)	Inclination (degrees)	Re-entered	Remarks
1976–125	Dec 12, 1976	Titan III D	533 – 247	96.9	Jan 28, 1979	Orbited 770 days.
1978–60	Jun 14, 1978	Titan III D	509 – 223	96.9	Aug 23, 1981	Orbited 1,165 days.
1980–10	Feb 7, 1980	Titan III D	500 – 229	96.9	Oct 30, 1982	Orbited 996 days.
1981–85	Sep 3, 1981	Titan III D	525 – 242	96.9	Nov 23, 1984	Orbited 1,175 days.
1982–111	Nov 17, 1982	Titan III D	518 – 281	96.9	Aug 13, 1985	Orbited 1,000 days.
USA-6	Dec 3, 1984	Titan 34 D	520 – 250	97.0	Still in orbit.	
None	Aug 28, 1985	Titan 34 D	–	–	–	Launch failure; believed to be a KH-11.

Soviet Reconnaissance Satellites

Name	Date	Launch Site*	Booster	Orbit (km)	Inclination (degrees)	Life-Time (days)	Remarks
				1961			
None	?	T	A-1	–	–	–	Launch failure in the 4th quarter.
				1962			
Cosmos 4	Apr 26	T	A-1	330 – 298	65.0	3	1st generation, low-resolution Vostok reconnaissance satellite. Radiation measurements.
Cosmos 7	Jul 28	T	A-1	369 – 210	65.0	4	1st generation, low-resolution. Radiation measurements.
Cosmos 9	Sep 27	T	A-1	358 – 301	65.0	4	1st generation, low-resolution.
Cosmos 10	Oct 17	T	A-1	380 – 210	65.0	4	1st generation, low-resolution.
Cosmos 12	Dec 22	T	A-1	405 – 211	65.0	8	1st generation, low-resolution.
				1963			
Cosmos 13	Mar 21	T	A-1	337 – 205	65.0	8	1st generation, low-resolution.
Cosmos 15	Apr 22	T	A-1	371 – 173	65.0	5	1st generation, low-resolution. Weather test equipment.
Cosmos 16	Apr 28	T	A-1	401 – 207	65.0	10	1st generation, low-resolution.
Cosmos 18	May 24	T	A-1	301 – 209	65.0	9	1st generation, low-resolution.
Cosmos 20	Oct 18	T	A-1	311 – 206	65.0	8	1st generation, low-resolution.
Cosmos 22	Nov 16	T	A-2	394 – 205	64.9	6	First 2nd generation, high-resolution.
Cosmos 24	Dec 19	T	A-1	408 – 211	65.0	9	1st generation, low-resolution.
None	?	T	?	–	–	–	Two reconnaissance satellite launches. Failed in 4th quarter, 1963
None	?	T	?	–	–	–	
				1964			
Cosmos 28	Apr 4	T	A-1	395 – 209	65.0	8	1st generation, low-resolution.
Cosmos 29	Apr 25	T	A-1	309 – 204	65.1	8	1st generation, low-resolution.
Cosmos 30	May 18	T	A-2	383 – 207	64.9	8	2nd generation, high-resolution.
Cosmos 32	Jun 10	T	A-1	333 – 209	51.3	8	1st generation, low-resolution.
Cosmos 33	Jun 23	T	A-1	293 – 209	65.0	8	1st generation, low-resolution.
Cosmos 34	Jul 1	T	A-2	360 – 205	65.0	8	2nd generation, high resolution.
Cosmos 35	Jul 15	T	A-1	268 – 217	51.3	8	1st generation, low-resolution.
Cosmos 37	Aug 14	T	A-1	300 – 205	65.0	8	1st generation, low-resolution.
Cosmos 45	Sep 13	T	A-2	337 – 206	64.9	5	2nd generation, high-resolution. Meteorological tests.
Cosmos 46	Sep 24	T	A-1	271 – 215	51.3	8	1st generation, low-resolution.
Cosmos 48	Oct 14	T	A-1	295 – 203	65.1	6	1st generation, low-resolution.
Cosmos 50	Oct 28	T	A-1	241 – 196	51.3	8	1st generation, low-resolution. Exploded.
				1965			
Cosmos 52	Jan 11	T	A-1	304 – 205	65.0	8	1st generation, low-resolution.
Cosmos 59	Mar 7	T	A-2	339 – 209	65.0	8	2nd generation, high-resolution.
Cosmos 64	Mar 25	T	A-1	271 – 206	65.0	8	1st generation, low-resolution.
Cosmos 65	Apr 17	T	A-2	342 – 210	65.0	8	2nd generation, high-resolution. Meteorological tests.
Cosmos 66	May 7	T	A-1	291 – 197	65.0	8	1st generation, low-resolution.
Cosmos 67	May 25	T	A-2	350 – 207	51.8	8	2nd generation, high-resolution.

* T = Tyuratam
  P = Plesetsk

Name	Date	Launch Site	Booster	Orbit (km)	Inclination (degrees)	Life-Time (days)	Remarks	
Cosmos 68	Jun 15	T	A-1	334 – 205	65.0	8	1st generation, low-resolution.	
Cosmos 69	Jun 25	T	A-2	332 – 211	65.0	8	2nd generation, high-resolution.	
Cosmos 77	Aug 3	T	A-2	291 – 200	51.8	8	2nd generation, high-resolution.	
Cosmos 78	Aug 14	T	A-1	329 – 206	69.0	8	1st generation, low-resolution.	
Cosmos 79	Aug 25	T	A-2	359 – 211	64.9	8	2nd generation, high-resolution.	
Cosmos 85	Sep 9	T	A-2	319 – 212	65.0	8	2nd generation, high-resolution.	
Cosmos 91	Sep 23	T	A-2	342 – 212	65.0	8	2nd generation, high-resolution.	
Cosmos 92	Oct 16	T	A-2	353 – 212	65.0	8	2nd generation, high-resolution. Meteorological tests and biological experiments.	
Cosmos 94	Oct 28	T	A-2	293 – 211	65.0	8	2nd generation, high-resolution. Biological experiments.	
Cosmos 98	Nov 27	T	A-1	570 – 216	65.0	8	1st generation, low-resolution.	
Cosmos 99	Dec 10	T	A-1	320 – 199	65.0	8	1st generation, low-resolution.	

1966

Name	Date	Launch Site	Booster	Orbit (km)	Inclination (degrees)	Life-Time (days)	Remarks	
Cosmos 104	Jan 7	T	A-1	401 – 204	65.0	8	1st generation, low-resolution.	
Cosmos 105	Jan 22	T	A-1	324 – 204	65.0	8	1st generation, low-resolution. Had new pressurized propulsion system and attitude control system to provide stability during thruster firing.	
Cosmos 107	Feb 10	T	A-1	322 – 204	65.0	8	1st generation, low-resolution.	
Cosmos 109	Feb 19	T	A-2	309 – 209	65.0	8	2nd generation, high-resolution. Biological experiments.	
Cosmos 112	Mar 17	P	A-1	565 – 214	72.0	8	1st generation, low-resolution. First Plesetsk launch.	
Cosmos 113	Mar 21	T	A-2	327 – 210	65.0	8	2nd generation, high-resolution.	
Cosmos 114	Apr 6	P	A-2	374 – 210	73.0	8	2nd generation, high-resolution.	TK*
Cosmos 115	Apr 20	T	A-1	294 – 190	65.0	8	1st generation, high-resolution.	
Cosmos 117	May 6	T	A-1	308 – 207	65.0	8	1st generation, high-resolution.	
Cosmos 120	Jun 8	T	A-2	300 – 200	51.8	8	2nd generation, low-resolution. (first of type)	
Cosmos 121	Jun 17	P	A-2	354 – 210	72.9	8	2nd generation, high-resolution.	
Cosmos 124	Jul 14	T	A-2	303 – 208	51.8	8	2nd generation, low-resolution.	
Cosmos 126	Jul 28	T	A-2	359 – 212	51.8	9	2nd generation, high-resolution.	TK
Cosmos 127	Aug 8	T	A-2	279 – 204	51.9	8	2nd generation, high-resolution.	TK
Cosmos 128	Aug 27	T	A-2	364 – 212	65.0	8	2nd generation, high-resolution.	
Cosmos 129	Oct 14	P	A-1	307 – 202	65.0	9	1st generation, low-resolution.	
Cosmos 130	Oct 20	T	A-2	340 – 211	65.0	8	2nd generation, high-resolution.	
Cosmos 131	Nov 12	P	A-2	360 – 205	72.9	8	2nd generation, high-resolution.	
Cosmos 132	Nov 19	T	A-1	280 – 207	65.0	8	1st generation, low-resolution.	
Cosmos 134	Dec 3	T	A-2	319 – 214	65.0	8	2nd generation, high-resolution. Scientific experiments.	
Cosmos 136	Dec 19	P	A-1	305 – 198	64.6	8	1st generation, high-resolution. Scientific experiments.	

1967

Name	Date	Launch Site	Booster	Orbit (km)	Inclination (degrees)	Life-Time (days)	Remarks	
Cosmos 138	Jan 19	P	A-1	293 – 193	65.0	8	1st generation, low-resolution.	
Cosmos 141	Feb 8	P	A-2	345 – 210	72.9	8	2nd generation, high-resolution.	TK
Cosmos 143	Feb 27	T	A-1	302 – 204	65.0	8	1st generation, low-resolution. Scientific experiments.	
Cosmos 147	Mar 13	P	A-1	317 – 198	65.0	8	1st generation, low-resolution.	
Cosmos 150	Mar 22	P	A-2	373 – 206	65.7	8	2nd generation, high-resolution.	TK
Cosmos 153	Apr 4	P	A-1	291 – 202	64.6	8	1st generation, low-resolution.	
Cosmos 155	Apr 12	T	A-2	296 – 203	51.8	8	2nd generation, high-resolution.	TK
Cosmos 157	May 12	T	A-1	296 – 202	51.3	8	Last 1st generation, low-resolution.	
Cosmos 161	May 22	P	A-2	343 – 205	65.7	8	2nd generation, high-resolution.	TK
Cosmos 162	Jun 1	T	A-2	280 – 201	51.8	8	2nd generation, low-resolution.	

*The TK, TF, TL and other two-letter groups indicate the recovery beacon signals that guide the ground team to the capsule.

Name	Date	Launch Site	Booster	Orbit (km)	Inclination (degrees)	Life-Time (days)	Remarks	
Cosmos 164	Jun 8	P	A-2	320 – 202	65.7	6	2nd generation, low-resolution.	
Cosmos 168	Jul 4	T	A-2	268 – 199	51.8	8	2nd generation, low-resolution.	
Cosmos 172	Aug 9	T	A-2	301 – 202	51.8	8	2nd generation, high-resolution.	
Cosmos 175	Sep 11	P	A-2	386 – 210	72.9	8	2nd generation, high-resolution.	TK
Cosmos 177	Sep 16	T	A-2	292 – 202	51.8	8	2nd generation, low-resolution.	
Cosmos 180	Sep 26	P	A-2	370 – 212	72.9	8	2nd generation, low-resolution.	
Cosmos 181	Oct 11	P	A-2	344 – 200	65.6	8	2nd generation, low-resolution.	
Cosmos 182	Oct 16	T	A-2	355 – 210	65.0	8	2nd generation, high-resolution.	TK
Cosmos 190	Nov 3	P	A-2	347 – 201	65.7	8	2nd generation, high-resolution.	TK
Cosmos 193	Nov 25	P	A-2	354 – 203	65.7	8	2nd generation, low-resolution.	
Cosmos 194	Dec 3	P	A-2	333 – 205	65.7	8	2nd generation, high-resolution.	TK
Cosmos 195	Dec 16	P	A-2	375 – 211	65.7	8	2nd generation, low-resolution.	

1968

Name	Date	Launch Site	Booster	Orbit (km)	Inclination (degrees)	Life-Time (days)	Remarks	
Cosmos 199	Jan 16	P	A-2	386 – 204	65.7	8	2nd generation, low-resolution. Exploded.	
Cosmos 201	Feb 6	T	A-2	355 – 210	65.0	8	2nd generation, high-resolution.	TK
Cosmos 205	Mar 5	P	A-2	310 – 201	65.7	8	2nd generation, low-resolution.	
Cosmos 207	Mar 16	P	A-2	342 – 210	65.6	8	2nd generation, high-resolution.	TK
Cosmos 208	Mar 21	T	A-2	305 – 207	65.0	12	First 3rd generation, low resolution, PDM TM-Science. Separated a capsule with high-energy gamma ray instruments.	TK
Cosmos 210	Apr 3	P	A-2	395 – 217	81.2	8	2nd generation, low-resolution.	
Cosmos 214	Apr 18	P	A-2	403 – 211	81.4	8	2nd generation, high-resolution.	
Cosmos 216	Apr 20	T	A-2	277 – 199	51.8	8	2nd generation, low-resolution.	
Cosmos 223	Jun 1	P	A-2	374 – 212	72.9	8	2nd generation, low-resolution.	
Cosmos 224	Jun 4	T	A-2	270 – 200	51.8	8	2nd generation, high-resolution. Measured atmospheric composition.	
Cosmos 227	Jun 18	T	A-2	281 – 194	51.8	8	2nd generation, high-resolution.	TK
Cosmos 228	Jun 21	T	A-2	259 – 206	51.6	12	3rd generation, low-resolution, PDM-Science. Separated a capsule for cosmic ray studies.	TK
Cosmos 229	Jun 26	P	A-2	354 – 210	72.8	8	2nd generation, high-resolution.	
Cosmos 231	Jul 10	T	A-2	330 – 211	65.0	8	2nd generation, low-resolution.	
Cosmos 232	Jul 16	P	A-2	352 – 202	65.0	8	2nd generation, high-resolution. Meteorological tests.	TK
Cosmos 234	Jul 30	T	A-2	310 – 210	51.8	6	2nd generation, high-resolution.	TK
Cosmos 235	Aug 9	T	A-2	303 – 207	51.8	8	2nd generation, low-resolution.	
Cosmos 237	Aug 27	P	A-2	343 – 201	65.4	8	2nd generation, high-resolution.	TK
Cosmos 239	Sep 5	T	A-2	282 – 202	51.8	8	2nd generation, high-resolution.	
Cosmos 240	Sep 14	T	A-2	293 – 197	51.8	7	2nd generation, low-resolution.	
Cosmos 241	Sep 16	P	A-2	343 – 201	65.4	8	2nd generation, high-resolution.	TK
Cosmos 243	Sep 23	T	A-2	319 – 210	71.3	11	3rd generation, low-resolution, PDM-Science. Separated a capsule with radio telescope for measurement of passive microwaves.	
Cosmos 246	Oct 7	P	A-2	348 – 147	65.4	5	2nd generation, low resolution. Observed Apollo 7 launch. (?)	TK
Cosmos 247	Oct 11	P	A-2	362 – 205	65.4	8	2nd generation, low-resolution.	
Cosmos 251	Oct 31	T	A-2	270 – 198	65.0	12	First 3rd generation, high-resolution, Morse-Maneuverable. Separated a capsule with radio astronomy and gamma ray experiments.	
Cosmos 253	Nov 13	P	A-2	355 – 206	65.4	5	2nd generation, low-resolution.	
Cosmos 254	Nov 21	P	A-2	350 – 203	65.4	8	2nd generation, high-resolution.	
Cosmos 255	Nov 29	P	A-2	336 – 201	65.4	8	2nd generation, low-resolution.	
Cosmos 258	Dec 10	T	A-2	325 – 210	65.0	8	2nd generation, low-resolution.	

Name	Date	Launch Site	Booster	Orbit (km)	Inclination (degrees)	Life-Time (days)	Remarks	
					1969			
Cosmos 263	Jan 12	P	A-2	346 – 205	65.4	8	2nd generation, low-resolution.	
Cosmos 264	Jan 23	T	A-2	330 – 219	70.0	13	3rd generation, high-resolution, Morse-Maneuverable. Separated an engine module. Carried radio astronomy and gamma ray package.	
Cosmos 266	Feb 25	P	A-2	358 – 208	72.9	8	2nd generation, low-resolution. [China crisis]	
Cosmos 267	Feb 26	T	A-2	346 – 210	65.0	8	2nd generation, high-resolution. [China crisis]	TK
Cosmos 270	Mar 6	P	A-2	350 – 205	65.4	8	2nd generation, high-resolution. [China crisis]	TK
Cosmos 271	Mar 15	P	A-2	342 – 200	65.4	8	2nd generation, high-resolution. [China crisis]	TK
Cosmos 273	Mar 22	P	A-2	356 – 205	65.4	8	2nd generation, low-resolution. [China crisis]	
Cosmos 274	Mar 24	T	A-2	323 – 213	65.0	8	2nd generation, high-resolution. Scientific experiments. [China crisis]	TK
Cosmos 276	Apr 4	P	A-2	410 – 214	81.4	7	2nd generation, high-resolution. [China crisis]	TK
Cosmos 278	Apr 9	P	A-2	338 – 203	65.0	8	2nd generation, low-resolution. [China crisis]	TG
Cosmos 279	Apr 15	T	A-2	280 – 194	51.8	8	2nd generation, high-resolution. [China crisis]	TK
Cosmos 280	Apr 23	T	A-2	272 – 206	51.6	13	3rd generation, high-resolution, Morse-Maneuverable. Separated an engine module. Carried weather experiments. [China crisis]	TK
Cosmos 281	May 13	P	A-2	317 – 194	65.4	8	2nd generation, low-resolution.	
Cosmos 282	May 20	P	A-2	343 – 209	65.4	8	2nd generation, high-resolution.	TK
Cosmos 284	May 29	T	A-2	308 – 207	51.8	8	2nd generation, high-resolution.	TK
Cosmos 286	Jun 15	P	A-2	349 – 206	65.4	8	2nd generation, high-resolution.	TK
Cosmos 287	Jun 24	T	A-2	268 – 190	51.8	8	2nd generation, low-resolution.	TK
Cosmos 288	Jun 27	T	A-2	281 – 201	51.8	8	2nd generation, high-resolution.	TK
Cosmos 289	Jul 10	P	A-2	350 – 200	65.4	5	2nd generation, high-resolution.	
Cosmos 290	Jul 22	P	A-2	352 – 200	65.4	8	2nd generation, low-resolution.	TG
Cosmos 293	Aug 16	T	A-2	270 – 208	51.8	12	3rd generation, low-resolution, PDM-Science. [China crisis]	TK
Cosmos 294	Aug 19	P	A-2	348 – 202	65.4	8	2nd generation, high-resolution. [China crisis]	TK
Cosmos 296	Aug 29	T	A-2	322 – 211	65.0	8	2nd generation, high-resolution. [China crisis]	TK
Cosmos 297	Sep 2	P	A-2	334 – 211	72.9	8	2nd generation, high-resolution. [China crisis]	
Cosmos 299	Sep 18	T	A-2	311 – 214	65.0	4	2nd generation, high-resolution. [China crisis]	TK
Cosmos 301	Sep 24	P	A-2	307 – 197	65.4	8	2nd generation, low-resolution. [China crisis]	
Cosmos 302	Oct 17	P	A-2	340 – 202	65.4	8	2nd generation, high-resolution.	TK
Cosmos 306	Oct 24	T	A-2	332 – 208	65.0	12	First 3rd generation, low-resolution, PDM.	TK
Cosmos 309	Nov 12	P	A-2	384 – 203	65.4	8	3rd generation, low-resolution, PDM-Science. Separated a capsule.	
Cosmos 310	Nov 15	T	A-2	347 – 208	65.0	8	2nd generation, high-resolution.	
Cosmos 313	Dec 3	P	A-2	276 – 204	65.4	12	3rd generation, low-resolution, PDM.	TK
Cosmos 317	Dec 23	P	A-2	302 – 209	65.4	13	First 3rd generation, high-resolution 2 tone-Maneuverable. Separated an engine module. Charged particle studies.	TF?

Name	Date	Launch Site	Booster	Orbit (km)	Inclination (degrees)	Life-Time (days)	Remarks	
					1970			
Cosmos 318	Jan 9	T	A-2	295 – 204	65.0	12	3rd generation, low-resolution, PDM.	TK
Cosmos 322	Jan 21	P	A-2	337 – 200	65.5	8	2nd generation, high-resolution.	TK
Cosmos 323	Feb 10	P	A-2	333 – 206	65.4	8	2nd generation, high-resolution.	TK
Cosmos 325	Mar 4	P	A-2	348 – 207	65.4	8	2nd generation, low-resolution.	TG
Cosmos 326	Mar 13	P	A-2	393 – 212	81.4	8	2nd generation, low-resolution.	
Cosmos 328	Mar 27	P	A-2	340 – 213	72.9	13	3rd generation, high-resolution, Morse-Maneuverable. Maneuvering engine was not separated.	TK
Cosmos 329	Apr 3	P	A-2	240 – 202	81.3	12	3rd generation, low-resolution, PDM.	TU?
Cosmos 331	Apr 8	T	A-2	347 – 213	65.0	8	2nd generation, high-resolution.	
Cosmos 333	Apr 15	P	A-2	265 – 217	81.4	13	3rd generation, high-resolution, Morse-Maneuverable. Separated an engine module.	TK
Cosmos 344	May 12	P	A-2	347 – 206	72.9	8	Last 2nd generation, low-resolution.	TG
Cosmos 345	May 20	T	A-2	276 – 193	51.8	8	2nd generation, high-resolution.	TK
Cosmos 346	Jun 10	T	A-2	289 – 201	51.8	7	2nd generation, high-resolution.	
Cosmos 349	Jun 17	P	A-2	350 – 203	65.4	8	2nd generation, high-resolution.	TK
Cosmos 350	Jun 26	T	A-2	267 – 204	51.8	12	3rd generation, low-resolution, PDM.	TG
Cosmos 352	Jul 7	T	A-2	309 – 205	51.8	8	2nd generation, high-resolution.	TK
Cosmos 353	Jul 9	P	A-2	309 – 211	65.4	12	3rd generation, low-resolution, PDM.	
Cosmos 355	Aug 7	P	A-2	342 – 202	66.4	8	Last 2nd generation, high-resolution.	TK
Cosmos 360	Aug 29	T	A-2	318 – 209	65.0	10	3rd generation, high-resolution, Morse-Maneuverable. Separated a capsule.	
Cosmos 361	Sep 8	P	A-2	326 – 207	72.9	13	3rd generation, high-resolution, Morse-Maneuverable. Separated an engine module.	
Cosmos 363	Sep 17	T	A-2	324 – 210	65.0	12	3rd generation, low-resolution, PDM.	
Cosmos 364	Sep 22	P	A-2	330 – 211	65.4	12	3rd generation, high-resolution, 2 tone-Maneuverable. Separated an engine.	
Cosmos 366	Oct 1	T	A-2	310 – 206	65.0	12	3rd generation, low-resolution, PDM.	
Cosmos 368	Oct 8	T	A-2	421 – 212	65.0	6	3rd generation, low-resolution, PDM-Science. Separated a capsule.	
Cosmos 370	Oct 9	T	A-2	307 – 208	65.0	13	3rd generation, high-resolution, Morse-Maneuverable. Separated an engine module.	TK
Cosmos 376	Oct 30	P	A-2	311 – 216	65.4	13	3rd generation, high-resolution, Morse-Maneuverable. Separated an engine module.	TK
Cosmos 377	Nov 11	T	A-2	305 – 208	65.0	12	3rd generation, low-resolution, PDM.	
Cosmos 383	Dec 3	P	A-2	293 – 208	65.4	13	3rd generation, high-resolution, 2 tone-Maneuverable. Maneuvered, but did not separate an engine module.	TF
Cosmos 384	Dec 10	P	A-2	314 – 212	72.9	12	3rd generation, low-resolution, PDM-Science. Separated a capsule.	TG
Cosmos 386	Dec 15	T	A-2	275 – 207	65.0	13	3rd generation, high-resolution, Morse-Maneuverable.	

Name	Date	Launch Site	Booster	Orbit (km)	Inclination (degrees)	Life-Time (days)	Remarks	
					1971			
Cosmos 390	Jan 12	T	A-2	296 – 208	65.0	13	3rd generation, high-resolution, 2 tone-Maneuverable. (?) Separated an engine module.	
Cosmos 392	Jan 21	T	A-2	300 – 207	65.0	12	3rd generation, low-resolution, PDM.	
Cosmos 396	Feb 18	P	A-2	310 – 211	65.4	13	3rd generation, high-resolution, Morse-Maneuverable. Separated an engine module.	TK
Cosmos 399	Mar 3	T	A-2	310 – 209	65.0	14	3rd generation, high-resolution, Morse-Maneuverable. Separated an engine module.	
Cosmos 401	Mar 27	P	A-2	322 – 216	72.9	13	3rd generation, high-resolution, Morse-Maneuverable. Separated an engine module.	TK
Cosmos 403	Apr 2	P	A-2	251 – 216	81.4	12	3rd generation, low-resolution, PDM.	
Cosmos 406	Apr 14	P	A-2	264 – 223	81.3	10	3rd generation, high-resolution, Morse-Maneuverable. Separated an engine module.	
Cosmos 410	May 6	T	A-2	300 – 207	65.0	12	3rd generation, low-resolution, PDM-Science. Separated a capsule.	
Cosmos 420	May 18	T	A-2	242 – 200	51.8	11	3rd generation, high-resolution, Morse-Maneuverable. Separated an engine module.	TK
Cosmos 424	May 28	P	A-2	309 – 214	65.4	13	3rd generation, high-resolution, Morse-Maneuverable. (?) Separated an engine module.	
Cosmos 427	Jun 11	P	A-2	337 – 211	72.9	12	3rd generation, high-resolution, 2 tone-Maneuverable. Separated an engine module.	
Cosmos 428	Jun 24	T	A-2	271 – 208	51.8	13	3rd generation, low-resolution, PDM-Science. Separated a capsule. Electron and gamma flux studies.	TG
Cosmos 429	Jul 20	T	A-2	260 – 204	51.8	13	3rd generation, low-resolution, Morse-Maneuverable. Separated an engine module.	
Cosmos 430	Jul 23	P	A-2	322 – 206	65.4	13	3rd generation, high-resolution, Morse-Maneuverable. Separated an engine module.	
Cosmos 431	Jul 30	T	A-2	262 – 202	51.8	12	3rd generation, low-resolution, PDM.	TG
Cosmos 432	Aug 5	T	A-2	262 – 209	51.8	13	3rd generation, high-resolution, Morse-Maneuverable. Separated an engine module.	TK
Cosmos 438	Sep 14	P	A-2	321 – 212	65.4	13	3rd generation, high-resolution, 2 tone-Maneuverable. Separated an engine module.	
Cosmos 439	Sep 21	P	A-2	308 – 219	65.4	11	3rd generation, low-resolution, PDM.	
Cosmos 441	Sep 28	T	A-2	228 – 209	65.0	12	3rd generation, high-resolution, Morse-Maneuverable. (?) Separated an engine module.	
Cosmos 442	Sep 29	P	A-2	321 – 211	72.9	13	3rd generation, high-resolution, 2 tone-Maneuverable. (?) Separated an engine module.	
Cosmos 443	Oct 7	P	A-2	325 – 211	65.4	12	3rd generation, low-resolution, PDM-Science. Separated a capsule.	

Name	Date	Launch Site	Booster	Orbit (km)	Inclination (degrees)	Life-Time (days)	Remarks
Cosmos 452	Oct 14	T	A-2	270 – 201	65.0	13	3rd generation, high-resolution, Morse-Maneuverable. (?) Separated a capsule.
Cosmos 454	Nov 2	P	A-2	284 – 210	66.4	14	3rd generation, high-resolution, Morse-Maneuverable. (?) Separated an engine module.
Cosmos 456	Nov 19	P	A-2	328 – 218	72.9	13	3rd generation, high-resolution, Morse-Maneuverable. Separated an engine module.
Cosmos 463	Dec 6	T	A-2	307 – 215	65.0	5	3rd generation, high-resolution, Morse-Maneuverable. Observed Indo-Pakistani war. Separated an engine module.
Cosmos 464	Dec 10	P	A-2	405 – 206	72.9	6	3rd generation, high-resolution, Morse-Maneuverable. (?) Observed Indo-Pakistani war. Separated an engine module.
Cosmos 466	Dec 16	T	A-2	302 – 207	65.0	11	3rd generation, high-resolution, Morse-Maneuverable. Separated an engine module.
Cosmos 470	Dec 27	P	A-2	272 – 195	65.4	10	First 3rd generation, low-resolution, 2 tone-Science. Separated a capsule.

1972

Name	Date	Launch Site	Booster	Orbit (km)	Inclination (degrees)	Life-Time (days)	Remarks
Cosmos 471	Jan 12	T	A-2	323 – 202	65.0	13	3rd generation, high-resolution, Morse-Maneuverable. Separated an engine module.
Cosmos 473	Feb 3	T	A-2	333 – 209	65.0	12	3rd generation, low-resolution, PDM.
Cosmos 474	Feb 16	T	A-2	347 – 207	65.0	13	3rd generation, high-resolution, Morse-Maneuverable. Separated an engine module.
Cosmos 477	Mar 4	P	A-2	328 – 212	72.9	12	3rd generation, low-resolution, PDM-Science. Separated a capsule with particle flux and radiation experiments.
Cosmos 478	Mar 15	P	A-2	319 – 213	65.4	13	3rd generation, high-resolution, Morse-Maneuverable. Separated an engine module.
Cosmos 483	Apr 3	P	A-2	345 – 212	72.9	12	3rd generation, high-resolution, Morse-Maneuverable. Separated an engine module.
Cosmos 484	Apr 5	P	A-2	236 – 203	81.3	13	3rd generation, low-resolution, PDM-Science. Separated a capsule for cosmic ray studies.
Cosmos 486	Apr 14	P	A-2	267 – 214	81.4	13	3rd generation, high-resolution, Morse-Maneuverable. Separated an engine module.
Cosmos 488	May 5	P	A-2	319 – 211	65.4	13	3rd generation, high-resolution, 2 tone-Maneuverable. Separated an engine module.
Cosmos 490	May 17	P	A-2	310 – 212	65.4	12	3rd generation, low-resolution, PDM-Science. Separated a capsule with high-energy electron flux and cosmic ray studies.
Cosmos 491	May 25	T	A-2	303 – 210	65.0	14	3rd generation, high-resolution, Morse-Maneuverable. Separated an engine module.
Cosmos 492	Jun 9	T	A-2	342 – 209	65.0	13	3rd generation, high-resolution, Morse-Maneuverable. Separated an engine module.

TK

Name	Date	Launch Site	Booster	Orbit (km)	Inclination (degrees)	Life-Time (days)	Remarks	
Cosmos 493	Jun 21	T	A-2	308 – 313	65.0	12	3rd generation, low-resolution, PDM	TK
Cosmos 495	Jun 23	P	A-2	298 – 206	65.4	13	3rd generation, high-resolution, Morse-Maneuverable. (?) Separated an engine module.	
Cosmos 499	Jul 6	T	A-2	283 – 209	51.8	9	3rd generation, high-resolution, Morse-Maneuverable. Separated an engine module.	TK
Cosmos 502	Jul 13	P	A-2	284 – 206	65.4	12	3rd generation, low-resolution, 2 tone-Science. Separated an engine module.	
Cosmos 503	Jul 19	P	A-2	304 – 208	65.4	13	3rd generation, high-resolution, Morse-Maneuverable. Separated an engine module.	TK
Cosmos 512	Jul 28	P	A-2	294 – 207	65.4	12	3rd generation, low-resolution, PDM.	
Cosmos 513	Aug 2	T	A-2	340 – 209	65.0	13	3rd generation, high-resolution, Morse-Maneuverable. Separated an engine module.	TK?
Cosmos 515	Aug 18	P	A-2	300 – 203	72.9	13	3rd generation, high-resolution, 2 tone-Maneuverable. Separated an engine module.	
Cosmos 517	Aug 30	T	A-2	305 – 207	65.0	12	3rd generation, low-resolution, PDM.	
Cosmos 518	Sep 15	P	A-2	330 – 208	72.9	9	3rd generation, low-resolution, PDM-Science. Separated a capsule.	
Cosmos 519	Sep 16	T	A-2	343 – 210	71.3	10	3rd generation, high-resolution, Morse-Maneuverable. Separated an engine module.	TK
Cosmos 522	Oct 4	P	A-2	342 – 214	72.9	13	3rd generation, high-resolution, Morse-Maneuverable. Separated an engine module.	
Cosmos 525	Oct 18	P	A-2	292 – 208	65.4	11	3rd generation, low-resolution, PDM-Science. Separated a capsule.	
Cosmos 527	Oct 31	P	A-2	330 – 214	65.4	13	3rd generation, high-resolution, 2 tone-Maneuverable. Separated an engine module.	
Cosmos 537	Nov 25	T	A-2	324 – 207	65.0	12	3rd generation, low-resolution, PDM.	
Cosmos 538	Dec 14	P	A-2	305 – 212	65.4	13	3rd generation, high-resolution, Morse-Maneuverable. Separated an engine module.	
Cosmos 541	Dec 27	P	A-2	371 – 242	81.4	12	3rd generation, low-resolution, 2 tone-Science. Separated a capsule.	TL

1973

Name	Date	Launch Site	Booster	Orbit (km)	Inclination (degrees)	Life-Time (days)	Remarks	
Cosmos 543	Jan 11	T	A-2	333 – 211	65.0	13	3rd generation, high-resolution, Morse-Maneuverable. Separated an engine module.	
Cosmos 547	Feb 1	T	A-2	330 – 208	65.0	12	3rd generation, low-resolution, PDM.	
Cosmos 548	Feb 8	P	A-2	322 – 214	65.4	13	3rd generation, high-resolution, Morse-Maneuverable. Separated an engine module.	TK
Cosmos 550	Mar 1	P	A-2	325 – 217	65.4	10	3rd generation, high-resolution, 2 tone-Maneuverable. Separated an engine module.	
Cosmos 551	Mar 6	T	A-2	316 – 210	65.0	14	3rd generation, high-resolution, Morse-Maneuverable. Separated an engine module.	

Name	Date	Launch Site	Booster	Orbit (km)	Inclination (degrees)	Life-Time (days)	Remarks	
Cosmos 552	Mar 22	P	A-2	337 – 211	72.9	12	3rd generation, low-resolution, PDM-Science. Separated a capsule.	
Cosmos 554	Apr 19	P	A-2	318 – 212	72.9	16	3rd generation, high-resolution, 2 tone-Maneuverable. Exploded after separation of maneuvering engine.	
Cosmos 555	Apr 25	P	A-2	253 – 216	81.3	12	3rd generation, low-resolution, PDM-Science. Separated a capsule.	
Cosmos 556	May 5	P	A-2	252 – 209	81.3	9	3rd generation, high-resolution, 2 tone-Maneuverable. Separated an engine module.	
Cosmos 559	May 18	P	A-2	345 – 217	65.4	5	3rd generation, high-resolution, 2 tone-Maneuverable. Separated an engine module.	
Cosmos 560	May 23	P	A-2	336 – 211	72.9	13	3rd generation, high-resolution, Morse-Maneuverable. Separated an engine module.	
Cosmos 561	May 25	P	A-2	317 – 215	65.4	12	3rd generation, low-resolution, PDM-Science. Separated a capsule with gamma ray telescope.	TK?
Cosmos 563	Jun 6	P	A-2	320 – 213	65.4	12	3rd generation, high-resolution, Morse-Maneuverable. Separated an engine module.	
Cosmos 572	Jun 10	T	A-2	294 – 211	51.7	13	3rd generation, high-resolution, Morse-Maneuverable. Separated an engine module.	TK
Cosmos 575	Jun 21	P	A-2	299 – 208	65.4	12	3rd generation, low-resolution, PDM. Separated a capsule.	
Cosmos 576	Jun 27	P	A-2	356 – 212	72.9	12	3rd generation, low-resolution, 2 tone-Science. Separated a capsule.	TL
Cosmos 577	Jul 25	P	A-2	312 – 209	65.4	13	3rd generation, high-resolution, Morse-Maneuverable. Separated an engine module.	
Cosmos 578	Aug 1	P	A-2	308 – 207	65.4	12	3rd generation, low-resolution, PDM.	
Cosmos 579	Aug 21	P	A-2	315 – 209	65.4	13	3rd generation, high-resolution, Morse-Maneuverable. Separated an engine module.	TK?
Cosmos 581	Aug 24	T	A-2	303 – 211	51.6	13	3rd generation, high-resolution, Morse-Maneuverable. Separated an engine module.	TK
Cosmos 583	Aug 30	T	A-2	316 – 208	65.0	13	3rd generation, low-resolution, PDM. Separated a capsule.	
Cosmos 584	Sep 6	P	A-2	360 – 213	72.9	14	3rd generation, high-resolution, Morse-Maneuverable. Separated an engine module.	TK
Cosmos 587	Sep 21	P	A-2	330 – 215	65.4	13	3rd generation, high-resolution, 2 tone-Maneuverable. Separated an engine module.	TF
Cosmos 596	Oct 3	P	A-2	310 – 211	65.4	6	3rd generation, low-resolution, PDM. Separated a capsule. Photographed Mid-East war.	TK
Cosmos 597	Oct 6	P	A-2	312 – 212	65.4	6	3rd generation, high-resolution, 2 tone-Maneuverable. Separated an engine module. Photographed Mid-East war.	

Name	Date	Launch Site	Booster	Orbit (km)	Inclination (degrees)	Life-Time (days)	Remarks	
Cosmos 598	Oct 10	P	A-2	360 – 213	72.9	6	3rd generation, high-resolution, Morse-Maneuverable. Separated an engine module. Photographed Mid-East war.	
Cosmos 599	Oct 15	T	A-2	294 – 206	65.0	13	3rd generation, low-resolution, PDM. Photographed Mid-East war.	
Cosmos 600	Oct 16	P	A-2	366 – 215	72.9	7	3rd generation, high-resolution, Morse-Maneuverable. Separated an engine module. Photographed Mid-East war.	
Cosmos 602	Oct 20	P	A-2	365 – 213	72.9	9	3rd generation, high-resolution, 2 tone-Maneuverable. Separated an engine module. Photographed Mid-East war.	
Cosmos 603	Oct 27	P	A-2	380 – 214	72.9	13	3rd generation, high-resolution, Morse-Maneuverable. Separated an engine module. Photographed Mid-East war.	
Cosmos 607	Nov 10	P	A-2	364 – 214	72.9	12	3rd generation, high-resolution, 2 tone-Maneuverable. (?) Separated an engine module.	
Cosmos 609	Nov 21	T	A-2	370 – 207	70.0	13	3rd generation, high-resolution, Morse-Maneuverable. Separated an engine module.	TK
Cosmos 612	Nov 28	P	A-2	371 – 214	72.9	13	3rd generation, high-resolution, 2 tone-Maneuverable. Separated an engine module.	
Cosmos 616	Dec 17	P	A-2	355 – 214	72.9	11	3rd generation, low-resolution, 2 tone-Science. Separated a capsule.	
Cosmos 625	Dec 21	P	A-2	346 – 214	72.8	13	3rd generation, high-resolution, 2 tone-Maneuverable. Separated an engine module.	

1974

Name	Date	Launch Site	Booster	Orbit (km)	Inclination (degrees)	Life-Time (days)	Remarks	
Cosmos 629	Jan 24	P	A-2	315 – 202	62.8	12	3rd generation, low-resolution, PDM-Science. Separated a capsule.	TG
Cosmos 630	Jan 30	P	A-2	367 – 213	72.9	14	3rd generation, high-resolution, 2 tone-Maneuverable. Separated an engine module.	
Cosmos 632	Feb 12	T	A-2	333 – 184	65.0	14	3rd generation, high-resolution, Morse-Maneuverable. Separated an engine module.	
Cosmos 635	Mar 14	P	A-2	350 – 212	72.9	12	3rd generation, low-resolution, PDM-Science. Separated a capsule.	TG
Cosmos 636	Mar 20	T	A-2	409 – 174	65.0	14	3rd generation, high-resolution, 2 tone-Maneuverable. Separated an engine module.	
Cosmos 639	Apr 4	P	A-2	238 – 209	81.3	11	3rd generation, high-resolution, 2 tone-Maneuverable. Separated an engine module.	
Cosmos 640	Apr 11	P	A-2	236 – 205	81.3	12	3rd generation, low-resolution, PDM. Possibly separated a capsule.	TG
Cosmos 649	Apr 29	P	A-2	320 – 189	62.8	12	3rd generation, high-resolution, 2 tone-Maneuverable. Separated an engine module.	
Cosmos 652	May 15	T	A-2	362 – 180	51.8	8	3rd generation, high-resolution, 2 tone-Maneuverable. Separated an engine module.	

Name	Date	Launch Site	Booster	Orbit (km)	Inclination (degrees)	Life-Time (days)	Remarks	
Cosmos 653	May 15	P	A-2	309 – 196	62.8	12	3rd generation, low-resolution, PDM. (Note two launches in one day.)	TG
Cosmos 657	May 30	P	A-2	317 – 182	62.8	14	3rd generation, high-resolution, 2 tone-Maneuverable. Separated an engine module.	TF
Cosmos 658	Jun 6	T	A-2	304 – 206	65.0	12	3rd generation, low-resolution, PDM.	TG
Cosmos 659	Jun 13	P	A-2	360 – 190	62.8	13	3rd generation, high-resolution, 2 tone-Maneuverable. Separated an engine module.	TF
Cosmos 664	Jun 29	P	A-2	364 – 212	72.9	12	3rd generation, low-resolution, 2 tone-Science. Separated a capsule.	TL
Cosmos 666	Jul 12	P	A-2	351 – 191	62.8	13	3rd generation, high-resolution, 2 tone-Maneuverable. Separated an engine module.	
Cosmos 667	Jul 25	T	A-2	342 – 182	65.0	13	Last 3rd generation, high-resolution, Morse-Maneuverable. (?) Separated an engine module.	
Cosmos 669	Jul 26	P	A-2	244 – 210	81.3	13	3rd generation, low-resolution, PDM-Science. Separated a capsule.	
Cosmos 671	Aug 7	P	A-2	369 – 191	67.8	13	3rd generation, high-resolution, 2 tone-Maneuverable. Separated an engine module.	
Cosmos 674	Aug 29	T	A-2	343 – 182	65.0	9	3rd generation, high-resolution, 2 tone-Maneuverable. Separated an engine module.	
Cosmos 685	Sep 20	T	A-2	303 – 208	65.0	12	3rd generation, low-resolution, PDM.	TG
Cosmos 688	Oct 18	P	A-2	371 – 188	62.8	12	3rd generation, high-resolution, 2 tone-Maneuverable. Separated an engine module.	TF
Cosmos 691	Oct 25	T	A-2	352 – 180	65.0	12	3rd generation, high-resolution, 2 tone-Maneuverable. Separated an engine module.	TF
Cosmos 692	Nov 1	P	A-2	315 – 201	62.8	12	3rd generation, low-resolution, PDM-Science. Separated a capsule.	
Cosmos 693	Nov 4	P	A-2	271 – 215	81.3	12	3rd generation, low-resolution, 2 tone-Science. Separated a capsule.	
Cosmos 694	Nov 16	P	A-2	344 – 213	72.9	13	3rd generation, high-resolution, 2 tone-Maneuverable. Separated an engine module.	
Cosmos 696	Nov 27	P	A-2	345 – 212	72.9	12	3rd generation, low-resolution, PDM.	TG
Cosmos 697	Dec 13	T	A-2	415 – 182	62.8	12	3rd generation, low-resolution, PDM-Science. (?) Separated a capsule.	
Cosmos 701	Dec 27	T	A-2	339 – 210	71.4	13	3rd generation, high-resolution, 2 tone-Maneuverable. Separated an engine module.	TF

1975

Name	Date	Launch Site	Booster	Orbit (km)	Inclination (degrees)	Life-Time (days)	Remarks	
Cosmos 702	Jan 17	T	A-2	334 – 210	71.4	12	3rd generation, low-resolution, PDM.	TG
Cosmos 704	Jan 23	P	A-2	329 – 213	72.9	14	3rd generation, high-resolution, 2 tone-Maneuverable. Separated an engine module.	TF
Cosmos 709	Feb 12	P	A-2	333 – 188	62.8	13	3rd generation, high-resolution, 2 tone-Maneuverable. Separated an engine module.	

Name	Date	Launch Site	Booster	Orbit (km)	Inclination (degrees)	Life-Time (days)	Remarks	
Cosmos 710	Feb 26	T	A-2	355 – 180	65.0	14	3rd generation, high-resolution, 2 tone-Maneuverable. Separated an engine module.	TF
Cosmos 719	Mar 12	T	A-2	329 – 182	65.0	13	3rd generation, high-resolution, 2 tone-Maneuverable. Separated an engine module.	TF
Cosmos 720	Mar 21	P	A-2	283 – 223	62.8	11	3rd generation, low-resolution, 2 tone-Science. Separated a capsule.	
Cosmos 721	Mar 26	P	A-2	241 – 210	81.3	12	3rd generation, low-resolution, PDM-Science. Separated a capsule.	
Cosmos 722	Mar 27	T	A-2	359 – 210	71.4	13	3rd generation, high-resolution, 2 tone-Maneuverable. Separated an engine module.	
Cosmos 727	Apr 16	T	A-2	358 – 180	65.0	12	3rd generation, high-resolution, 2 tone-Maneuverable. Separated an engine module.	
Cosmos 728	Apr 18	P	A-2	350 – 211	72.8	11	3rd generation, low-resolution, PDM-Science. Separated a capsule.	TG
Cosmos 730	Apr 24	P	A-2	251 – 212	81.3	12	3rd generation, high-resolution, 2 tone-Maneuverable. Separated an engine module.	
Cosmos 731	May 21	T	A-2	313 – 207	65.0	12	3rd generation, low-resolution, PDM-Science. Separated a capsule.	
Cosmos 740	May 28	T	A-2	347 – 181	65.0	13	3rd generation, high-resolution, 2 tone-Maneuverable. Separated an engine module.	
Cosmos 741	May 30	P	A-2	246 – 210	81.4	12	3rd generation, low-resolution, PDM.	
Cosmos 742	Jun 3	P	A-2	375 – 189	62.8	12	3rd generation, high-resolution, 2 tone-Maneuverable. Separated an engine module.	TF
Cosmos 743	Jun 12	P	A-2	355 – 190	62.8	13	3rd generation, high-resolution, 2 tone-Maneuverable. Separated an engine module.	
Cosmos 746	Jun 25	P	A-2	346 – 188	62.8	13	3rd generation, high-resolution, 2 tone-Maneuverable. Separated an engine module.	
Cosmos 747	Jun 27	P	A-2	309 – 197	62.8	12	3rd generation, low-resolution, PDM-Science. Separated a capsule.	
Cosmos 748	Jul 3	P	A-2	339 – 184	62.8	13	3rd generation, high-resolution, 2 tone-Maneuverable. Separated an engine module.	TF
Cosmos 751	Jul 23	P	A-2	335 – 203	62.8	12	3rd generation, low-resolution, PDM.	TG
Cosmos 753	Jul 31	P	A-2	351 – 189	62.8	13	3rd generation, high-resolution, 2 tone-Maneuverable. Separated an engine module.	
Cosmos 754	Aug 13	T	A-2	345 – 210	71.4	13	3rd generation, high-resolution, 2 tone-Maneuverable. Separated an engine module.	TF
Cosmos 757	Aug 27	P	A-2	337 – 190	62.8	13	3rd generation, high-resolution, 2 tone-Maneuverable. Separated an engine module.	
Cosmos 758	Sep 5	P	A-2	351 – 181	67.2	20	First 4th generation, high-resolution, Soyuz reconnaissance satellite. Exploded.	
Cosmos 759	Sep 12	P	A-2	281 – 234	62.8	11	3rd generation, low-resolution, 2 tone-Science. Separated a capsule. Photographed Diego Garcia.	

Name	Date	Launch Site	Booster	Orbit (km)	Inclination (degrees)	Life-Time (days)	Remarks	
Cosmos 760	Sep 16	T	A-2	355 – 181	65.0	14	3rd generation, high-resolution, 2 tone-Maneuverable. Separated an engine module.	
Cosmos 769	Sep 23	P	A-2	331 – 211	72.9	12	3rd generation, low-resolution, PDM-Science. Separated a capsule.	
Cosmos 771	Sep 25	P	A-2 .	247 – 219	81.3	13	3rd generation, high-resolution, 2 tone-Maneuverable. Separated an engine module.	TK
Cosmos 774	Oct 1	T	A-2	333 – 212	71.4	14	3rd generation, high-resolution, 2 tone-Maneuverable. Separated an engine module.	TF
Cosmos 776	Oct 17	P	A-2	310 – 203	62.8	12	3rd generation, low-resolution, PDM-Science. Separated a capsule.	TG
Cosmos 779	Nov 4	P	A-2	334 – 188	62.8	14	3rd generation, high-resolution, 2 tone-Maneuverable. Separated an engine module.	TF
Cosmos 780	Nov 21	T	A-2	298 – 206	65.0	12	3rd generation, low-resolution, PDM-Science. Separated a capsule.	
Cosmos 784	Dec 3	P	A-2	252 – 216	81.3	12	3rd generation, low-resolution, PDM-Science. Separated a capsule.	
Cosmos 786	Dec 16	T	A-2	346 – 180	65.0	13	3rd generation, high-resolution, 2 tone-Maneuverable. Separated an engine module.	

1976

Name	Date	Launch Site	Booster	Orbit (km)	Inclination (degrees)	Life-Time (days)	Remarks	
Cosmos 788	Jan 7	P	A-2	343 – 191	62.8	13	3rd generation, high-resolution, 2 tone-Maneuverable.	TF
Cosmos 799	Jan 29	T	A-2	328 – 210	71.4	12	3rd generation, low-resolution, PDM.	
Cosmos 802	Feb 11	T	A-2	355 – 180	65.0	14	3rd generation, high-resolution, 2 tone-Maneuverable.	TF
Cosmos 805	Feb 20	P	A-2	372 – 181	67.2	20	4th generation, high-resolution.	
Cosmos 806	Mar 10	T	A-2	353 – 182	71.4	13	3rd generation, high-resolution, 2 tone-Maneuverable.	TF
Cosmos 809	Mar 18	T	A-2	322 – 210	65.0	12	3rd generation, low-resolution, PDM.	TG
Cosmos 810	Mar 26	P	A-2	358 – 188	62.8	13	3rd generation, high-resolution, 2 tone-Maneuverable.	
Cosmos 811	Mar 31	P	A-2	361 – 212	72.9	12	3rd generation, low-resolution, 2 tone. Mapping, geodesy, earth resources. Separated a capsule.	
Cosmos 813	Apr 9	P	A-2	250 – 212	81.3	12	3rd generation, low-resolution, PDM.	
Cosmos 815	Apr 28	P	A-2	254 – 218	81.3	13	3rd generation, high-resolution, 2 tone-Maneuverable.	
Cosmos 817	May 4	T	A-2	347 – 178	65.0	13	3rd generation, high-resolution, 2 tone-Maneuverable.	
Cosmos 819	May 20	T	A-2	307 – 204	65.0	12	3rd generation, low-resolution, PDM.	
Cosmos 820	May 21	P	A-2	236 – 214	81.4	12	3rd generation, high-resolution, 2 tone-Maneuverable. Earth resources studies.	
Cosmos 821	May 26	P	A-2	338 – 212	72.8	13	3rd generation, high-resolution, 2 tone-Maneuverable.	
Cosmos 824	Jun 8	T	A-2	345 – 209	71.4	13	3rd generation, high-resolution, 2 tone-Maneuverable.	
Cosmos 833	Jun 16	P	A-2	335 – 189	62.8	13	3rd generation, high-resolution, 2 tone-Maneuverable.	

Name	Date	Launch Site	Booster	Orbit (km)	Inclination (degrees)	Life-Time (days)	Remarks	
Cosmos 834	Jun 24	P	A-2	263 – 223	81.4	12	3rd generation, low-resolution, PDM.	
Cosmos 835	Jun 29	T	A-2	338 – 180	65.0	13	3rd generation, high-resolution, 2 tone-Maneuverable.	
Cosmos 840	Jul 14	P	A-2	343 – 212	72.9	12	3rd generation, low-resolution, PDM.	
Cosmos 844	Jul 22	P	A-2	385 – 181	67.1	39	4th generation, high-resolution. Exploded.	
Cosmos 847	Aug 4	P	A-2	342 – 189	62.8	13	3rd generation, high-resolution, 2 tone-Maneuverable.	
Cosmos 848	Aug 12	P	A-2	325 – 214	62.8	13	3rd generation, low-resolution, PDM-Science. Separated a capsule.	
Cosmos 852	Aug 28	T	A-2	354 – 179	65.0	13	3rd generation, high-resolution, 2 tone-Maneuverable.	
Cosmos 854	Sep 3	P	A-2	337 – 177	81.4	13	3rd generation, high-resolution, 2 tone-Maneuverable.	
Cosmos 855	Sep 21	P	A-2	366 – 212	72.9	12	3rd generation, low-resolution, 2 tone. Mapping, geodesy, earth resources. Separated a capsule.	
Cosmos 856	Sep 22	T	A-2	322 – 210	65.0	13	3rd generation, low-resolution, PDM-Science. Separated a capsule.	TG
Cosmos 857	Sep 24	P	A-2	346 – 185	62.8	13	3rd generation, high-resolution, 2 tone-Maneuverable.	
Cosmos 859	Oct 10	T	A-2	360 – 180	65.0	11	3rd generation, high-resolution, 2 tone-Maneuverable. Earth resources studies.	
Cosmos 863	Oct 25	P	A-2	370 – 187	62.8	11	3rd generation, high-resolution, 2 tone-Maneuverable.	
Cosmos 865	Nov 1	P	A-2	350 – 212	72.9	12	3rd generation, low-resolution, PDM-Science. Separated a capsule.	
Cosmos 866	Nov 11	T	A-2	306 – 182	65.0	12	3rd generation, high-resolution, 2 tone-Maneuverable.	
Cosmos 867	Nov 24	P	A-2	418 – 258	62.8	13	First 3rd generation, medium-resolution.	TF
Cosmos 879	Dec 9	P	A-2	241 – 217	81.4	13	3rd generation, low-resolution, PDM.	TG
Cosmos 884	Dec 17	T	A-2	346 – 178	65.0	12	3rd generation, high-resolution, 2 tone-Maneuverable.	

1977

Name	Date	Launch Site	Booster	Orbit (km)	Inclination (degrees)	Life-Time (days)	Remarks	
Cosmos 888	Jan 6	T	A-2	346 – 178	65.0	13	3rd generation, high-resolution, 2 tone-Maneuverable.	
Cosmos 889	Jan 20	T	A-2	353 – 210	71.4	12	3rd generation, low-resolution, PDM.	TG
Cosmos 892	Feb 9	P	A-2	454 – 170	72.9	13	3rd generation, high-resolution, 2 tone-Maneuverable.	
Cosmos 896	Mar 3	P	A-2	216 – 194	72.9	13	3rd generation, high-resolution, 2 tone-Maneuverable.	TF
Cosmos 897	Mar 10	P	A-2	371 – 182	72.9	13	3rd generation, high-resolution, 2 tone-Maneuverable.	TF
Cosmos 898	Mar 17	P	A-2	258 – 222	81.4	13	3rd generation, low-resolution, PDM-Science. Separated a capsule.	TG
Cosmos 902	Apr 7	P	A-2	307 – 179	81.4	13	3rd generation, high-resolution, 2 tone-Maneuverable.	
Cosmos 904	Apr 20	T	A-2	350 – 210	71.4	14	3rd generation, low-resolution, PDM.	
Cosmos 905	Apr 26	P	A-2	366 – 179	67.1	30	4th generation, high-resolution.	

Name	Date	Launch Site	Booster	Orbit (km)	Inclination (degrees)	Life-Time (days)	Remarks	
Cosmos 907	May 5	P	A-2	388 – 187	62.8	11	3rd generation, high-resolution, 2 tone-Maneuverable.	
Cosmos 908	May 17	T	A-2	307 – 180	51.8	14	3rd generation, high-resolution, 2 tone-Maneuverable.	TF
Cosmos 912	May 26	P	A-2	257 – 219	81.4	13	3rd generation, high-resolution, 2 tone-Maneuverable. Earth resources studies. Separated an engine module.	
Cosmos 914	May 31	T	A-2	327 – 210	65.0	13	3rd generation, low-resolution, PDM-Science. Separated a capsule.	
Cosmos 915	Jun 8	P	A-2	306 – 182	62.8	13	3rd generation, high-resolution, 2 tone-Maneuverable.	
Cosmos 916	Jun 10	P	A-2	307 – 250	62.8	12	3rd generation, low-resolution, 2 tone. Mapping, geodesy, earth resources. Separated a capsule.	
Cosmos 920	Jun 22	T	A-2	364 – 180	65.0	13	3rd generation, high-resolution, 2 tone-Maneuverable.	
Cosmos 922	Jun 30	P	A-2	323 – 212	62.8	13	3rd generation, low-resolution, PDM.	
Cosmos 927	Jul 12	P	A-2	403 – 178	72.9	13	3rd generation, high-resolution, 2 tone-Maneuverable.	
Cosmos 932	Jul 20	T	A-2	342 – 180	65.0	13	3rd generation, high-resolution, 2 tone-Maneuverable.	
Cosmos 934	Jul 27	P	A-2	264 – 238	62.8	13	3rd generation, high-resolution, 2 tone-Maneuverable.	TF
Cosmos 935	Jul 29	P	A-2	276 – 225	81.3	13	3rd generation, low-resolution, PDM.	
Cosmos 938	Aug 24	P	A-2	365 – 189	62.8	13	3rd generation, high-resolution, 2 tone-Maneuverable.	TF
Cosmos 947	Aug 27	P	A-2	346 – 211	72.8	13	3rd generation, low-resolution,	TG
Cosmos 948	Sep 2	P	A-2	265 – 217	81.4	13	3rd generation, high-resolution, 2 tone-Maneuverable. Earth resources studies. Separated an engine module.	
Cosmos 949	Sep 6	P	A-2	348 – 184	62.8	30	4th generation, high-resolution.	
Cosmos 950	Sep 13	P	A-2	305 – 213	62.8	14	3rd generation, low-resolution, 2 tone.	
Cosmos 953	Sep 16	P	A-2	354 – 188	62.8	13	3rd generation, high-resolution, 2 tone-Maneuverable.	TF
Cosmos 957	Sep 30	T	A-2	381 – 181	65.0	13	3rd generation, medium-resolution.	
Cosmos 958	Oct 11	P	A-2	369 – 265	62.8	13	3rd generation, medium-resolution.	
Cosmos 964	Dec 4	P	A-2	391 – 180	72.9	13	3rd generation, high-resolution, 2 tone. (?)	
Cosmos 966	Dec 12	T	A-2	316 – 210	65.0	12	3rd generation, low-resolution, PDM-Science. Separated a capsule.	
Cosmos 969	Dec 20	P	A-2	340 – 188	62.8	14	3rd generation, high-resolution, 2 tone-Maneuverable.	
Cosmos 973	Dec 27	T	A-2	348 – 210	71.4	13	3rd generation, low-resolution, PDM-Science. Separated a capsule.	

1978

Name	Date	Launch Site	Booster	Orbit (km)	Inclination (degrees)	Life-Time (days)	Remarks	
Cosmos 974	Jan 6	P	A-2	356 – 188	62.8	13	3rd generation, high-resolution, 2 tone-Maneuverable.	
Cosmos 984	Jan 13	P	A-2	313 – 215	62.8	9	3rd generation, low-resolution, PDM.	TF
Cosmos 986	Jan 24	T	A-2	341 – 179	65.0	14	3rd generation, high-resolution, 2 tone-Maneuverable.	TF
Cosmos 987	Jan 31	T	A-2	359 – 183	62.8	14	3rd generation, high-resolution, 2 tone-Maneuverable.	TF

Name	Date	Launch Site	Booster	Orbit (km)	Inclination (degrees)	Life-Time (days)	Remarks	
Cosmos 988	Feb 8	P	A-2	363 – 210	72.8	12	3rd generation, low-resolution, 2 tone-Maneuverable. Mapping, geodesy, earth resources.	?
Cosmos 989	Feb 14	T	A-2	354 – 178	65.0	14	3rd generation, high-resolution, 2 tone-Maneuverable.	
Cosmos 992	Mar 4	T	A-2	346 – 210	71.4	13	3rd generation, low-resolution, PDM.	
Cosmos 993	Mar 10	P	A-2	368 – 182	72.9	13	3rd generation, high-resolution, 2 tone-Maneuverable.	TF
Cosmos 995	Mar 17	P	A-2	262 – 221	81.4	13	3rd generation, low-resolution, PDM.	
Cosmos 999	Mar 30	T	A-2	376 – 180	71.4	13	3rd generation, high-resolution, 2 tone-Maneuverable.	TF
Cosmos 1002	Apr 6	T	A-2	305 – 209	65.0	13	3rd generation, low-resolution, 2 tone.	
Cosmos 1003	Apr 20	P	A-2	349 – 185	62.8	14	3rd generation, high-resolution, 2 tone-Maneuverable.	TF
Cosmos 1004	May 5	P	A-2	311 – 213	62.8	13	3rd generation, low-resolution, PDM-Science. Separated a capsule.	TG
Cosmos 1007	May 16	P	A-2	384 – 180	72.9	13	3rd generation, high-resolution, 2 tone-Maneuverable. Earth resources studies.	TF
Cosmos 1010	May 23	P	A-2	257 – 218	81.4	13	3rd generation, high-resolution, 2 tone-Maneuverable.	
Cosmos 1012	May 25	P	A-2	280 – 214	62.8	13	3rd generation, low-resolution, 2 tone.	
Cosmos 1021	Jun 10	T	A-2	336 – 180	65.0	13	3rd generation, high-resolution, 2 tone-Maneuverable.	TF
Cosmos 1022	Jun 12	P	A-2	374 – 182	72.9	13	3rd generation, high-resolution, 2 tone-Maneuverable.	
Cosmos 1026	Jul 2	T	A-2	261 – 209	51.8	4	3rd generation, low-resolution, 2 tone. Mission purpose in doubt. (Mapping, geodesy, earth resources?)	
Cosmos 1028	Aug 5	P	A-2	272 – 182	67.1	30	4th generation, high-resolution.	
Cosmos 1029	Aug 29	P	A-2	353 – 186	62.8	10	3rd generation, high-resolution, 2 tone-Maneuverable.	
Cosmos 1031	Sep 12	P	A-2	351 – 191	62.8	13	3rd generation, high-resolution, 2 tone-Maneuverable.	TF
Cosmos 1032	Sep 19	P	A-2	249 – 218	81.4	13	3rd generation, low-resolution, 2 tone. Separated a capsule.	TG
Cosmos 1033	Oct 3	P	A-2	268 – 223	81.4	13	3rd generation, high-resolution, 2 tone-Maneuverable. Earth resources studies. Separated an engine module.	TK
Cosmos 1042	Oct 6	P	A-2	326 – 187	62.8	13	3rd generation, high-resolution, 2 tone-Maneuverable.	TF
Cosmos 1044	Oct 17	P	A-2	315 – 211	62.8	13	3rd generation, low-resolution, 2 tone.	
Cosmos 1046	Nov 1	P	A-2	353 – 212	72.9	12	3rd generation, low-resolution, 2 tone. Mapping, geodesy, earth resources. Separated a capsule.	TL
Cosmos 1047	Nov 15	P	A-2	378 – 182	72.9	13	3rd generation, high-resolution, 2 tone-Maneuverable.	TF
Cosmos 1049	Nov 21	P	A-2	375 – 183	72.9	13	3rd generation, high-resolution, 2 tone-Maneuverable.	TF
Cosmos 1050	Nov 28	P	A-2	298 – 258	62.8	14	3rd generation, high-resolution, 2 tone-Maneuverable.	TF
Cosmos 1059	Dec 7	P	A-2	360 – 188	62.8	13	3rd generation, high-resolution, 2 tone-Maneuverable.	TF

Name	Date	Launch Site	Booster	Orbit (km)	Inclination (degrees)	Life-Time (days)	Remarks	
Cosmos 1060	Dec 8	T	A-2	311 – 209	65.0	13	3rd generation, low-resolution, 2 tone.	
Cosmos 1061	Dec 14	P	A-2	333 – 211	62.8	13	3rd generation, low-resolution, 2 tone. Separated a capsule.	TG
Cosmos 1068	Dec 26	P	A-2	401 – 187	62.8	13	3rd generation, high-resolution, 2 tone-Maneuverable.	
Cosmos 1069	Dec 28	P	A-2	290 – 244	62.8	13	3rd generation, low-resolution, 2 tone. Mapping, geodesy, earth resources. Separated a capsule.	

1979

Name	Date	Launch Site	Booster	Orbit (km)	Inclination (degrees)	Life-Time (days)	Remarks	
Cosmos 1070	Jan 11	P	A-2	316 – 214	62.8	9	3rd generation, low-resolution, 2 tone.	
Cosmos 1071	Jan 13	P	A-2	360 – 190	62.8	13	3rd generation, high-resolution, 2 tone-Maneuverable.	
Cosmos 1073	Jan 30	P	A-2	350 – 187	62.8	13	3rd generation, high-resolution, 2 tone-Maneuverable.	
Cosmos 1078	Feb 22	P	A-2	306 – 160	72.9	8	3rd generation, high-resolution, 2 tone-Maneuverable.	
Cosmos 1079	Feb 27	P	A-2	359 – 179	67.1	12	4th generation, high-resolution. May have malfunctioned.	
Cosmos 1080	Mar 14	P	A-2	230 – 180	72.9	14	3rd generation, high-resolution, 2 tone-Maneuverable.	TF
Cosmos 1090	Mar 31	P	A-2	354 – 212	72.9	13	3rd generation, low-resolution, 2 tone.	TG
Cosmos 1095	Apr 20	P	A-2	404 – 209	72.9	14	3rd generation, medium-resolution.	TF
Cosmos 1097	Apr 27	P	A-2	357 – 180	62.8	30	4th generation, high-resolution.	
Cosmos 1098	May 15	P	A-2	382 – 180	72.9	13	3rd generation, high-resolution, 2 tone. Earth resources studies?	TF
Cosmos 1099	May 17	P	A-2	274 – 224	81.4	13	3rd generation, high-resolution, 2 tone-Maneuverable. Earth resources studies. Separated an engine module.	
Cosmos 1102	May 25	P	A-2	288 – 222	81.4	13	3rd generation, low-resolution, 2 tone. Earth resources studies. Separated a capsule.	
Cosmos 1103	May 31	P	A-2	396 – 264	62.8	14	3rd generation, medium-resolution.	
Cosmos 1105	Jun 8	P	A-2	281 – 223	81.4	13	3rd generation, high-resolution, 2 tone-Maneuverable. Earth resources studies. Separated an engine module.	TK
Cosmos 1106	Jun 12	P	A-2	264 – 222	81.4	13	3rd generation, low-resolution, 2 tone. Earth resources studies. Separated a capsule.	TG
Cosmos 1107	Jun 15	P	A-2	328 – 209	72.9	14	3rd generation, high-resolution, 2 tone-Maneuverable. Separated an engine module.	
Cosmos 1108	Jun 22	P	A-2	272 – 224	81.3	13	3rd generation, high-resolution, 2 tone-Maneuverable. Earth resources studies. Separated an engine module.	TK
Cosmos 1111	Jun 29	P	A-2	353 – 264	63.0	15	3rd generation, medium-resolution.	TF
Cosmos 1113	Jul 10	T	A-2	350 – 180	65.0	13	3rd generation, high-resolution, 2 tone-Maneuverable.	TF
Cosmos 1115	Jul 13	P	A-2	263 – 222	81.4	13	3rd generation, high-resolution, 2 tone-Maneuverable. Earth resources studies. Separated an engine module.	TK

Name	Date	Launch Site	Booster	Orbit (km)	Inclination (degrees)	Life-Time (days)	Remarks	
Cosmos 1117	Jul 25	P	A-2	349 – 187	62.8	13	3rd generation, high-resolution, 2 tone-Maneuverable.	TF
Cosmos 1118	Jul 27	P	A-2	273 – 222	81.4	13	3rd generation, low-resolution, 2 tone. Earth resources studies. Separated a capsule.	
Cosmos 1119	Aug 3	P	A-2	267 – 222	81.3	12	3rd generation, low-resolution, 2 tone. Separated a capsule.	TL
Cosmos 1120	Aug 11	T	A-2	376 – 181	70.4	13	3rd generation, high-resolution, 2 tone-Maneuverable.	TF
Cosmos 1121	Aug 14	P	A-2	375 – 180	67.2	30	4th generation, high-resolution.	
Cosmos 1122	Aug 17	P	A-2	260 – 218	81.4	13	3rd generation, low-resolution, 2 tone. Earth resources studies. Separated a capsule.	
Cosmos 1123	Aug 21	P	A-2	266 – 221	81.4	13	3rd generation, high-resolution, 2 tone-Maneuverable. Earth resources studies. Separated an engine module.	TK
Cosmos 1126	Aug 31	P	A-2	421 – 208	72.9	14	3rd generation, medium-resolution.	TF
Cosmos 1127	Sep 5	P	A-2	300 – 226	81.4	13	3rd generation, high-resolution, 2 tone-Maneuverable. Earth resources studies. Separated an engine module.	TF
Cosmos 1128	Sep 14	P	A-2	354 – 184	62.8	13	3rd generation, high-resolution, 2 tone-Maneuverable.	TF
Cosmos 1138	Sep 28	P	A-2	398 – 210	72.9	14	3rd generation, medium-resolution.	TF
Cosmos 1139	Oct 5	P	A-2	357 – 212	72.9	13	3rd generation, low-resolution, 2 tone. Mapping, geodesy, earth resources. Separated a capsule.	TL
Cosmos 1142	Oct 22	P	A-2	403 – 203	72.9	13	3rd generation, medium-resolution.	TF
Cosmos 1144	Nov 2	P	A-2	378 – 179	67.2	32	4th generation, high-resolution.	
Cosmos 1147	Dec 12	P	A-2	407 – 207	72.9	14	3rd generation, medium-resolution.	TF
Cosmos 1148	Dec 28	P	A-2	367 – 180	67.1	13	3rd generation, high-resolution, 2 tone-Maneuverable.	

1980

Name	Date	Launch Site	Booster	Orbit (km)	Inclination (degrees)	Life-Time (days)	Remarks	
Cosmos 1149	Jan 9	P	A-2	414 – 208	72.9	14	3rd generation, medium-resolution.	TF
Cosmos 1152	Jan 24	P	A-2	370 – 181	67.1	13	4th generation, high-resolution.	
Cosmos 1155	Feb 7	P	A-2	422 – 206	72.9	14	3rd generation, medium-resolution.	
Cosmos 1165	Feb 21	P	A-2	379 – 182	72.9	13	3rd generation, medium-resolution.	TF
Cosmos 1166	Mar 4	P	A-2	406 – 208	72.9	14	3rd generation, medium-resolution.	TF
Cosmos 1170	Apr 1	T	A-2	386 – 181	70.4	11	3rd generation, high-resolution, 2 tone-Maneuverable.	
Cosmos 1173	Apr 17	T	A-2	379 – 180	70.3	11	3rd generation, high-resolution, 2 tone-Maneuverable. May have separated an engine module.	
Cosmos 1177	Apr 29	P	A-2	365 – 181	67.2	44	4th generation, high-resolution.	
Cosmos 1178	May 7	P	A-2	417 – 207	72.9	15	3rd generation, medium-resolution.	
Cosmos 1180	May 15	P	A-2	269 – 240	62.8	12	3rd generation, low-resolution, 2 tone. Mapping, geodesy, earth resources. Separated a capsule.	TL
Cosmos 1182	May 23	P	A-2*	278 – 221	82.3	13	3rd generation, high-resolution, 2 tone-Maneuverable. (?) Earth resources studies.	

* These flights were once thought to have been launched by the F-2 booster. It is now clear they were standard A-2 launches.

Name	Date	Launch Site	Booster	Orbit (km)	Inclination (degrees)	Life-Time (days)	Remarks	
Cosmos 1183	May 28	P	A-2	414 – 208	72.9	14	3rd generation, medium-resolution.	TF
Cosmos 1185	Jun 6	P	A-2*	308 – 226	82.3	14	3rd generation, high-resolution, 2 tone-Maneuverable. (?) Earth resources studies.	TF
Cosmos 1187	Jun 12	P	A-2	332 – 210	72.9	14	3rd generation, high-resolution, 2 tone-Maneuverable.	TF
Cosmos 1189	Jun 26	P	A-2	330 – 209	72.9	14	3rd generation, high-resolution, 2 tone-Maneuverable.	TF
Cosmos 1200	Jul 9	P	A-2	332 – 209	72.9	14	3rd generation, high-resolution, 2 tone-Maneuverable.	TF
Cosmos 1201	Jul 15	P	A-2*	274 – 220	82.3	13	3rd generation, high-resolution, 2 tone-Maneuverable. Earth resources studies. Separated an engine module.	TK
Cosmos 1202	Jul 24	P	A-2	333 – 209	72.9	14	3rd generation, high-resolution, 2 tone-Maneuverable.	
Cosmos 1203	Jul 31	P	A-2*	303 – 227	82.3	14	3rd generation, high-resolution, 2 tone-Maneuverable. (?) Earth resources studies.	TF
Cosmos 1205	Aug 12	P	A-2	332 – 208	72.8	14	3rd generation, high-resolution, 2 tone-Maneuverable.	TF
Cosmos 1207	Aug 22	P	A-2*	282 – 218	82.3	13	3rd generation, high-resolution, 2 tone-Maneuverable. Earth resources studies. Separated an engine module.	TK
Cosmos 1208	Aug 26	P	A-2	362 – 181	67.1	29	4th generation, high-resolution.	
Cosmos 1209	Sep 3	P	A-2*	306 – 222	82.3	14	3rd generation, high-resolution, 2 tone-Maneuverable. Earth resources studies.	TF
Cosmos 1210	Sep 19	P	A-2*	268 – 195	82.3	14	3rd generation, high-resolution, 2 tone.	TF
Cosmos 1211	Sep 23	P	A-2*	261 – 215	82.4	11	3rd generation, low-resolution, 2 tone. Mapping, geodesy, earth resources. Separated a capsule.	TL
Cosmos 1212	Sep 26	P	A-2*	275 – 216	82.3	14	3rd generation, high-resolution, 2 tone-Maneuverable. Earth resources studies. Separated a capsule.	TK
Cosmos 1213	Oct 3	P	A-2	343 – 207	72.8	14	3rd generation, high-resolution, 2 tone-Maneuverable.	TF
Cosmos 1214	Oct 10	P	A-2	368 – 181	67.2	13	3rd generation, high-resolution, 2 tone-Maneuverable.	
Cosmos 1216	Oct 16	P	A-2	404 – 209	72.9	14	3rd generation, medium-resolution.	TF
Cosmos 1218	Oct 30	T	A-2	374 – 178	64.9	43	4th generation, high-resolution. Photographed Rapid Deployment Force exercises in Egypt, as well as Iran/Iraqi war.	
Cosmos 1219	Oct 31	P	A-2	353 – 205	72.9	13	3rd generation, high-resolution, 2 tone-Maneuverable.	
Cosmos 1221	Nov 12	P	A-2	424 – 207	72.9	14	3rd generation, medium-resolution. Photographed Rapid Deployment Force exercises in Egypt.	TF
Cosmos 1224	Dec 1	P	A-2	403 – 209	72.9	14	3rd generation, medium-resolution.	TF
Cosmos 1227	Dec 16	P	A-2	325 – 209	72.9	12	3rd generation, high-resolution, 2 tone-Maneuverable. (?)	
Cosmos 1236	Dec 26	P	A-2	388 – 180	67.1	26	4th generation, high-resolution.	

1981

Name	Date	Launch Site	Booster	Orbit (km)	Inclination (degrees)	Life-Time (days)	Remarks	
Cosmos 1237	Jan 6	P	A-2	385 – 194	72.8	14	3rd generation, medium-resolution.	
Cosmos 1239	Jan 16	P	A-2*	234 – 213	82.3	12	3rd generation, low-resolution, 2 tone. Mapping and geodesy.	

Name	Date	Launch Site	Booster	Orbit (km)	Inclination (degrees)	Life-Time (days)	Remarks	
Cosmos 1240	Jan 20	T	A-2	361 – 167	64.9	28	4th generation, high-resolution.	
Cosmos 1245	Feb 13	P	A-2	377 – 196	72.8	14	3rd generation, medium-resolution.	
Cosmos 1246	Feb 18	T	A-2	265 – 195	64.9	23	4th generation, high-resolution.	
Cosmos 1248	Mar 5	P	A-2	345 – 171	67.1	30	4th generation, high-resolution.	
Cosmos 1259	Mar 17	T	A-2	382 – 206	70.3	14	3rd generation, medium-resolution.	TF
Cosmos 1262	Apr 7	P	A-2	392 – 196	72.9	14	3rd generation, high-resolution, 2 tone-Maneuverable.	
Cosmos 1264	Apr 15	T	A-2	385 – 208	70.3	14	3rd generation, medium-resolution.	TF
Cosmos 1265	Apr 16	P	A-2	287 – 198	72.9	12	3rd generation, high-resolution, 2 tone-Maneuverable.	TF
Cosmos 1268	Apr 28	T	A-2	368 – 208	70.4	14	3rd generation, high-resolution, 2 tone-Maneuverable.	TF
Cosmos 1270	May 18	T	A-2	348 – 173	64.8	30	4th generation, high-resolution.	
Cosmos 1272	May 21	P	A-2	379 – 208	70.3	14	3rd generation, medium-resolution.	
Cosmos 1273	May 22	P	A-2*	249 – 211	82.3	13	3rd generation, high-resolution, 2 tone-Maneuverable. Earth resources studies.	
Cosmos 1274	Jun 3	P	A-2	353 – 170	67.1	30	4th generation, high-resolution.	
Cosmos 1276	Jun 16	P	A-2*	237 – 214	82.4	13	3rd generation, high-resolution, 2 tone-Maneuverable. Earth resources studies.	TK
Cosmos 1277	Jun 17	T	A-2	377 – 207	70.4	14	3rd generation, medium-resolution.	
Cosmos 1279	Jul 1	T	A-2	362 – 210	70.4	14	3rd generation, high-resolution, 2 tone-Maneuverable.	TF
Cosmos 1280	Jul 2	P	A-2*	273 – 256	82.3	13	3rd generation, high-resolution, 2 tone-Maneuverable. Earth resources studies.	
Cosmos 1281	Jul 7	P	A-2	414 – 357	72.8	14	3rd generation, medium-resolution.	TF
Cosmos 1282	Jul 15	T	A-2	334 – 172	64.9	30	4th generation, high-resolution.	
Cosmos 1283	Jul 17	P	A-2*	260 – 181	82.3	14	3rd generation, medium-resolution. Earth resources studies.	TF
Cosmos 1284	Jul 29	P	A-2*	229 – 192	82.3	14	3rd generation, medium-resolution. Earth resources studies.	TF
Cosmos 1296	Aug 13	P	A-2	353 – 171	67.1	30	4th generation, high-resolution.	
Cosmos 1297	Aug 18	P	A-2	363 – 197	72.8	12	3rd generation, high-resolution, 2 tone-Maneuverable.	
Cosmos 1298	Aug 21	T	A-2	330 – 172	64.9	42	4th generation, high-resolution.	
Cosmos 1301	Aug 27	P	A-2*	271 – 213	82.3	14	3rd generation, high-resolution, 2 tone-Maneuverable. Earth resources studies.	
Cosmos 1303	Sep 4	T	A-2	375 – 206	70.4	14	3rd generation, medium-resolution.	
Cosmos 1307	Sep 15	P	A-2	393 – 196	72.9	14	3rd generation, medium-resolution.	TF
Cosmos 1309	Sep 18	P	A-2*	256 – 211	82.3	13	3rd generation, low-resolution. Mapping and geodesy.	TL
Cosmos 1313	Oct 1	T	A-2	279 – 231	70.3	14	3rd generation, high-resolution, 2 tone-Maneuverable.	TF
Cosmos 1314	Oct 9	P	A-2*	235 – 212	82.3	13	3rd generation, high-resolution, 2 tone-Maneuverable. Earth resources studies.	TK
Cosmos 1316	Oct 15	T	A-2	278 – 232	70.3	14	3rd generation, high-resolution, 2 tone-Maneuverable.	TF
Cosmos 1318	Nov 3	P	A-2	352 – 171	67.1	30	4th generation, high-resolution.	
Cosmos 1319	Nov 13	T	A-2	337 – 207	70.3	14	3rd generation, medium-resolution.	
Cosmos 1329	Dec 4	T	A-2	261 – 233	65.0	14	3rd generation, high-resolution, 2 tone-Maneuverable.	TF
Cosmos 1330	Dec 19	T	A-2	379 – 167	70.3	31	4th generation, high-resolution.	

1982

Name	Date	Launch Site	Booster	Orbit (km)	Inclination (degrees)	Life-Time (days)	Remarks	
Cosmos 1332	Jan 12	P	A-2*	248 – 210	82.3	13	3rd generation, low-resolution, 2 tone. Mapping and geodesy.	TL
Cosmos 1334	Jan 20	P	A-2	289 – 226	72.8	14	3rd generation, high-resolution, 2 tone-Maneuverable.	

Name	Date	Launch Site	Booster	Orbit (km)	Inclination (degrees)	Life-Time (days)	Remarks	
Cosmos 1336	Jan 30	T	A-2	356 – 169	70.3	27	4th generation, high-resolution.	
Cosmos 1338	Feb 16	P	A-2	362 – 200	72.9	14	3rd generation, medium-resolution.	
Cosmos 1342	Mar 5	P	A-2	299 – 194	72.8	14	3rd generation, high-resolution, 2 tone-Maneuverable.	
Cosmos 1343	Mar 17	P	A-2	287 – 197	72.8	14	3rd generation, high-resolution, 2 tone-Maneuverable.	TF
Cosmos 1347	Apr 2	T	A-2	340 – 172	70.3	50	4th generation, high-resolution.	
Cosmos 1350	Apr 15	P	A-2	357 – 171	67.1	31	4th generation, high-resolution.	
Cosmos 1352	Apr 21	T	A-2	360 – 208	70.3	14	3rd generation, medium-resolution.	
Cosmos 1353	Apr 23	P	A-2*	241 – 211	82.3	13	3rd generation, high-resolution, 2 tone-Maneuverable. Earth resources studies.	TK
Cosmos 1368	May 21	T	A-2	341 – 211	70.3	14	3rd generation, high-resolution, 2 tone-Maneuverable.	TF
Cosmos 1369	May 25	P	A-2	275 – 197	64.8	44	4th generation, high-resolution.	
Cosmos 1373	Jun 2	T	A-2	347 – 210	70.3	14	3rd generation, medium-resolution.	TF
Cosmos 1376	Jun 8	P	A-2*	274 – 261	82.3	14	3rd generation, high-resolution, 2 tone-Maneuverable. Earth resources studies.	TF
Cosmos 1381	Jun 18	T	A-2	450 – 372	70.3	13	3rd generation, medium-resolution.	TF
Cosmos 1384	Jun 30	T	A-2	354 – 170	67.1	30	4th generation, high-resolution.	
Cosmos 1385	Jul 6	P	A-2*	236 – 186	82.3	14	3rd generation, medium-resolution. Earth resources studies.	
Cosmos 1387	Jul 13	P	A-2*	234 – 214	82.3	13	3rd generation, high-resolution, 2 tone-Maneuverable. Earth resources studies.	
Cosmos 1396	Jul 27	P	A-2	298 – 198	72.86	14	3rd generation, high-resolution, 2 tone-Maneuverable.	TF
Cosmos 1398	Aug 3	P	A-2*	234 – 216	82.3	10	3rd generation, low-resolution, 2 tone. Mapping and geodesy.	TL
Cosmos 1399	Aug 4	T	A-2	345 – 170	64.9	43	4th generation, high-resolution.	
Cosmos 1401	Aug 20	P	A-2*	274 – 261	82.3	14	3rd generation, high-resolution, 2 tone-Maneuverable. Earth resources studies.	
Cosmos 1403	Sep 1	T	A-2	416 – 354	70.3	14	3rd generation, medium-resolution.	TF
Cosmos 1404	Sep 1	P	A-2	414 – 358	72.8	14	3rd generation, medium-resolution. Note 2 launches on same day.	TF
Cosmos 1406	Sep 8	P	A-2*	220 – 212	82.3	13	3rd generation, high-resolution, 2 tone-Maneuverable. Earth resources studies.	
Cosmos 1407	Sep 15	P	A-2	340 – 174	67.1	31	4th generation, high-resolution.	
Cosmos 1411	Sep 30	P	A-2	357 – 198	72.8	14	3rd generation, high-resolution, 2 tone-Maneuverable.	
Cosmos 1416	Oct 14	T	A-2	278 – 231	70.3	14	3rd generation, high-resolution, 2 tone-Maneuverable.	TF
Cosmos 1419	Nov 2	T	A-2	285 – 228	70.3	14	3rd generation, high-resolution, 2 tone-Maneuverable.	TF
Cosmos 1421	Nov 18	T	A-2	280 – 230	70.3	14	3rd generation, high-resolution, 2 tone-Maneuverable.	
Cosmos 1422	Dec 3	P	A-2	288 – 228	72.8	14	3rd generation, high-resolution, 2 tone-Maneuverable.	
Cosmos 1424	Dec 16	T	A-2	349 – 171	64.9	43	4th generation, high-resolution.	
Cosmos 1425	Dec 23	T	A-2	415 – 348	69.9	14	3rd generation, medium-resolution.	
Cosmos 1426	Dec 28	T	A-2	351 – 205	50.5	67	First 5th generation, digital imagery satellite.	

1983

Name	Date	Launch Site	Booster	Orbit (km)	Inclination (degrees)	Life-Time (days)	Remarks	
Cosmos 1438	Jan 27	T	A-2	293 – 175	70.4	11	3rd generation, high-resolution, 2 tone-Maneuverable.	
Cosmos 1439	Feb 6	T	A-2	295 – 160	70.3	16	4th generation, high-resolution. Unusually short mission.	
Cosmos 1440	Feb 10	P	A-2*	275 – 260	82.3	14	3rd generation, high-resolution, 2 tone-Maneuverable. Earth resources studies.	

Name	Date	Launch Site	Booster	Orbit (km)	Inclination (degrees)	Life-Time (days)	Remarks	
Cosmos 1442	Feb 25	P	A-2	360 – 169	67.1	45	4th generation, high-resolution.	
Cosmos 1444	Mar 2	P	A-2	416 – 358	72.8	14	3rd generation, medium-resolution.	
Cosmos 1446	Mar 16	T	A-2	241 – 222	69.9	14	3rd generation, high-resolution, 2 tone-Maneuverable. Observed Iran/Iraqi war.	
Cosmos 1449	Mar 31	P	A-2	417 – 356	72.8	15	3rd generation, medium-resolution.	
Cosmos 1451	Apr 8	P	A-2*	323 – 227	82.3	14	3rd generation, high-resolution, 2 tone-Maneuverable.	
Cosmos 1454	Apr 22	P	A-2	343 – 170	67.1	30	4th generation, high-resolution. Observed Lebanon.	
Cosmos 1457	Apr 26	T	A-2	350 – 171	70.4	43	4th generation, high-resolution. Observed Lebanon.	
Cosmos 1458	Apr 28	P	A-2*	245 – 212	82.3	13	3rd generation, high-resolution, 2 tone-Maneuverable. Earth resources studies.	
Cosmos 1460	May 6	T	A-2	417 – 350	70.3	14	3rd generation, medium-resolution.	TF
Cosmos 1462	May 17	P	A-2*	277 – 259	82.3	14	3rd generation, high-resolution, 2 tone-Maneuverable. Earth resources studies.	TF
Cosmos 1466	May 26	T	A-2	345 – 174	64.8	41	4th generation, high-resolution.	
Cosmos 1467	May 31	P	A-2	417 – 357	72.8	12	3rd generation, medium-resolution.	
Cosmos 1468	Jun 7	P	A-2*	277 – 252	82.3	14	3rd generation, high-resolution, 2 tone-Maneuverable. Earth resources studies.	TF
Cosmos 1469	Jun 14	P	A-2	342 – 231	72.8	10	3rd generation, high-resolution, 2 tone-Maneuverable.	
Cosmos 1471	Jun 28	P	A-2	344 – 185	67.1	30	4th generation, high-resolution. Observed El Salvador.	
Cosmos 1472	Jul 5	P	A-2*	362 – 338	82.3	14	3rd generation, medium-resolution. Earth resources studies.	TF
Cosmos 1482	Jul 13	T	A-2	413 – 352	69.9	14	3rd generation, medium-resolution.	
Cosmos 1484	Jul 20	P	A-2*	275 – 260	82.3	14	3rd generation, medium-resolution. Earth resources studies.	TF
Cosmos 1485	Jul 26	P	A-2	415 – 358	72.8	14	3rd generation, medium-resolution.	TF
Cosmos 1487	Aug 5	P	A-2*	275 – 261	82.3	14	3rd generation, high-resolution, 2 tone-Maneuverable.	TF
Cosmos 1488	Aug 9	P	A-2	415 – 356	72.8	14	3rd generation, medium-resolution.	?
Cosmos 1489	Aug 10	T	A-2	365 – 171	64.7	44	4th generation, high-resolution. Observed Lebanon.	
Cosmos 1493	Aug 23	P	A-2	414 – 360	72.9	14	3rd generation, medium-resolution.	
Cosmos 1495	Sep 3	P	A-2*	236 – 215	82.3	13	3rd generation, high-resolution, 2 tone-Maneuverable. Earth resources studies.	TK
Cosmos 1496	Sep 7	P	A-2	341 – 170	67.1	42	4th generation, high-resolution. Observed El Salvador.	
Cosmos 1497	Sep 9	P	A-2	416 – 357	72.8	14	3rd generation, medium-resolution.	
Cosmos 1498	Sep 14	P	A-2*	275 – 261	82.3	14	3rd generation, high-resolution, 2 tone-Maneuverable. Earth resources studies.	
Cosmos 1499	Sep 17	P	A-2	416 – 357	72.8	14	3rd generation, medium-resolution.	
Cosmos 1504	Oct 14	T	A-2	306 – 171	64.8	23	4th generation, high-resolution. Observed Lebanon and Grenada.	
Cosmos 1505	Oct 21	P	A-2	414 – 356	72.8	14	3rd generation, medium-resolution.	
Cosmos 1509	Nov 17	P	A-2	290 – 225	72.8	14	3rd generation, high-resolution, 2 tone-Maneuverable.	
Cosmos 1511	Nov 30	P	A-2	325 – 183	67.0	44	4th generation, high-resolution.	
Cosmos 1512	Dec 7	P	A-2	416 – 355	72.8	14	3rd generation, medium-resolution.	
Cosmos 1516	Dec 27	T	A-2	276 – 196	64.8	44	4th generation, high-resolution.	

1984

Name	Date	Launch Site	Booster	Orbit (km)	Inclination (degrees)	Life-Time (days)	Remarks	
Cosmos 1530	Jan 11	P	A-2	415 – 356	72.8	14	3rd generation, medium-resolution.	
Cosmos 1532	Jan 14	P	A-2	355 – 167	67.1	44	4th generation, high-resolution.	
Cosmos 1533	Jan 26	P	A-2	414 – 348	70.3	14	3rd generation, medium-resolution.	
Cosmos 1537	Feb 16	P	A-2	273 – 259	82.3	14	3rd generation, high-resolution, 2 tone-Maneuverable. Earth resources studies.	

Name	Date	Launch Site	Booster	Orbit (km)	Inclination (degrees)	Life-Time (days)	Remarks
Cosmos 1539	Feb 28	P	A-2	241 – 169	67.1	41	4th generation, high-resolution.
Cosmos 1542	Mar 7	T	A-2	414 – 348	70.3	14	3rd generation, medium-resolution.
Cosmos 1543	Mar 10	P	A-2	394 – 216	62.8	26	3rd generation, medium-resolution. (?)
Cosmos 1545	Mar 21	P	A-2	415 – 356	72.81	15	3rd generation, medium-resolution.
Cosmos 1548	Apr 10	P	A-2	334 – 167	67.1	45	4th generation, high-resolite.
Cosmos 1549	Apr 19	P	A-2	415 – 356	72.9	14	3rd generation, medium-resolution.
Cosmos 1551	May 11	P	A-2	279 – 196	72.8	12	3rd generation, high-resolution, 2 tone-Maneuverable.
Cosmos 1552	May 14	T	A-2	322 – 182	64.9	173	5th generation, digital imagery satellite.
Cosmos 1557	May 22	P	A-2	247 – 211	82.3	13	3rd generation, low-resolution, 2 tone. Mapping and geodesy.
Cosmos 1558	May 25	P	A-2	294 – 168	67.1	44	4th generation, high-resolution.
Cosmos 1568	Jun 1	P	A-2	414 – 356	72.8	13	3rd generation, medium-resolution.
Cosmos 1571	Jun 11	T	A-2	415 – 348	70.0	15	3rd generation, medium-resolution.
Cosmos 1572	Jun 15	P	A-2	272 – 259	82.3	14	3rd generation, high-resolution, 2 tone-Maneuverable. Earth resources studies.
Cosmos 1573	Jun 19	P	A-2	309 – 231	72.8	9	3rd generation, high-resolution, 2 tone-Maneuverable.
Cosmos 1575	Jun 22	P	A-2	275 – 261	82.3	15	3rd generation, high-resolution, 2 tone-Maneuverable.
Cosmos 1576	Jun 26	P	A-2	351 – 170	67.1	59	5th generation, digital imagery satellite.
Cosmos 1580	Jun 29	P	A-2	347 – 243	62.8	14	3rd generation, medium-resolution.
Cosmos 1582	Jul 19	P	A-2	279 – 213	82.3	14	3rd generation, high-resolution, 2 tone-Maneuverable. Earth resources studies.
Cosmos 1583	Jul 24	P	A-2	416 – 356	72.8	15	3rd generation, medium-resolution.
Cosmos 1584	Jul 27	P	A-2	365 – 180	82.3	14	3rd generation, medium-resolution. (?) Earth resources studies. (May have gone into wrong orbit.)
Cosmos 1585	Jul 31	T	A-2	302 – 174	64.7	58	5th generation, digital imagery satellite.
Cosmos 1587	Aug 6	P	A-2	368 – 197	72.8	25	In-space storage of a 3rd generation, medium-resolution satellite.
Cosmos 1590	Aug 16	P	A-2	266 – 210	82.3	14	3rd generation, high-resolution, 2 tone-Maneuverable.
Cosmos 1591	Aug 30	P	A-2	263 – 209	82.3	14	3rd generation, high-resolution, 2 tone-Maneuverable.
Cosmos 1592	Sep 4	P	A-2	287 – 225	72.8	14	3rd generation, high-resolution, 2 tone-Maneuverable.
Cosmos 1597	Sep 12	P	A-2	244 – 211	82.3	13	3rd generation, high-resolution, 2 tone-Maneuverable. Earth resources studies.
Cosmos 1599	Sep 25	P	A-2	327 – 180	67.1	61	5th generation, digital imagery satellite. (?)
Cosmos 1600	Sep 27	T	A-2	416 – 349	68.9	14	3rd generation, medium-resolution.
Cosmos 1608	Nov 14	T	A-2	250 – 195	69.9	33	4th generation, high-resolution.
Cosmos 1609	Nov 14	P	A-2	414 – 356	72.8	14	3rd generation, medium-resolution. After a long interval, two launches on same day.
Cosmos 1611	Nov 21	T	A-2	351 – 173	64.7	51	5th generation, high-resolution satellite. (?)
Cosmos 1613	Nov 29	P	A-2	356 – 197	72.8	25	In-space storage of a 3rd generation, medium-resolution satellite.

1985

Name	Date	Launch Site	Booster	Orbit (km)	Inclination (degrees)	Life-Time (days)	Remarks
Cosmos 1616	Jan 9	T	A-2	358 – 173	64.9	54	5th generation, digital imagery satellite. (?)
Cosmos 1623	Jan 16	T	A-2	415 – 349	69.9	14	3rd generation, medium-resolution.

Name	Date	Launch Site	Booster	Orbit (km)	Inclination (degrees)	Life-Time (days)	Remarks
Cosmos 1628	Feb 6	P	A-2	415 – 355	72.8	14	3rd generation, medium-resolution.
Cosmos 1630	Feb 27	T	A-2	334 – 174	64.8	55	5th generation, digital imagery satellite.
Cosmos 1632	Mar 1	P	A-2	268 – 211	72.8	14	3rd generation, high-resolution, 2 tone-Maneuverable.
Cosmos 1643	Mar 25	T	A-2	276 – 183	64.7	207	5th generation, digital imagery satellite.
Cosmos 1644	Apr 3	T	A-2	415 – 349	70.3	14	3rd generation, medium-resolution.
Cosmos 1647	Apr 18	P	A-2	323 – 169	67.1	53	4th generation, high-resolution.
Cosmos 1648	Apr 25	P	A-2	327 – 229	82.3	11	3rd generation, medium-resolution. (?)
Cosmos 1649	May 15	P	A-2	415 – 356	72.8	14	3rd generation, medium-resolution.
Cosmos 1653	May 22	P	A-2	273 – 259	82.3	14	3rd generation, high-resolution, 2 tone-Maneuverable. Earth resources studies.
Cosmos 1654	May 23	T	A-2	343 – 172	64.8	–	5th generation, digital imagery satellite. Exploded.
Cosmos 1657	Jun 7	P	A-2	284 – 182	82.2	14	3rd generation, high-resolution, 2 tone-Maneuverable. (?)
Cosmos 1659	Jun 13	P	A-2	415 – 357	72.9	14	3rd generation, medium-resolution.
Cosmos 1663	Jun 21	P	A-2	274 – 258	82.3	14	3rd generation, high-resolution, 2 tone-Maneuverable. Earth resources studies.
Cosmos 1664	Jun 21	P	A-2	379 – 224	72.8	9	3rd generation, high-resolution, 2 tone-Maneuverable. Two launches on same day.
Cosmos 1665	Jul 3	P	A-2	290 – 225	72.8	14	3rd generation, high-resolution, 2 tone-Maneuverable.
Cosmos 1668	Jul 15	T	A-2	281 – 230	70.3	14	3rd generation, high-resolution, 2 tone-Maneuverable.
Cosmos 1671	Aug 2	P	A-2	365 – 230	72.8	14	3rd generation, high-resolution, 2 tone-Maneuverable.
Cosmos 1672	Aug 7	P	A-2	262 – 186	82.3	14	3rd generation, high-resolution, 2 tone-Maneuverable.
Cosmos 1673	Aug 8	T	A-2	272 – 198	64.7	42	5th generation, digital imagery satellite. (?)
Cosmos 1676	Aug 16	P	A-2	345 – 167	67.1	59	4th generation, high-resolution.
Cosmos 1678	Aug 28	P	A-2	272 – 258	82.3	14	3rd generation, high-resolution, 2 tone-Maneuverable. Earth resources studies.
Cosmos 1679	Aug 29	T	A-2	342 – 172	64.8	50	5th generation, digital imagery satellite.
Cosmos 1681	Sep 6	P	A-2	226 – 219	82.3	13	3rd generation, high-resolution, 2 tone-Maneuverable. Earth resources studies.
Cosmos 1683	Sep 19	P	A-2	414 – 356	72.8	15	3rd generation, medium-resolution.
Cosmos 1685	Sep 26	P	A-2	416 – 356	72.8	14	3rd generation, medium-resolution.
Cosmos 1696	Oct 16	T	A-2	281 – 230	70.3	14	3rd generation, high-resolution, 2 tone-Maneuverable.
Cosmos 1699	Oct 25	P	A-2	338 – 168	67.1	59	4th generation, high-resolution.
Cosmos 1702	Nov 13	P	A-2	414 – 356	72.8	14	3rd generation, medium-resolution.
Cosmos 1705	Dec 3	P	A-2	415 – 356	72.8	14	3rd generation, medium-resolution.
Cosmos 1706	Dec 11	P	A-2	340 – 162	67.1	60	4th generation, high-resolution.
Cosmos 1708	Dec 13	P	A-2	273 – 257	82.2	14	3rd generation, high-resolution, 2 tone-Maneuverable. Earth resources studies.

Notes: Third-generation reconnaissance satellite varieties were identified using published accounts and the author's analysis.

Phillip Clark, an English writer on Soviet space activity, has developed the following chart to identify 3rd generation varieties in the 1980s:

Type	Inclination (degrees)		Orbit (km)				
Low resolution	82.3	Initial:	245	(7) – 212	(2)		
Medium resolution	62.8	Initial:	370	(20) – 255	(2)		
		Later:	413	(8) – 339	(13)		
	70.4	Initial:	377	(7) – 208	(1)		
		Later:	416	(1) – 357	(5)		
	72.9	Initial:	382	(10) – 196	(1)		
		Later:	413	(6) – 356	(8)		
	82.3	Initial:	246	(5) – 181	(1)		
		Later:	371	(1) – 325	(0)		
High resolution	70.4	Initial:	376	(9) – 208	(1)		
		Later:	278	(3) – 228	(3)		
	72.9	Initial:	301	(14) – 195	(2)		
		Later:	245	(2) – 218	(5)		
			260	(6) – 213	(1)		
			302	(15) – 226	(7)		
			357	(2) – 229	(5)		
	82.3	Initial:	235	(2) – 215	(1)		
		Later:	235	(7) – 212	(4)		
		Initial:	274	(8) – 210	(1)		
		Later:	275	(3) – 263	(5)		

These are average orbits with standard deviations in parentheses. Since this chart appeared in the April 1983 *Journal of the British Interplanetary Society,* a new profile has appeared:

72.8	Initial:	379 – 224	
	Later:	268 – 211	

These are considered to be high-resolution satellites. The 4th and 5th generation launches are separated by lifetime, 5th generation having lifetimes greater than 50 days. Also, the inclinations differ:

4th Generation	5th Generation
62.8°	50.5°
64.9	67.1
67.1	64.8
70.4	

It can be speculated that both types are Soyuz-based.

Third generation reconnaissance satellites are decided both by high versus low resolution, and according to the telemetry format used. The three types are:

Pulse-Duration Modulation (PDM) – This is a string of "bleeps." The position of each represents a particular measurement. The duration of these bleeps changes as consumables, such as film or attitude control fuel, is used up or a system switches to another mode of operation. "Word 13," for instance, represents the film usage.

Morse code – The satellite transmits 12 groups of 3 letters in morse code. These actually represent binary code; a dot is a binary 0 while a dash is a binary 1. On high-resolution missions, the 6th letter group is MWI; when the maneuvering engine is separated, it changes to MRI. This gives 24 hours' warning of satellite recovery. Group 7 represents film supply.

2-tone – This simply means the satellite alternately transmits a brief signal on two different frequencies. This is a tracking beacon rather than telemetry.

Chinese Reconnaissance Satellites

Name	Date	Booster	Orbit (km)	Inclination (degrees)	Life-Time (days)	Remarks
China 3	Jul 26, 1975	FB-1	464 – 186	69.0	50	Radio transmission satellite.
China 4	Nov 26, 1975	FB-1	484 – 173	63.0	6	Film return satellite.
China 5	Dec 16, 1975	FB-1	387 – 186	69.0	42	Radio transmission.
China 7	Dec 7, 1976	FB-1	479 – 172	59.4	2	Film return.
China 8	Jan 26, 1978	FB-1	479 – 162	57.0	5	Film return.
China 12	Sep 9, 1982	FB-1	384 – 175	62.9	4	Film return.
China 13	Aug 19, 1983	FB-1	382 – 173	63.3	5	Film return.
China 16	Sep 12, 1984	FB-1	400 – 174	67.9	5	Film return.
China 17	Oct 21, 1985	FB-1	393 – 171	67.9	5	Film return.

U.S. Heavy ELINT Satellites

Name	Date	Booster	Orbit (km)	Inclination (degrees)	Remarks
1962 Omega	Jun 18, 1962	Thor Agena B	411 – 370	82.1	First heavy ELINT.
1963–3	Jan 16, 1963	Thor Agena D	533 – 459	81.9	
1963–27	Jun 29, 1963	TAT Agena D	536 – 484	82.3	
1964–11	Feb 28, 1964	TAT Agena D	520 – 479	82.0	
1964–35	Jul 2, 1964	TAT Agena D	529 – 501	82.0	
1964–72	Nov 4, 1964	TAT Agena D	526 – 512	82.0	
1965–55	Jul 17, 1965	TAT Agena D	512 – 471	70.1	
1966–9	Feb 9, 1966	TAT Agena D	512 – 508	82.0	
1966–118	Dec 29, 1966	TAT Agena D	496 – 486	75.0	
1967–71	Jul 25, 1967	TAT Agena D	513 – 458	75.0	
1968–4	Jan 17, 1968	TAT Agena D	546 – 450	75.1	
1968–86	Oct 5, 1968	LTTAT	511 – 483	74.9	New larger booster.
1969–65	Jul 31, 1969	LTTAT	541 – 462	75.0	
1970–66	Aug 26, 1970	LTTAT	504 – 484	74.9	
1971–60	Jul 16, 1971	LTTAT	508 – 488	75.0	Last heavy ELINT.

U.S. ELINT Sub-satellites

Name	Date	Booster	Orbit (km)	Inclination (degrees)	Remarks
1963–35	Aug 29, 1963	Thor Agena D	431 – 310	81.8	First radar ELINT subsatellite.
1963–42	Oct 29, 1963	TAT Agena D	585 – 285	89.9	Radar ELINT.
1963–55	Dec 21, 1963	TAT Agena D	388 – 321	64.5	Radar ELINT.
1964–36	Jul 6, 1964	Atlas Agena D	377 – 297	92.9	Radar ELINT. Shift to new launch vehicle.
1964–68	Oct 23, 1964	Atlas Agena D	336 – 323	95.5	Radar ELINT.
1965–31	Apr 28, 1965	Atlas Agena D	559 – 490	95.2	Radar ELINT.
1965–50	Jun 25, 1965	Atlas Agena D	510 – 496	107.6	Radar ELINT.
1965–62	Aug 3, 1965	Atlas Agena D	515 – 501	107.3	Radar ELINT.
1966–39	May 14, 1966	Atlas Agena D	559 – 517	109.9	Radar ELINT.
1966–74	Aug 16, 1966	Atlas Agena D	524 – 510	93.1	Radar ELINT.
1966–83	Sep 16, 1966	Atlas Agena D	501 – 460	94.0	Radar ELINT.
1967–43	May 9, 1967	LTTAT	809 – 555	85.1	Radar ELINT. New launch vehicle.
1967–62	Jun 16, 1967	LTTAT	517 – 501	80.2	Radar ELINT.
1967–109	Nov 2, 1967	LTTAT	524 – 455	81.6	Radar ELINT.
1968–8	Jan 24, 1968	LTTAT	542 – 473	81.6	Radar ELINT.
1968–20	Mar 14, 1968	LTTAT	522 – 481	83.0	Radar ELINT.
1968–52	Jun 20, 1968	LTTAT	519 – 437	85.1	Radar ELINT.
1968–78	Sep 18, 1968	LTTAT	514 – 500	83.2	Radar ELINT.
1968–112	Dec 12, 1968	LTTAT	1,468 – 1,391	80.3	First ABM ELINT sub-satellite.
1969–10	Feb 5, 1969	LTTAT	1,441 – 1,396	80.4	ABM ELINT.
1969–26	Mar 19, 1969	LTTAT	513 – 504	83.0	Radar ELINT.
1969–41	May 2, 1969	LTTAT	473 – 401	65.7	Radar ELINT.
1969–79	Sep 22, 1969	LTTAT	496 – 490	85.1	Radar ELINT.
1970–16	Mar 4, 1970	LTTAT	514 – 442	88.1	Radar ELINT.
1970–40	May 20, 1970	LTTAT	503 – 491	83.1	Radar ELINT.
1970–98	Nov 18, 1970	LTTAT	511 – 487	83.1	Radar ELINT.
1971–76	Sep 10, 1971	LTTAT	507 – 492	75.0	Radar ELINT.
1972–2D	Jan 20, 1972	Titan III D	549 – 472	96.6	Radar ELINT. Shift to new launch vehicle.
1972–52C	Jul 7, 1972	Titan III D	504 – 497	96.1	Radar ELINT.
1972–79C	Oct 10, 1972	Titan III D	1,469 – 1,423	95.6	ABM ELINT.
1973–88B	Nov 10, 1973	Titan III D	508 – 486	96.3	Radar ELINT.
1973–88D			1,458 – 1,419	96.9	ABM ELINT. (Both launched on same Big Bird.)
1974–20C	Apr 10, 1974	Titan III D	531 – 503	94.0	Radar ELINT.
1974–85B	Oct 29, 1974	Titan III D	535 – 520	96.0	Radar ELINT.
1976–65C	Jul 8, 1976	Titan III D	632 – 628	97.5	Radar ELINT. (New higher orbit.)
1978–29B	Mar 16, 1978	Titan III D	645 – 639	95.8	Radar ELINT.
1979–25B	Mar 16, 1979	Titan III D	628 – 621	95.7	Radar ELINT.
1980–52C	Jun 18, 1980	Titan III D	1,333 – 1,331	96.6	ABM ELINT.
1982–41C	May 11, 1982	Titan III D	707 – 701	96.0	Radar ELINT (Ferret-D).
1983–60C	Jun 20, 1983	Titan 34 D	1,290 – 1,290	96.6	ABM ELINT.
USA-3	Jun 25, 1984	Titan 34 D	700 circular	96.5	Radar ELINT.

U.S. Geosynchronous ELINT Satellites

Name	Date	Booster	Orbit (km)	Inclination (degrees)	Remarks
1973–13	Mar 6, 1973	Atlas Agena D	35,855 – 35,679	0.2	First Rhyolite stationed above the Horn of Africa.
1977–38	May 23, 1977	Atlas Agena D	Classified	?	Over Borneo.
1977–114	Dec 11, 1977	Atlas Agena D	Classified	?	In-orbit spare.
1978–38	Apr 7, 1978	Atlas Agena D	Classified	?	In-orbit spare.
USA-8	Jan 24, 1985	Shuttle IUS	Classified	?	First Aquacade.

Soviet Small ELINT Satellites

Name	Date	Launch Site[*]	Booster	Orbit (km)	Inclination (degrees)	Remarks
1965						
Cosmos 103	Dec 28	T	C-1	600 – 600	56.0	Test mission?
		No launches in 1966.				
1967						
Cosmos 151	Mar 24	T	C-1	630 – 630	56.0	Test mission?
Cosmos 189	Oct 30	P	C-1	600 – 535	74.0	First ELINT satellite.
1968						
Cosmos 200	Jan 19	P	C-1	536 – 536	74.0	
Cosmos 236	Aug 27	T	C-1	655 – 600	56.0	Test mission?
Cosmos 250	Oct 30	P	C-1	556 – 523	74.0	
1969						
Cosmos 269	Mar 5	P	C-1	558 – 526	74.0	
Cosmos 315	Dec 20	P	C-1	556 – 521	74.0	
1970						
Cosmos 330	Apr 7	P	C-1	548 – 514	74.1	Replaced Cosmos 250.
Cosmos 358	Aug 21	P	C-1	549 – 517	74.0	
Cosmos 387	Dec 16	P	C-1	560 – 528	74.0	Replaced Cosmos 269.
1971						
Cosmos 395	Feb 18	P	C-1	570 – 534	74.0	
Cosmos 425	May 29	P	C-1	556 – 511	74.0	
Cosmos 436	Sep 7	P	C-1	550 – 514	74.0	Booster exploded.
Cosmos 437	Sep 10	P	C-1	558 – 523	74.0	Replacement launch for Cosmos 436.
Cosmos 460	Nov 30	P	C-1	553 – 520	74.0	
1972						
Cosmos 479	Mar 21	P	C-1	549 – 517	74.0	
Cosmos 500	Jul 10	P	C-1	554 – 509	74.0	Replaced Cosmos 425.
Cosmos 536	Nov 3	P	C-1	555 – 514	74.0	
1973						
Cosmos 544	Jan 20	P	C-1	561 – 513	74.0	Replaced Cosmos 460.
Cosmos 549	Feb 28	P	C-1	556 – 513	74.0	
Cosmos 582	Aug 28	P	C-1	559 – 521	74.0	
Cosmos 609	Nov 27	P	C-1	560 – 515	74.0	
1974						
Cosmos 631	Feb 6	P	C-1	565 – 522	74.0	Replaced Cosmos 479.
Cosmos 655	May 21	P	C-1	549 – 520	74.0	Replaced Cosmos 536.
Cosmos 661	Jun 21	P	C-1	555 – 513	74.0	
Cosmos 698	Dec 18	P	C-1	566 – 515	74.0	Replaced Cosmos 549.

[*] T = Tyuratam
P = Plesetsk

Name	Date	Launch Site	Booster	Orbit (km)	Inclination (degrees)	Remarks
				1975		
Cosmos 707	Feb 5	P	C-1	550 – 505	74.0	
Cosmos 749	Jul 4	P	C-1	557 – 511	74.0	Replaced Cosmos 610.
Cosmos 781	Nov 21	P	C-1	557 – 508	74.0	Replaced Cosmos 698.
				1976		
Cosmos 787	Jan 6	P	C-1	564 – 519	74.0	Replaced Cosmos 655.
Cosmos 790	Jan 23	P	C-1	559 – 513	74.0	Replaced Cosmos 631.
Cosmos 812	Apr 6	P	C-1	558 – 504	74.1	Replaced Cosmos 544.
Cosmos 845	Jul 27	P	C-1	557 – 505	74.0	Replaced Cosmos 812.
Cosmos 870	Dec 2	P	C-1	560 – 511	74.0	Replaced Cosmos 500.
				1977		
Cosmos 899	Mar 24	P	C-1	552 – 505	74.1	Replaced Cosmos 437.
Cosmos 924	Jul 4	P	C-1	560 – 514	74.0	Replaced Cosmos 436.
Cosmos 960	Oct 25	P	C-1	549 – 505	74.0	Replaced Cosmos 870.
				1978		
Cosmos 1008	May 17	P	C-1	551 – 501	74.0	Replaced Cosmos 845.
Cosmos 1062	Dec 15	P	C-1	548 – 508	74.0	Replaced Cosmos 899.
				1979		
Cosmos 1114	Jul 11	P	C-1	558 – 507	74.0	Last launch. Replaced Cosmos 924.

Soviet Heavy ELINT Satellites

Name	Date	Launch Site[*]	Booster	Orbit (km)	Inclination (degrees)	Remarks
1970						
Cosmos 389	Dec 18	P	A-1	699 – 655	81.0	First heavy ELINT.
1971						
Cosmos 405	Apr 7	P	A-1	706 – 676	81.3	
1972						
Cosmos 476	Mar 1	P	A-1	651 – 618	81.2	
Cosmos 542	Dec 28	P	A-1	653 – 554	81.2	
1973						
Cosmos 604	Oct 29	P	A-1	647 – 624	81.2	
1974						
Cosmos 673	Aug 16	P	A-1	648 – 620	81.2	
1975						
Cosmos 744	Jun 20	P	A-1	650 – 612	81.2	
Cosmos 756	Aug 22	P	A-1	649 – 627	81.2	
1976						
Cosmos 808	Mar 16	P	A-1	647 – 618	81.3	
Cosmos 851	Aug 27	P	A-1	649 – 592	81.0	
1977						
Cosmos 895	Feb 27	P	A-1	648 – 613	81.2	
Cosmos 925	Jul 7	P	A-1	645 – 622	81.2	
Cosmos 955	Sep 20	P	A-1	664 – 631	81.2	Started UFO panic.
1978						
Cosmos 975	Jan 10	P	A-1	680 – 637	81.2	
Cosmos 1005	May 12	P	A-1	672 – 626	81.2	
Cosmos 1025	Jun 28	P	F-2	680 – 649	82.5	Test mission?
Cosmos 1043	Oct 10	P	A-1	650 – 625	81.1	
Cosmos 1063	Dec 19	P	A-1	661 – 632	81.2	
1979						
Cosmos 1076	Feb 12	P	F-2	678 – 647	82.5	Test mission?
Cosmos 1077	Feb 14	P	A-1	651 – 629	81.2	
Cosmos 1093	Apr 14	P	A-1	650 – 625	81.3	
Cosmos 1116	Jul 20	P	A-1	649 – 608	81.2	
Cosmos 1143	Oct 26	P	A-1	665 – 625	81.2	
Cosmos 1145	Nov 27	P	A-1	652 – 629	81.2	

[*] P = Plesetsk

Name	Date	Launch Site	Booster	Orbit (km)	Inclination (degrees)	Remarks
1980						
Cosmos 1151	Jan 23	P	F-2	678 – 650	82.5	Test mission?
Cosmos 1154	Jan 30	P	A-1	671 – 634	81.3	
Cosmos 1184	Jun 4	P	A-1	662 – 621	81.3	
Cosmos 1206	Aug 15	P	A-1	659 – 624	81.2	
Cosmos 1222	Nov 21	P	A-1	659 – 624	81.2	
1981						
Cosmos 1242	Jan 27	P	A-1	655 – 625	81.1	
Cosmos 1271	May 19	P	A-1	649 – 626	81.2	
Cosmos 1300	Aug 24	P	F-2	664 – 636	82.5	First operational F-2 HEAVY ELINT satellite.
Cosmos 1315	Oct 13	P	A-1	665 – 625	81.1	
Cosmos 1328	Dec 3	P	F-2	664 – 636	82.5	
1982						
Cosmos 1340	Feb 19	P	A-1	638 – 621	81.2	
Cosmos 1346	Mar 31	P	A-1	660 – 621	81.1	
Cosmos 1356	May 5	P	A-1	671 – 632	81.1	
Cosmos 1378	Jun 10	P	F-2	663 – 634	82.5	
Cosmos 1400	Aug 5	P	A-1	654 – 630	81.1	
Cosmos 1408	Sep 16	P	F-2	669 – 635	82.5	
1983						
Cosmos 1437	Jan 20	P	A-1	657 – 630	81.1	
Cosmos 1441	Feb 16	P	A-1	633 – 624	81.1	Last A-1 launch.
Cosmos 1455	Apr 23	P	F-2	665 – 635	82.5	
Cosmos 1470	Jun 23	P	F-2	670 – 635	82.5	
Cosmos 1500	Sep 28	P	F-2	665 – 633	82.5	
Cosmos 1515	Dec 15	P	F-2	663 – 636	82.5	
1984						
Cosmos 1536	Feb 8	P	F-2	665 – 634	82.5	
Cosmos 1544	Mar 15	P	F-2	664 – 634	82.5	
Cosmos 1602	Sep 28	P	F-2	667 – 634	82.5	
Cosmos 1606	Oct 18	P	F-2	655 – 631	82.5	
1985						
Cosmos 1626	Jan 24	P	F-2	664 – 630	82.5	
Cosmos 1633	Mar 5	P	F-2	660 – 634	82.5	
Cosmos 1666	Jul 3	P	F-2	666 – 633	82.5	
Cosmos 1674	Aug 8	P	F-2	664 – 632	82.5	
Cosmos 1703	Nov 22	P	F-2	666 – 635	82.5	
Cosmos 1707	Dec 12	P	F-2	655 – 634	82.5	

Soviet Very Heavy ELINT Satellites

Name	Date	Launch Site[*]	Booster	Orbit (km)	Inclination (degrees)	Remarks
Cosmos 1603	Sep 28, 1984	T	D-1-e	856 – 850	71.0	First launch.
Cosmos 1656	May 30, 1985	T	D-1-e	861 – 806	71.1	Positioned 45° away from Cosmos 1603.
Cosmos 1697	Oct 23, 1985	T	?	854 – 849	71.0	
Cosmos 1714	Dec 28, 1985	T	?	443 – 853	70.9	Booster malfunctioned, leaving satellite in incorrect orbit.

[*] T = Tyuratam

Notes: It has been suspected that Cosmos 1697 and Cosmos 1714 used the Soviet new SL-16/J-1 medium-lift booster. This is unconfirmed.

Soviet Nuclear Ocean Surveillance Satellites

Name	Date	Launch Site[*]	Booster	Orbit (km)	Inclination (degrees)	Life-Time (days)	Remarks
Cosmos 198	Dec 27, 1967	T	F-1-m	281 – 265	65.1	2	
Cosmos 209	Mar 22, 1968	T	F-1-m	282 – 250	65.1	6	
None	Jan 24/25, 1969	T	F-1-m	–	–	–	Launch failure.
Cosmos 367	Oct 3, 1970	T	F-1-m	280 – 250	65.2	0	Failed immediately after reaching orbit.
Cosmos 402	Apr 1, 1971	T	F-1-m	279 – 261	65.0	8	
Cosmos 469	Dec 25, 1971	T	F-1-m	276 – 259	65.0	10	
Cosmos 516	Aug 21, 1972	T	F-1-m	277 – 256	65.0	31	
None	Apr 25, 1973	T	F-1-m	–	–	–	Launch failure.
Cosmos 626	Dec 27, 1973	T	F-1-m	280 – 257	65.0	46	
Cosmos 651	May 15, 1974	T	F-1-m	276 – 256	65.0	71	
Cosmos 654	May 17, 1974	T	F-1-m	277 – 261	65.0	74	
Cosmos 723	Apr 2, 1975	T	F-1-m	277 – 256	65.0	46	
Cosmos 724	Apr 7, 1975	T	F-1-m	276 – 258	65.0	65	
Cosmos 785	Dec 12, 1975	T	F-1-m	278 – 259	65.0	0.68	Failed after reaching orbit.
Cosmos 860	Oct 17, 1976	T	F-1-m	278 – 260	65.0	24	
Cosmos 861	Oct 21, 1976	T	F-1-m	280 – 256	65.0	60	
Cosmos 952	Sep 16, 1977	T	F-1-m	278 – 258	65.0	22	
Cosmos 954	Sep 18, 1977	T	F-1-m	277 – 251	65.0	–	Re-entered over Canada.
Cosmos 1176	Apr 29, 1980	T	F-1-m	265 – 260	65.0	134	Test flight.
Cosmos 1249	Mar 5, 1981	T	F-1-m	264 – 251	64.9	106	
Cosmos 1266	Apr 21, 1981	T	F-1-m	267 – 248	64.9	8	Failed early.
Cosmos 1299	Aug 24, 1981	T	F-1-m	266 – 247	65.0	13	Failed early.
Cosmos 1365	May 14, 1982	T	F-1-m	264 – 252	65.0	136	
Cosmos 1372	Jun 2, 1982	T	F-1-m	270 – 246	64.9	71	
Cosmos 1402	Aug 30, 1982	T	F-1-m	264 – 251	65.0	–	Re-entered over Indian Ocean and South Atlantic.
Cosmos 1412	Oct 2, 1982	T	F-1-m	266 – 251	65.0	39	
Cosmos 1579	Jun 29, 1984	T	F-1-m	264 – 249	65.0	125	(Exact lifetime not available.)
Cosmos 1607	Oct 31, 1984	T	F-1-m	264 – 250	65.0	93	
Cosmos 1670	Aug 1, 1985	T	F-1-m	264 – 252	65.0	83	
Cosmos 1677	Aug 23, 1985	T	F-1-m	263 – 251	65.0	61	Both Cosmos 1670 and Cosmos 1677 were boosted on Oct 23.

[*] T = Tyuratam

Soviet ELINT Ocean Surveillance Satellites

Name	Date	Launch Site[*]	Booster	Orbit (km)	Inclination (degrees)	Remarks
Cosmos 699	Dec 24, 1974	T	F-1-m	454 – 436	65.0	Exploded Apr 17, 1975.
Cosmos 777	Oct 29, 1975	T	F-1-m	456 – 437	65.0	Exploded January, 1976.
Cosmos 838	Jul 2, 1976	T	F-1-m	456 – 438	65.0	Exploded Jun/Jul 1977.
Cosmos 868	Nov 26, 1976	T	F-1-m	457 – 438	65.0	
Cosmos 937	Aug 24, 1977	T	F-1-m	457 – 438	65.0	
Cosmos 1094	Apr 18, 1979	T	F-1-m	457 – 437	65.0	
Cosmos 1096	Apr 25, 1979	T	F-1-m	457 – 439	65.0	First pair of ELINT ocean surveillance satellites.
Cosmos 1167	May 14, 1980	T	F-1-m	457 – 433	65.0	
Cosmos 1220	Nov 4, 1980	T	F-1-m	454 – 432	65.0	
Cosmos 1260	Mar 20, 1981	T	F-1-m	447 – 428	65.0	
Cosmos 1286	Aug 4, 1981	T	F-1-m	444 – 431	65.0	Worked with Cosmos 1260.
Cosmos 1306	Sep 14, 1981	T	F-1-m	424 – 171	64.9	Booster problem placed satellite into incorrect orbit. Orbit raised over 8 days. Replaced Cosmos 1260.
Cosmos 1337	Feb 11, 1982	T	F-1-m	446 – 428	65.0	Worked with Cosmos 1286 and Cosmos 1306.
Cosmos 1355	Apr 29, 1982	T	F-1-m	446 – 428	65.0	
Cosmos 1405	Sep 4, 1982	T	F-1-m	444 – 430	65.0	
Cosmos 1461	May 7, 1983	T	F-1-m	444 – 429	65.0	
Cosmos 1507	Oct 29, 1983	T	F-1-m	442 – 433	65.0	
Cosmos 1567	May 30, 1984	T	F-1-m	442 – 432	65.0	
Cosmos 1588	Aug 7, 1984	T	F-1-m	446 – 426	65.0	
Cosmos 1625	Jan 23, 1985	T	F-1-m	370 – 116	65.0	Upper stage failed to re-start. Satellite re-entered after a few hours.
Cosmos 1646	Apr 18, 1985	T	F-1-m	443 – 429	65.0	
Cosmos 1682	Sep 19, 1985	T	F-1-m	443 – 429	65.0	

[*] T = Tyuratam

U.S. Ocean Surveillance Satellites

Name	Date	Booster	Orbit (km)	Inclination (degrees)	Remarks
1971–110	Dec 14, 1971	LTTAT	999 – 983	70.0	Test flight.
1976–38	Apr 3, 1976	Atlas F	1,130 – 1,093	63.4	First White Cloud.
1977–112	Dec 8, 1977	Atlas F	1,069 – 1,054	63.4	
1980–19	Mar 3, 1980	Atlas F	1,150 – 1,035	63.0	
None	Dec 9, 1980	Atlas F	–	–	Failed to orbit. Meant to replace first White Cloud.
1983–8	Feb 9, 1983	Atlas F	1,170 – 1,150	63.4	
1983–56	Jun 10, 1983	Atlas F	1,165 – 1,045	63.3	
1984–12	Feb 5, 1984	Atlas F	Not available	–	

Notes: Three payloads were launched on a single booster. Orbit listed is the average.

U.S. Early Warning Satellites

Name	Date	Booster	Orbit (km)	Inclination (degrees)	Remarks
Midas 1	Feb 26, 1960	Atlas Agena A	–	–	Launch failure.
Midas 2	May 24, 1960	Atlas Agena A	511 – 484	33.0	
Midas 3	Jul 12, 1961	Atlas Agena B	3,534 – 3,358	91.2	
Midas 4	Oct 21, 1961	Atlas Agena B	3,756 – 2,496	95.8	
1962-Kappa	Apr 9, 1962	Atlas Agena B	3,382 – 2,814	86.6	
None	Dec 17, 1962	Atlas Agena B	–	–	Launch failure.
1963–14	May 9, 1963	Atlas Agena B	3,680 – 3,604	87.4	
None	Jun 12, 1963	Atlas Agena B	–	–	Launch failure.
1963–30	Jul 19, 1963	Atlas Agena B	3,727 – 3,670	88.4	
1966–51	Jun 9, 1966	Atlas Agena D	3,616 – 174	90.0	Left in transfer orbit.
1966–77	Aug 19, 1966	Atlas Agena D	3,700 – 3,680	90.0	
1966–89	Oct 5, 1966	Atlas Agena D	3,702 – 3,682	90.2	
1968–63	Aug 6, 1968	Atlas Agena D	39,860 – 31,680	9.90	First Program 949 launch.
1969–36	Apr 13, 1969	Atlas Agena D	39,270 – 32,670	9.90	Program 949.
1970–46	Jun 19, 1970	Atlas Agena D	33,685 – 178	28.2	Program 949; left in transfer orbit.
1970–69	Sep 1, 1970	Atlas Agena D	39,855 – 31,947	10.3	
1970–93	Nov 6, 1970	Titan III C	35,886 – 26,050	7.80	First Program 647 launch; transtage failure.
1971–39	May 5, 1971	Titan III C	35,840 – 35,651	0.87	Program 647.
None	Dec 4, 1971	Atlas Agena D	–	–	Program 949 satellite; failed to orbit.
1972–10	Mar 1, 1972	Titan III C	35,962 – 35,416	0.20	Program 647.
1972–101	Dec 20, 1972	Atlas Agena D	40,728 – 31,012	9.70	Program 949.
1973–40	Jun 12, 1973	Titan III C	35,786 – 35,777	0.30	Program 647.
1975–55	Jun 18, 1975	Titan III C	40,800 – 30,200	9.00	
1975–118	Dec 14, 1975	Titan III C	35,785 – 35,671	3.00	
1976–59	Jun 26, 1976	Titan III C	35,860 – 35,620	0.50	
1977–07	Feb 6, 1977	Titan III C	35,755 – 35,532	0.10	
1978–58	Jun 10, 1978	Titan III C	42,039 – 29,929	12.0	
1979–53	Jun 10, 1979	Titan III C	35,854 – 35,712	1.80	
1979–86	Oct 1, 1979	Titan III C	41,497 – 30,443	7.50	
1981–25	Mar 16, 1981	Titan III C	35,800 – 35,776	?	
1981–107	Oct 31, 1981	Titan III C	35,788 – 35,771	?	
1982–19	Mar 6, 1982	Titan III C	35,600 – 35,521	1.90	

Soviet Early Warning Satellites

Name	Date	Launch Site[*]	Booster	Orbit (km)	Inclination (degrees)	Remarks
1972						
Cosmos 520	Sep 19	P	A-2-e	39,319 – 625	62.8	First known Early Warning launch.
1973						
Cosmos 606	Nov 2	P	A-2-e	39,360 – 625	62.8	Replaced Cosmos 520.
Cosmos 665	Jun 29	P	A-2-e	39,384 – 633	62.9	
1975						
Cosmos 706	Jan 30	P	A-2-e	39,812 – 625	62.8	
1976						
Cosmos 862	Oct 22	P	A-2-e	39,300 – 610	62.8	
1977						
Cosmos 903	Apr 11	P	A-2-e	40,170 – 630	62.8	
Cosmos 917	Jun 16	P	A-2-e	40,150 – 625	62.5	
Cosmos 931	Jul 20	P	A-2-e	40,180 – 600	62.8	Replaced Cosmos 665.
1978						
Cosmos 1024	Jun 24	P	A-2-e	40,000 – 630	62.8	Replaced Cosmos 931.
Cosmos 1030	Sep 6	P	A-2-e	40,100 – 650	62.8	
1979						
Cosmos 1109	Jun 27	P	A-2-e	40,130 – 626	62.8	
Cosmos 1124	Aug 28	P	A-2-e	40,070 – 620	62.8	Replaced Cosmos 1030.
1980						
Cosmos 1164	Feb 12	P	A-2-e	640 – 220	62.8	Booster failure left satellite in parking orbit. Satellite meant to replace Cosmos 1109.
Cosmos 1172	Apr 12	P	A-2-e	40,160 – 637	62.8	Replaced Cosmos 1109.
Cosmos 1188	Jun 14	P	A-2-e	40,165 – 628	62.8	Replaced Cosmos 1024.
Cosmos 1191	Jul 2	P	A-2-e	40,165 – 646	62.8	Replaced Cosmos 1124.
Cosmos 1217	Oct 24	P	A-2-e	40,165 – 642	62.8	Replaced Cosmos 1188.
Cosmos 1223	Nov 27	P	A-2-e	40,165 – 614	62.8	Replaced Cosmos 903.
1981						
Cosmos 1247	Feb 19	P	A-2-e	39,734 – 621	62.9	
Cosmos 1261	Mar 31	P	A-2-e	39,756 – 593	62.9	
Cosmos 1278	Jun 19	P	A-2-e	40,214 – 615	62.8	
Cosmos 1285	Aug 4	P	A-2-e	40,429 – 587	62.9	Failed to stabilize ground track.
Cosmos 1317	Oct 31	P	A-2-e	40,162 – 584	62.8	

[*] P = Plesetsk

Name	Date	Launch Site	Booster	Orbit (km)	Inclination (degrees)	Remarks
1982						
Cosmos 1341	Mar 3	P	A-2-e	39,684 – 630	62.9	
Cosmos 1348	Apr 7	P	A-2-e	39,751 – 596	62.8	Replaced Cosmos 1172.
Cosmos 1367	May 20	P	A-2-e	39,264 – 581	62.8	
Cosmos 1382	Jun 25	P	A-2-e	39,436 – 590	62.8	
Cosmos 1409	Sep 22	P	A-2-e	39,690 – 612	63.0	
1983						
Cosmos 1456	Apr 25	P	A-2-e	39,387 – 620	62.9	
Cosmos 1481	Jul 8	P	A-2-e	39,200 – 643	62.9	
Cosmos 1518	Dec 28	P	A-2-e	39,345 – 584	62.9	
1984						
Cosmos 1541	Mar 6	P	A-2-e	39,650 – 706	62.9	
Cosmos 1547	Apr 4	P	A-2-e	39,221 – 582	62.9	
Cosmos 1569	Jun 6	P	A-2-e	39,834 – 588	62.9	
Cosmos 1581	Jul 3	P	A-2-e	39,707 – 639	62.9	
Cosmos 1586	Aug 2	P	A-2-e	39,737 – 609	62.9	
Cosmos 1596	Sep 7	P	A-2-e	39,248 – 604	62.9	
Cosmos 1604	Oct 4	P	A-2-e	39,716 – 606	62.9	
1985						
Cosmos 1658	Jun 11	P	A-2-e	39,745 – 580	62.8	
Cosmos 1661	Jun 18	P	A-2-e	39,743 – 607	62.9	
Cosmos 1675	Aug 12	P	A-2-e	39,729 – 596	62.8	
Cosmos 1684	Sep 24	P	A-2-e	39,762 – 580	62.9	
Cosmos 1687	Sep 30	P	A-2-e	39,732 – 610	62.9	
Cosmos 1698	Oct 22	P	A-2-e	39,725 – 603	62.9	
Cosmos 1701	Nov 3	P	A-2-e	39,719 – 619	63.0	

U.S. Vela Nuclear Detection Satellites

Name	Date	Booster	Orbit (km)	Inclination (degrees)	Remarks
Vela 1	Oct 16, 1963	Atlas Agena D	110,888 – 101,851	38.3	First launch.
Vela 2			117,415 – 101,055	38.0	
Vela 3	Jul 17, 1964	Atlas Agena D	104,516 – 102,078	39.5	
Vela 4			111,737 – 94,371	40.8	
Vela 5	Jul 20, 1965	Atlas Agena D	95,967 – 88,355	35.2	
Vela 6			121,452 – 101,589	34.9	
Vela 7	Apr 28, 1967	Titan III C	112,585 – 109,089	32.1	First advanced Vela launch.
Vela 8			114,587 – 107,489	33.1	
Vela 9	May 23, 1969	Titan III C	111,644 – 110,463	32.7	
Vela 10			112,009 – 110,657	32.8	
Vela 11	Apr 8, 1970	Titan III C	110,739 – 110,091	37.7	
Vela 12			111,379 – 111,111	32.9	

Notes: All launches from the Cape. Satellites were orbited in pairs aboard a single booster.

Sources for Appendix B

This list of military satellite launches was assembled from a variety of sources: the monthly satellite digests in *Spaceflight Magazine;* the Library of Congress reports, *Soviet Space Programs* 1971–75, Vol. 1, and *Soviet Space Programs* 1976–80, Part 1 and 3; the Goddard Spaceflight Center's "Satellite Situation Report"; *American Space Exploration—the First Decade* by William Roy Shelton (Boston: Little Brown, 1967); and various articles in *Aviation Week & Space Technology, Spaceflight,* and the *Journal of the British Interplanetary Society.*

Bibliographic Essay

The literature on orbital reconnaissance, unlike more conventional espionage, is limited. For the reader seeking further information, the author has found several sources to be of particular value. First is Philip J. Klass' *Secret Sentries in Space* (New York: Random House, 1971). This was the first book on reconnaissance satellites and despite being written in 1971 (before Big Bird), it remains an essential starting point.

Details of soviet military space activities can be found in the Library of Congress series *Soviet Space Programs* (Library of Congress, 1976). Books that look at the broader field of intelligence analysis, policy, and its use and abuse are Lawrence Freedman's *U.S. Intelligence and the Soviet Strategic Threat* (Boulder, Co.: Westview Press, 1977) and John Prados' *The Soviet Estimate* (New York: The Dial Press, 1982).

Information on the U-2 program, from the pilot's viewpoint, is in Francis Gary Powers' autobiography *Operation Overflight* (New York: Holt Rinehart Winston, 1970). Details on the aircraft itself are in Jay Miller's *Lockheed U-2* (Austin, Tx: Aerofax, 1983). Clarence "Kelly" Johnson provides the designer's viewpoint in *Kelly, More Than My Share of It All* (Washington, D.C.: Smithsonian Institution Press, 1985).

For news on space activities, both U.S. and Soviet (civil and military), three magazines are particularly important. *Aviation Week & Space Technology* provides in-depth coverage not found in other media. *Spaceflight* magazine and *The Journal of the British Interplanetary Society* are unparalleled. Their coverage and analysis of Soviet space activities, for instance, are of a quality found nowhere else.

Index